Land Use Effects on Streamflow and Water Quality in the Northeastern United States

Land Use Effects on Streamflow and Water Quality in the Northeastern United States

Avril L. de la Crétaz
Paul K. Barten

CRC Press
Taylor & Francis Group
Boca Raton London New York

CRC Press is an imprint of the
Taylor & Francis Group, an informa business

CRC Press
Taylor & Francis Group
6000 Broken Sound Parkway NW, Suite 300
Boca Raton, FL 33487-2742

© 2007 by Taylor & Francis Group, LLC
CRC Press is an imprint of Taylor & Francis Group, an Informa business

No claim to original U.S. Government works
Printed in the United States of America on acid-free paper
10 9 8 7 6 5 4 3 2 1

International Standard Book Number-13: 978-0-8493-9187-3 (Hardcover)

Visit the Taylor & Francis Web site at
http://www.taylorandfrancis.com

and the CRC Press Web site at
http://www.crcpress.com

Dedication

To my children and grandchildren
—A.L. dlC.

To Peter E. Black, Kenneth N. Brooks, and Arthur R. Eschner
Exceptional teachers and mentors
—P.K.B.

Table of Contents

Preface

The scope and scale of watershed protection and restoration efforts has grown exponentially since the 1980s. Current examples are numerous and varied. They include (1) regional efforts to protect and restore the Great Lakes, Chesapeake Bay, Lake Champlain, and many major rivers; (2) renewed emphasis on source protection in public water supplies; (3) riparian and aquatic ecosystem restoration for salmon and other endangered species; (4) urban and community forestry projects to reduce stormwater volume and improve its quality; and (5) the reconnection of urban areas to nearby creeks and rivers with greenway projects, to name just a few. The corresponding increase in the awareness, engagement, and expectations of the public, political leaders, and environmental regulators is both a cause and an effect of these efforts. The groundswell of support for this work is evident in the work of more than 3000 watershed associations across the United States and a noteworthy group of national and international conservation and environmental organizations.

During the past few decades, the watershed science and management literature also has grown exponentially in volume, complexity, diversity, *and* value. This growth has been fueled by advances in field sampling, laboratory equipment, spatial information technologies (e.g., remote sensing and geographic information systems), computer software for data analysis, graphics, and modeling, and the recognition that interdisciplinary education and research is needed to address pressing problems. The revolution in electronic communications and Internet resources has added to this embarrassment of riches and also to the prospect of information overload.

The broad scope of watershed management efforts can make the organization, synthesis, and delivery of scientific information a daunting challenge. Although watershed management programs and projects often share common goals (maintaining or improving water quality, restoring a "natural" flow regime, conserving biological diversity, etc.), it has long been recognized that unique patterns of historical and contemporary land use and sets or combinations of biophysical attributes (climate, soils, flora, fauna, etc.) require site-specific or place-based adaptation of management practices. These practices have been derived from a common set of scientific principles, ecological processes, and operational experience, yet it often can be difficult to see the forest for the trees.

The conditions and observations described above suggested to us that a comprehensive volume was needed. It could not nor should not be encyclopedic, nor could it be burdened with the usual caveats ("it is beyond the scope of this book to discuss") in relation to knowledge and information needed by our intended audience—watershed managers, scientists, upper division and graduate students, executive and legislative staff members, and citizen leaders.

This book began as a narrowly focused effort to review the literature on the effects of forest conversion (to residential, commercial, or industrial land use) on streamflow and water quality in the northeastern United States. While retaining the same geographic focus, it grew into a more comprehensive volume as we carefully considered the needs of watershed managers and scientists in the 21st century and sought the advice of colleagues. The need is ubiquitous—state-of-the-science information to inform the development and implementation of watershed management programs and projects. Furthermore, since a patchwork of land uses (forest, agricultural, residential, commercial, and industrial) is the norm for most watersheds, we thought they should be described and discussed individually, then cumulatively, just as they affect streamflow and water quality.

The two part structure of this volume is based on two hypotheses. First, the scarcest resource for scientists and practitioners is time. Therefore the opportunity to conduct exhaustive literature searches and thorough reviews is correspondingly limited. Second, few people have academic training or professional experience of sufficient breadth to fully access the salient literature in watershed science and management. (We needed the help of an expert panel, more than 1000 references, and several years to complete the task.) As a result, we opted for a review of key topics and concepts in Part I as the scientific foundation for the examples on the effects of forest management, agriculture, and urbanization contained in Part II. Time will tell if our goal was appropriate and how close we came to the mark.

We are indebted to the expert panel listed below for their knowledge of the literature, insightful review comments, editorial suggestions, encouragement, and dedication to the advancement of watershed science and management. This book would not have been possible without their help and guidance.

Dr. Derek B. Booth
University of Washington, Department of Civil Engineering, Seattle, Washington

Dr. John C. Clausen
University of Connecticut, Department of Natural Resources Management and Engineering, Storrs, Connecticut

Dr. Christopher Eagar
USDA Forest Service, Northern Research Station, Durham, New Hampshire

Mr. Thomas D. Kyker-Snowman
Massachusetts Division of Water Supply Protection, Quabbin Reservoir, Belchertown, Massachusetts

Dr. Andrew N. Sharpley
USDA Agricultural Research Service, Pasture Systems and Watershed Management Research Unit, Pennsylvania State University, State College, Pennsylvania

Mr. Albert H. Todd
USDA Forest Service, Northeastern Area State and Private Forestry, Annapolis, Maryland

Dr. Elon S. Verry
USDA Forest Service (retired), North Central Research Station, Grand Rapids, Minnesota

This project was supported by a grant from the USDA Forest Service, Northeastern Area State and Private Forestry. We are especially grateful to Al Todd, Robin Morgan, and Kathryn Maloney for their vision, patience, flexibility, and encouragement as a short technical report morphed into a book. The University of Massachusetts and Harvard University (through the Bullard Fellowship program at the Harvard Forest) supported Paul Barten's sabbatical leave in 2003–2004.

Ethan Nadeau did masterful work with the many graphs and diagrams used to illustrate this book. His scientific training and field experience in aquatic ecology coupled with his creativity, artistic skill, and attention to detail were great assets to this project. We thank the staff at the University of Massachusetts–Amherst libraries. We admire and value their persistence and professionalism. Finally, we thank our editors, Matt Lamoreaux, David Fausel, Mimi Williams, and all the staff at the Taylor & Francis Group for guiding us through the publication process.

Avril L. de la Crétaz and Paul K. Barten
Amherst, Massachusetts

About the Authors

Avril L. de la Crétaz is a research associate at the University of Massachusetts Amherst. She received her Ph.D. (2000) from the University of Massachusetts in forest ecology and a M.S. in environmental studies from Boston University (1993). In her dissertation research she studied the interaction of deer browsing and fern invasion in the 55,000 acre Quabbin Forest, the primary watershed for the metropolitan Boston water supply system, and tested strategies for restoration of the forest understory. Since 2001, Dr. de la Crétaz has worked in the field of forest hydrology and watershed management, specializing in the area of land use effects on water quality.

Paul K. Barten is associate professor of forest resources at the University of Massachusetts–Amherst (1997–present) and codirector of the USDA Forest Service–University of Massachusetts Amherst Watershed Partnership. His research and teaching focuses on watershed management and source water protection. He earned a Ph.D. (1988) and M.S. (1985) in forest hydrology and watershed management from the University of Minnesota and undergraduate degrees in forestry from the State University of New York (SUNY) College of Environmental Science and Forestry (B.S., 1983) and the New York State Ranger School (A.A.S., 1977). Dr. Barten served on the faculty of Yale University (1988–1997) and was a Bullard Fellow at Harvard University (2003–2004). He is the chair of a National Research Council committee (2005–2007) that is reviewing the hydrologic effects of forest management in the United States. He also serves as a scientist-at-large on the Research Committee of the Sustainable Forest Management Network in Canada.

About This Book

This book was designed and written in two parts that anticipate the needs and interests of watershed managers, foresters, landscape architects, planners, engineers, scientists, legislative and nongovernmental organization (NGO) staff members, students, teachers, and community leaders. The scope, scale, and complexity of watershed protection and restoration efforts—and the associated scientific literature—have grown exponentially since the 1980s. Most watersheds are a patchwork of different land uses (i.e., forests, farms, suburbs, urban areas, etc.) and present a wide range of management challenges and opportunities. This book provides a synthesis of and gateway to the diverse scientific literature that is urgently needed to solve contemporary problems.

Part I summarizes the scientific principles and processes that define and govern the interactions between activities on the land and conditions in streams, lakes, and estuaries. Chapter 2 reviews and explains the hydrologic processes and key watershed characteristics referred to throughout the book. Chapter 3 describes the chemical processes, transport mechanisms, and environmental effects of the most common pollutants: nutrients, pesticides, and metals. Chapter 4 describes freshwater ecosystems and explains how aquatic organisms can be used to evaluate stream health. Chapter 5 focuses on the structure and function of riparian areas and the effectiveness of riparian buffers. The importance of the physical environment (climate, topography, geology, and soils) and local conditions as a source of site-specific variation in streamflow and water quality is discussed throughout Part I.

Part II builds directly on the principles and processes discussed in Part I to examine the streamflow and water quality effects of specific land uses: forest management, agriculture, and urban development (Chapters 6, 7, and 8, respectively). A wide variety of published case studies are used to present and assess best management practices and innovative technologies for stormwater management, nonpoint source pollution mitigation, and aquatic ecosystem restoration. Chapter 9 summarizes large-scale studies of mixed land use watersheds. It also explores the spatial, temporal, and cumulative effects of land use. Key findings are summarized in Chapter 10 and are then used as the basis of widely applicable watershed management guidelines and recommendations.

Avril L. de la Crétaz and Paul K. Barten
University of Massachusetts Amherst

1 Introduction

With the disappearance of the forest, all is changed.

George Perkins Marsh, 1864, *Man and Nature*

The connection between the loss of forests, land use, streamflow, and water quality has long been recognized. More than 2000 years ago, Plato described a prehistoric Athens where rich and fertile lowland soils were surrounded by wooded hills and rains soaked into upland clay soils and supplied water for rivers and springs. Following the cutting of these mountain forests the springs dried up and floods carried the soil away to the sea "leaving the land nothing but skin and bone" (Plato [c. 427–347 B.C.]). Beginning in 1342, communities in Switzerland began to establish "protection" forests as a safeguard against avalanches. Between 1532 and 1777, 322 protection forests were established. An excerpt from a 1608 letter to the government of Venice directly attributes "the great and frequent inundations and the huge quantities of refuse and mud, which the mountain torrents are carrying and depositing at the present day in the lagoon, something unknown to antiquity" to the loss of forests (Kittredge, 1948). The author clearly understood the direct influence of mountain forests on soil stability and water quality and, as a result, their protective value to downstream communities.

In 1864 George Perkins Marsh published *Man and Nature* (later titled *The Earth as Modified by Human Action*), a seminal work on the impacts of people on the physical environment. Marsh focused on the devastating effects on wildlife, fisheries, and marine mammals caused by deforestation and the exploitive use of natural resources. He described soil erosion, flooding, and sedimentation caused by widespread forest clearing in Europe and the United States as follows:

> The soil is bared of its covering of leaves, broken, and loosened by the plow, deprived of the fibrous rootlets, which held it together, dried and pulverized by sun and wind... The face of the earth is no longer a sponge, but a dust-heap, and the floods, which the waters of the sky pour over it, hurry swiftly along its slopes, carrying in suspension vast quantities of earthy particles, which increase the abrading power and mechanical force of the current, and, augmented by the sand and gravel of falling banks, fill the beds of streams, divert them into new channels, and obstruct their outlets (Marsh, 1864).

Dr. Raphael Zon, the first director of the Lakes States Forest Experiment Station of the U.S. Forest Service began his study *Forests and Water in Light of Scientific Investigation* in the early 1900s. His synthesis of the literature from Russia, Europe, and North America led to the formulation of these generalizations and conclusions:

1. The total discharge of rivers depends upon climate.
2. Forests tend to equalize streamflow throughout the year.
3. Forests retard snowmelt.
4. Forests prevent erosion.
5. Forests cannot prevent floods produced by exceptional precipitation, but they can mitigate their destructiveness.

Zon's work (1927) marked the beginning of the rigorous study of forests and streamflow processes in the United States. Coweeta, Fernow, Hubbard Brook, Marcell, and many other U.S. Forest Service research forests built upon Zon's groundbreaking work.

Another book, entitled *Man's Role in Changing the Face of the Earth* (Thomas et al., 1956), and dedicated to the memory of George Perkins Marsh, reported the results of an international symposium that brought together more than 50 leading scientists of the day to discuss the effects of human civilization on the environment. Participants, including Luna Leopold, the eminent geologist and hydrologist, and Lewis Mumford, the internationally renowned expert on urban systems and land use planning, contributed papers discussing the detrimental effects of the loss of natural vegetation and urban development, among other topics. Recent advances in computing, mapping, and analytical technology have revealed new levels of complexity in watershed science and management. In many respects, a century of systematic research has quantitatively described phenomena that have been observed for thousands of years.

Over the last 50 years, population growth and increased rates of consumption have greatly increased the demand for water and other natural resources. In their comprehensive report entitled "Running Pure: The Importance of Forest Protected Areas to Drinking Water," Dudley and Stolton (2003) show the extent to which the world's drinking water (more 100 major cities) is linked to forested watersheds. They also emphasize the "clear link between forests and the quality of water coming out of a catchment." This is, or at least should be, a familiar refrain from the historical literature. They reinforce this key point by quoting the United Nations Food and Agriculture Organization: "The loss of forest cover and conversion to other land uses can adversely affect freshwater supplies, threatening the survival of millions of people and damaging the environment." While survival may not be at stake in developed countries, it is increasingly clear that forest conversion and inappropriate land and resource use jeopardizes ecological *and* public health. Water is a renewable natural resource, yet it also is a *finite* and essential resource that must be diligently protected and conserved.

Forests in the northeastern United States were cleared for agriculture and the development of towns and cities during the 1700s and 1800s. A suite of social, political, and economic changes led to the abandonment of agricultural land and natural regeneration of forests in the late 1800s and early 1900s. Today, forestland is being converted to other uses (residential, industrial, and commercial) throughout the region (Steele, 1999; Stein et al., 2005). The piecemeal nature of this change usually does not attract much attention until the cumulative effect of forest loss and new development becomes obvious and very costly to mitigate, offset, or reverse. In addition, industrialization and increased energy consumption have introduced water and air pollutants on a scale—both in type and quantity—that would have astounded Marsh, Zon, and their contemporaries. Today, scientists and managers concerned with protecting water supplies and aquatic habitat must quantify pollution from nutrients and pesticides from both agricultural and urban land uses. In addition, metals, and a host of newly introduced chemical compounds used in industry, transportation, and medicine are a matter of great concern. Pollutants are transported by water, sediment particles, and in atmospheric circulation. Characterizing cause-effect relationships and ecological patterns and processes is inherently complex. It also is the essential scientific foundation for watershed management.

1.1   THE GEOGRAPHIC AND SCIENTIFIC SCOPE OF THIS BOOK

Our intent in writing this book is to provide a single source that introduces the reader to the principles, processes, and recent scientific literature regarding forest conversion and land use effects on streamflow and water quality. The book was written primarily for watershed and resource management professionals. We hope to assist practitioners with management and regulatory responsibilities to make systematic and timely use of the science. Other readers may include researchers, teachers, upper division undergraduate and graduate students, and nongovernmental organization (NGO) and legislative staff members.

We have focused on peer-reviewed literature published since 1980, with some references to earlier papers and historical publications that are seminal or unique contributions. Our specific region of interest is the USDA Forest Service Northeastern Area. This area includes 20 states and the District of Columbia and can be divided into three major regions: (1) the New England states and northern New York; (2) the Mid-Atlantic, including southern New York and Long Island, New Jersey, Maryland, Delaware, and Pennsylvania; and (3) the Midwest, including the Great Lakes states (Ohio, Indiana, Michigan, Illinois, Wisconsin, Minnesota), and Iowa and Missouri. Studies from other areas, primarily the Pacific Northwest and the Southeast, are included when (in spite of regional variations in climate, topography, soils, and vegetation) they include key examples of ecosystem patterns and processes and management principles that are also relevant to the Northeastern Area. A general description of the physical environment and population growth and distribution in the Northeastern Area follows in this chapter.

1.2 THE STRUCTURE AND ORGANIZATION OF THIS BOOK

This book has two parts: Part I summarizes the scientific principles that define and govern interactions between activities on the land and conditions in streams, lakes, and estuaries. Chapter 2 provides an overview of hydrologic processes and key watershed characteristics. Chapter 3 describes the processes determining the movement and environmental effects of the most common pollutants: nutrients, pesticides, and metals. Chapter 4 describes freshwater ecosystems and explains how aquatic organisms can be used to monitor stream health. Chapter 5 focuses on definitions and functions of riparian areas and the functions and limitations of riparian buffers. Throughout these chapters we emphasize the importance of the physical environment (climate, topography, geology, and soils) as a source of variation in stream response to activities on the land. We also note how general principles operate within the context of local conditions.

In Part II we build upon the principles and processes discussed in Part I to examine the streamflow and water quality effects of specific land uses: forest management, agriculture, and urban development (Chapters 6, 7, and 8, respectively). Chapter 9 examines large-scale studies of mixed land use watersheds and examines the phenomenon of cumulative effects—the combined environmental results of multiple actions over space and time. Chapter 10 summarizes the conclusions of previous chapters and presents recommendations for watershed management programs and practices.

1.3 THE PHYSICAL ENVIRONMENT OF THE NORTHEASTERN UNITED STATES

The removal of natural vegetation (forests or grasslands) and the imposition of other land uses (agricultural, urban development) can increase the movement of sediment (from soil erosion) and pollutants from the land to streams. Sediment and pollutants are carried by water that moves through and over the soil. The rate of delivery (loading) of sediment and pollutants to streams depends on the soil characteristics, geology, topography, and climate of a particular region or location. In order to understand the effects of disturbance and human land use on streamflow and water quality in a particular setting, it is first necessary to know and understand the physical environment.

1.3.1 BEDROCK GEOLOGY AND TOPOGRAPHY

Differing patterns of bedrock geology and topography are the result of millions of years of geologic history. (Whitney [1994:39–52] provides a concise geologic history of the Northeast region. For an in-depth discussion of the New England–New York area, see Randall [2001]). Topographic regions have been mapped and categorized (Fenneman, 1938) into physiographic provinces that aid in identifying and describing the various regions and their geologic history (Figure 1.1 and Figure 1.2). The area now known as the Appalachian plateau formed the original eastern border of the North American land mass (an ancient continent named "Laurentia"). Beginning about 1.2 billion years ago, colliding continental plates compressed the bedrock at the edge of the North American land mass. This pushed

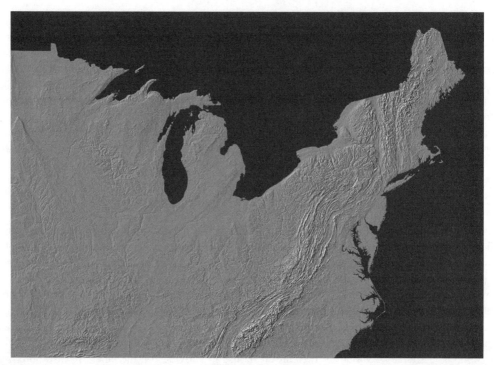

FIGURE 1.1 Topography of the northeastern United States. (Image courtesy of the U.S. Geological Survey, http://pubs.usgs.gov/imap/12206/usa_shade.tif; accessed July 2006.)

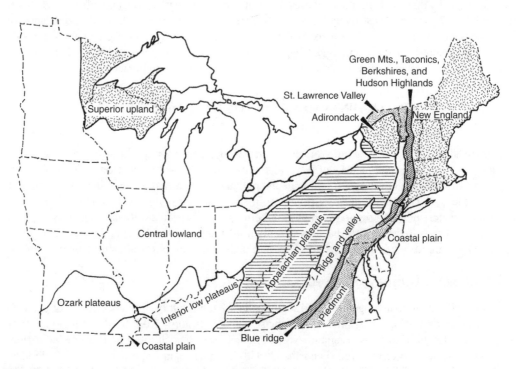

FIGURE 1.2 Major physiographic provinces and regions of the northeastern United States. (Whitney, 1994; reprinted with the permission of Cambridge University Press.)

sedimentary and metamorphic rocks upward in repeated episodes of mountain building, creating the Ridge and Valley province and the Appalachian plateau, mountainous terrain that separates the East Coast from the Central Lowlands region (Skehan, 2001). At the time of their formation, this eastern mountain chain was as high as the Rocky Mountains or the European Alps.

The Green Mountains, Taconics, Berkshires, and Hudson Highlands were formed when chains of volcanic islands collided with the ancient coast of North America in the Taconic mountain building event from 470 to 440 million years ago. Bedrock in this region includes a mix of sedimentary rock from the eastern edge of the continent, igneous rock from volcanic activity, and metamorphic rock created from the intense pressure of colliding tectonic plates. The New England physiographic province formed from a collection of fragments of other continents that adhered to the original North American land mass as tectonic plates collided (Skehan, 2001). There are numerous mountain ranges and individual peaks in the interior of New England. Elevations range from sea level along the coast to 1917 m at the peak of Mount Washington in New Hampshire (Breault and Harris, 1997). The bedrock in this region consists of metamorphic and igneous rocks, schist, gneiss, and granite, reflecting the ancient volcanic activity in much of the region. The Connecticut River Valley, the site of a continental rift, lies on sedimentary bedrock (Bell, 1985).

The Piedmont region or Appalachian foothills is formed on metamorphic bedrock that was once buried under the mountain chains at the edge of the continent. Millions of years of erosion wore down the mountains, exposing the original Piedmont bedrock (Plank and Schenck, 1998).

Cape Cod, Long Island, southern New Jersey, Delaware, and the eastern portion of Maryland lie in the Coastal Plain region. Sediment emanating from soil and channel erosion in the mountainous lands to the west have been deposited here for millions of years. In the Coastal Plain, terrestrial and marine sediments are subject to repeated cycles of erosion and deposition from wave action and sea level variation. These continuing processes create layers of sand, gravel, silt, and clay that form aquifers. Groundwater is stored in sand between confining clay layers (Watt, 2000). These sedimentary deposits can be very deep. The depth to bedrock at Cape May, on the southern tip of New Jersey, is more than 1900 m (Widmer, 1964).

The Midwestern states—Ohio, Indiana, Michigan, Illinois, southern Wisconsin, eastern Minnesota, Iowa, and northern Missouri—are predominantly located in the Central Lowlands region, an area not subject to the mountain building processes at work closer to the Atlantic Ocean. The Central Lowlands is a region of gently rolling hills and flatlands (Feldman et al., 1977; Myers et al., 2000; Whitney, 1994).

1.3.2 Glacial History

In much of the northern United States, surficial geology and soil type are a product of glacial processes (Figure 1.3). The retreat of the most recent ice sheet (16,000 to 7,000 years ago) reshaped the surface of the landscape (Pielou, 1991). Deposits of boulders, cobbles, gravel, sand, silt, and clay left by retreating glaciers were the parent material for most soils. These deposits are generally referred to as "glacial drift" and may be divided into two major categories: (1) material formed by the immense pressure of the glacier grinding on the bedrock in the region (glacial till) and (2) material sorted and deposited by glacial meltwater (outwash and lacustrine deposits) (Bennett and Glasser, 1996).

Glacial till is unstratified or poorly sorted, consisting of a jumbled mix of particle sizes. If the till develops on igneous and metamorphic bedrock, it contains a mixture of rocks and coarse fragments, sand, and gravel, with a relatively low proportion of finer particles (silts and clays). Till that develops on softer sedimentary bedrock with less resistance to the grinding action of the glaciers has a higher proportion of silts and clays and fewer stones. Loess is a term applied to particles of glacial till that have been picked up by the wind and redeposited many kilometers from their original location. Loess is primarily composed of silt, with some very fine sands and coarse clays. Loess deposits can be up to 8 m thick in some areas (Brady and Weil, 2002).

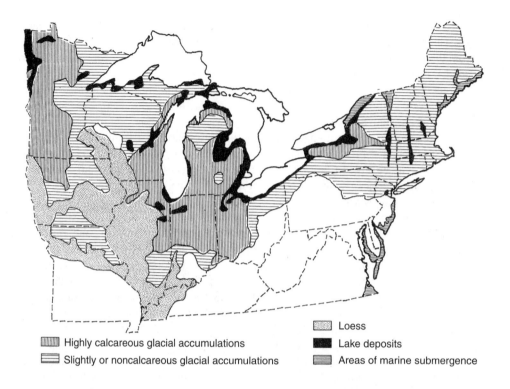

Highly calcareous glacial accumulations

Slightly or noncalcareous glacial accumulations

Loess

Lake deposits

Areas of marine submergence

FIGURE 1.3 Surficial geology of the northeastern United States showing the distribution of parent materials of soils. Unshaded regions are areas of consolidated rocks, south of the glacial border. (Whitney, 1994; reprinted with the permission of Cambridge University Press.)

Glacial outwash and lake deposits carried by melting water flowing from the glacier are well sorted (i.e., particles of similar size are deposited in layers or strata). Fast moving glacial meltwaters carried along sand, silt, and clay particles, leaving heavier rocks and coarse fragments behind. When glacial streams entered a lake they slowed, losing energy, and the larger, heavier particles dropped out, forming sandy outwash deposits at the edges of glacial lakes. Smaller particles, silts, and clays were carried further into the lakes and deposited on lake beds. When the lakes were drained by evaporation or the failure of natural dams, these deposits were left behind (Randall, 2001).

1.3.3 Soils

Differences in bedrock geology, glacial history, and climate have resulted in characteristic soil types in the various regions of the northeastern United States (Table 1.1). In much of New England, poorly sorted glacial till soils formed over igneous and metamorphic rocks in the uplands. Cold, wet climates and coniferous vegetation contribute to the formation of acidic, low-nutrient soils (spodosols) at higher elevations (Brady and Weil, 2002). Well-sorted sandy outwash and lacustrine and alluvial soils (with some fine textured till deposits) occur in river valleys (Bell, 1985; Randall, 2001). In the eastern Great Lakes region (Ohio, Michigan, Indiana), the advance and recession of the glaciers ground sedimentary bedrock into fine grain silt and clay till that formed nutrient-rich loamy soils interspersed with sandy outwash (Casey et al., 1997). In some areas, low-permeability till deposits have covered older layers of highly permeable sand and gravel, forming confined aquifers (Myers et al., 2000). Further west, in parts of Minnesota, Iowa, and Illinois, a cold, relatively dry continental climate resulted in the formation of nutrient-rich prairie (grassland) soils with high organic matter content (Brady and Weil, 2002). Large regions of loess deposits can be found in Illinois, Wisconsin, Minnesota, Iowa, and Missouri (Whitney, 1994).

TABLE 1.1
Dominant Soil Orders of the Northeastern United States (Brady and Weil, 2002)

Order name	Description	Primary locations	Uses
Alfisols	High-nutrient (calcium, magnesium) soils, high clay content, moderate to high cation exchange capacity (CEC)	Great Lakes region: Ohio, Indiana, Illinois, Wisconsin, southern Michigan, eastern Minnesota	Productive soils that result in good hardwood forest growth and crop yields
Entisols	New soils found in floodplains and Coatal Plains	River valleys and the Atlantic Coastal Plain	River valley floodplain soils are nutrient-rich and very productive agricultural soils; sandy floodplain soils have low natural fertility
Histosols	Wetland soils with accumulated partially decomposed organic matter	Occur in wetlands throughout the Northeast, large areas of histosols in northern Minnesota, Wisconsin, Michigan	Large areas have been drained for agriculture
Inceptisols	Young soils; low levels of organic matter and lower CECs and nutrient levels than alfisols	Mountainous regions, New York, Pennsylvania, southern New England, West Virginia, eastern Ohio	Forestry: relatively low productivity
Mollisols	Prairie soils, accumulated calcium-rich organic matter, formed under grasslands	Great Plains, northern Illinois, western Minnesota, Iowa	Highly productive croplands in regions with sufficient rainfall
Spodosols	Leached, acidic soils formed in cool climates under coniferous forests	Northern New England, mountainous regions of western Massachusetts, northern New York	Forestry: relatively low productivity
Ultisols	Old soils, leached and weathered clay soils that have formed in unglaciated regions	South of the boundary of maximum glacial extent, New Jersey, Delaware, Maryland, West Virginia, Virginia, Kentucky, southern Missouri	Less fertile than mollisols and alfisols, less acidic than spodosols; can be used for agriculture with good management, lime, and fertilizer; productive forest soils

South of the glacial boundary there are weathered, leached, low-nutrient clay soils in interior regions and deep sand deposits along the Coastal Plain (Figure 1.3) (Brady and Weil, 2002). Unglaciated soils or residual soils that have "developed in place" are found in the southern parts of New Jersey and large areas of Pennsylvania and Ohio, all of Delaware, Maryland, Virginia, West Virginia, and Kentucky, and much of Missouri. Frequently these soils contain a high proportion of leached, low-nutrient clays.

Soils can vary substantially in permeability and susceptibility to erosion as a result of differences in particle size. Fine textured soils are the least permeable to water, however, small particles (clays) and substances that adhere to fine textured soils (e.g., phosphorus and some pesticides) are carried farther by water than are coarse textured soils (sands). (Silt particles are intermediate in size between sands and clays.) Fine textured soils in areas with steep slopes, (e.g., the Driftless Area of Wisconsin and the Maryland Piedmont) are therefore highly likely to be vulnerable to erosion and to provide sources of sediment and adsorbed pollutants. Sandy soils are more permeable and have higher infiltration capacity than clay soils. Sandy soils, especially deep sandy soils in level terrain, are vulnerable to shallow groundwater contamination by soluble, mobile contaminants such as nitrate and volatile organic compounds (VOCs) (Ayers et al., 2000; McCobb and LeBlanc, 2002). Since overland flow occurs much less frequently on well-drained sandy soils, erosion rates are typically low. In contrast, sands are more susceptible to erosion in stream channels and in floodplains where silts, clays, and organic soils offer more structural resistance. The movement of water in soils and stream channels is discussed in much greater detail in Chapter 2.

1.3.4 CLIMATE AND WIND PATTERNS

The climate in the northeastern United States is temperate and seasonal. There is systematic variation within the region in relation to latitude, elevation, and proximity to large water bodies (the Atlantic Ocean or the Great Lakes), and proximity to the Appalachian Mountains (windward or leeward). Temperatures in eastern coastal regions are moderated by the ocean. Seasonal temperature variations are more extreme in interior locations. Cool, dry continental polar air masses meet warm, humid maritime tropical air masses in the New England and the Mid-Atlantic regions. Precipitation often occurs at the interface when warm air is lifted and cooled (Whitney, 1994).

In general, precipitation decreases from east to west. In central New England, mean annual precipitation averages 109 cm, compared to approximately 81 to 89 cm in Wisconsin and west-central Illinois (Groschen et al., 2000; Peters et al., 1998). The exception to this would be in areas receiving additional (lake effect) rain or snow from air masses that cross the Great Lakes. For example, mean annual precipitation between 1960 and 1995 was 111 cm at Erie, Pennsylvania, on the southern shore of Lake Erie (Casey et al., 1997).

There is substantial and systematic variation in rain and snow in mountainous areas. In New England, mean annual precipitation in the northern Connecticut River Valley averages 86 cm, compared to 165 cm on nearby mountain peaks (Zimmerman et al., 1996). This is noteworthy because greater precipitation—especially on soils with low clay content on the East Coast—can lead to a greater leaching of soluble nutrients to streams and groundwater (Binkley, 2001; Vitousek et al., 1979; Whitney, 1994).

Air masses generally traverse the region from the west to east. This strongly influences the pattern of atmospheric deposition of pollutants that degrade water quality in regions far from their point of origin. Clark et al. (2000) examined data from 85 relatively undeveloped sites (all U.S. Geological Survey [USGS] monitoring stations) across the United States and found that concentrations and yields of nitrate tended to be highest in New England and the Mid-Atlantic states. They attribute this trend to patterns of airflow and atmospheric deposition.

1.4 POPULATION GROWTH AND DISTRIBUTION

With the exception of West Virginia and the District of Columbia, the population of the Northeastern Area has grown steadily since 1950 (Table 1.2). This growth in population has been accompanied by an expansion of residential development into rural areas. An analysis of population data from 1990 to 2000 (Richardson and Gordon, 2004) for major metropolitan areas in New England, the Mid-Atlantic states, and the Midwest shows that, in general, suburban areas grew at a faster rate than core central cities. In the Philadelphia, Detroit–Ann Arbor–Flint, Cleveland–Akron, and St. Louis metropolitan areas, the population of the central cities decreased while the suburbs increased (Table 1.3). An Associated Press article published in June 2006 noted a continuation of this trend between 2001 and 2005: Detroit lost 65,000 people between 2000 and 2005, while Philadelphia lost 54,000 during the same period. These populations reflected movement both to outer suburbs and to western and southern states.

The forests in much of the Northeastern Area were cleared and converted to crop and pasture-lands during 18th- and 19th-century European settlement. Forest clearing began in the coastal areas of New England and the Mid-Atlantic states in the 1600s and proceeded westward until the forests and grasslands of Illinois, Wisconsin, and Minnesota were cleared or converted in the 1800s (Whitney, 1994). Beginning in about 1840, farm abandonment and emigration to the Ohio Valley,

TABLE 1.2
A Summary of Population Change, 1950 to 2000, in the Northeastern United States (U.S. Census)

Region	State	Area (km²)	Population in 1950	Population in 2000	Population density in 1950 (persons/km²)	Population density in 2000 (persons/km²)	Percent change in population
New England and New York	Connecticut	12,549	2,007,000	3,406,000	160	271	70
	Maine	79,940	914,000	1,277,000	11	16	39
	Massachusetts	20,300	4,691,000	6,349,000	231	313	35
	New Hampshire	23,230	533,000	1,236,000	23	53	132
	New York	122,310	14,830,000	18,976,000	121	155	28
	Rhode Island	2,707	792,000	1,048,000	293	387	32
	Vermont	23,955	378,000	609,000	16	25	61
Mid-Atlantic	Delaware	5,063	318,000	784,000	63	155	147
	Maryland	25,317	2,343,000	5,296,000	93	209	126
	New Jersey	19,215	4,835,000	8,414,000	252	438	74
	Pennsylvania	116,084	10,498,000	12,281,000	90	106	17
	West Virginia	62,385	2,006,000	1,808,000	32	29	−10
	District of Columbia	158	802,000	572,000	5,076	3,620	−29
Midwest	Illinois	143,986	8,712,000	12,419,000	61	86	43
	Indiana	92,903	3,934,000	6,080,000	42	65	55
	Iowa	144,716	2,621,000	2,926,000	18	20	12
	Michigan	147,135	6,372,000	9,938,000	43	68	56
	Minnesota	206,208	2,982,000	4,919,000	14	24	65
	Missouri	178,446	3,955,000	5,595,000	22	31	41
	Ohio	106,068	7,947,000	11,353,000	75	107	43
	Wisconsin	140,673	3,435,000	5,364,000	24	38	56

TABLE 1.3
Population Change (percent) for Selected Metropolitan Areas in the Northeastern United States (1990 to 2000): Core Cities versus Outlying Metropolitan Areas (modified from Richardson and Gordon, 2004)

Consolidated metropolitan statistical area (CMSA)	Entire CMSA	Core central city	All other core central cities (>100,000) in CMSA	Remainder of metropolitan area
New York, northern New Jersey–Long Island, NY–NJ–CT–PA CMSA	8.4	9.4	0.8	7.2
Chicago–Gary–Kenosha, IL–IN–WI CMSA	11.1	4.0	20.0	14.4
Philadelphia–Wilmington–Atlantic City, PA–NJ–DE–MD CMSA	5.0	4.3	NA	8.4
Boston–Worcester–Lawrence, MA–NH–ME–CT CMSA	6.7	2.6	3.8	7.5
Detroit–Ann Arbor–Flint, MI CMSA	5.2	7.5	4.5	9.1
Minneapolis–St. Paul, MN–WI CMSA	16.9	3.9	12.2	26.2
Cleveland–Akron, OH CMSA	3.0	5.4	2.7	5.6
St. Louis, MO–IL CMSA	4.5	12.2	NA	7.6

TABLE 1.4
Changing Patterns in Population versus Forest Area in Selected Northeastern States in 1900 and 2000 (U.S. Census, USDA Forest Service, Harvard Forest LTER, State GIS Agencies: MassGIS, NJDEP GIS, and PASDA)

State	Year	Population	Forest area (km²)	No. of people/km² of forest	Forest (hectares)/person
Connecticut	1900	910,000	8,560	106	0.9
	2000	3,282,031	17,835	184	0.5
Maine	1900	695,000	101,865	7	14.7
	2000	1,253,040	177,451	7	14.2
Massachusetts	1900	2,788,000	15,084	185	0.5
	2000	6,175,169	27,312	226	0.4
New Hampshire	1900	412,000	23,041	18	5.6
	2000	1,201,134	47,242	25	3.9
New Jersey	1900	1884,000	NA	NA	NA
	2000	8,414,350	9,510	885	0.1
New York	1900	7,283,000	24,463	298	0.3
	2000	18,196,601	74,698	244	0.4
Pennsylvania	1900	6,302,000	NA	NA	NA
	2000	12,281,054	68,796	179	0.6
Rhode Island	1900	430,000	1,953	220	0.5
	2000	990,819	3,468	286	0.4
Vermont	1900	344,000	14,675	23	4.3
	2000	593,740	46,959	13	7.9

coupled with the opening of the Erie Canal, expansion of railroad systems, and industrialization in eastern cities, led to natural reforestation in many parts of New England, New York, and Pennsylvania. This trend has been reversed in many states by population increases, expansion of highway

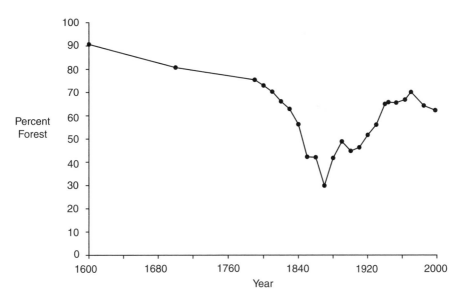

FIGURE 1.4 Changes in Massachusetts forest cover, 1600 to 2000 (Foster and Gould, 2005).

systems, and associated suburban development (Table 1.4). Second-growth forests are being cleared and farmland is being converted to residential, industrial, and commercial use at an alarming rate. The 300-year time series of forest cover in Massachusetts developed by scientists at the Harvard Forest clearly shows a trend that is typical for many other states (Figure 1.4).

1.5 SUMMARY

For thousands of years people have observed that the loss of forests and grasslands leads to overland flow, soil erosion, more frequent floods and droughts, and the degradation of water quality and aquatic ecosystems. Social, economic, and demographic changes during the 18th and 19th centuries led to dramatic changes in forest cover and land use in the northeastern United States. Population increases and accelerating urban and suburban sprawl from the 1950s to the present are reversing decades of natural reforestation and generating a familiar suite of ecological problems—as well as new challenges related to synthetic compounds and energy use. The formulation of effective management plans and implementation of mitigation and restoration practices depends upon a thorough understanding of the inherent complexity of watershed systems. Topography, geology, soils, climate, land use history, and contemporary resource use combine to influence the pathway and rate of water movement and the transport of sediment, nutrients, and contaminants.

In this book we attempt to build a bridge from the basic scientific information presented in Part I (Chapters 1 through 5) to the interdisciplinary research and research needs presented in Part II (Chapters 6 through 10). Decades of research and operational experience have substantially reduced, but can never eliminate, the uncertainty and complexity associated with watershed management. Nevertheless, this impressive body of literature should motivate 21st-century researchers and managers to emulate the examples of Marsh, Zon, Leopold, and others. They were willing to challenge conventional wisdom, apply deductive reasoning, and undertake empirical work that required great persistence and resourcefulness in order to understand, explain, and counteract the forces that were threatening the sustainability of forests, water, and, ultimately, human populations. Try to imagine the ecological condition of the northeastern United States had they not succeeded. Now, imagine the condition of the northeastern United States in 2105 and lead by example in research and management.

REFERENCES

Associated Press, Suburbs grow while big cities shrink, June 20, 2006.

Ayers, M.A., Kennen, J.G., and Stackelberg, P.E., Water quality in the Long Island–New Jersey coastal drainages, New York and New Jersey, 1996–98, Circular 1201, U.S. Geological Survey, Reston, VA, 2000.

Bell, M., The face of Connecticut: people, geology and the land, Bulletin 110, State Geological and Natural History Survey of Connecticut, Hartford, 1985.

Bennett, M.R. and Glasser, N.F., *Glacial Geology: Ice Sheets and Landforms*, John Wiley & Sons, Chichester, England, 1996.

Binkley, D., Patterns and processes of variation in nitrogen and phosphorus concentrations in forested streams, Technical Bulletin 836, National Council for Air and Stream Improvement, Research Triangle Park, NC, 2001.

Brady, N.C. and Weil, R.R., *The Nature and Properties of Soils*, 13th ed., Macmillan, New York, 2002.

Breault, R.F. and Harris, S.L., Geographical distribution and potential for adverse biological effects of selected trace elements and organic compounds in streambed sediment in the Connecticut, Housatonic, and Thames River basins, 1992–1994, Water Resources Investigations Report 97-4169, U.S. Geological Survey, Reston, VA, 1997.

Casey, G.D., Myers, D.N., Finnegan, D.P., and Wieczorek, M.E., National water quality assessment of the Lake Erie–Lake St. Clair basin, Michigan, Indiana, Ohio, Pennsylvania, and New York—environmental and hydrologic setting, Water Resources Investigations Report 97-4256, U.S. Geological Survey, Reston, VA, 1997.

Clark, G.M., Mueller, D.K., and Mast, M.A., Nutrient concentrations and yields in undeveloped stream basins of the United States, *J. Am. Water Resourc. Assoc.*, 36, 849–860, 2000.

Dudley, N. and Stolton, S., Running pure: the importance of forest protected areas to drinking water, World Bank/World Wildlife Fund Alliance for Forest Conservation and Sustainable Use, Washington, D.C., 2003.

Feldman, R.M., Coogan, A.H., and Heimlich, R.A., *Field Guide: Southern Great Lakes*, Kendall/Hunt, Dubuque, IA, 1977.

Fenneman, N.M., *Physiography of the Eastern United States*, McGraw-Hill, New York, 1938.

Foster, D.R. and Gould, E., HF013: Forest change and human populations in New England, Harvard Forest LTER data archive, 2005, http://harvardforest.fas.harvard.edu/data/archive.html; accessed July 2006.

Groschen, G.E., Harris, M.A., King, R.B., Terrio, P.J., and Warner, K.L., Water quality in the lower Illinois River basin, Illinois, 1995–98, Circular 1209, U.S. Geological Survey, Washington, D.C., 2000.

Harvard Forest LTER (Long-Term Ecological Research), http://harvardforest.fas.harvard.edu; accessed July 2006.

Kittredge, J., *Forest Influences*, McGraw-Hill, New York, 1948.

Marsh, G.P., *Man and Nature* (reprinted 1965), Lowenthal, D. (ed.), Harvard University Press, Cambridge, MA, 1864.

MassGIS, http//www.mass.gov/mgis; accessed July 2006.

McCobb, T.D. and LeBlanc, D.R., Detection of fresh ground water and a contaminant plume beneath Red Brook Harbor, Cape Cod, Massachusetts, 2000, Water Resources Investigations Report 02-4166, U.S. Geological Survey, Reston, VA, 2002.

Myers, D.N., Thomas, M.A., Frey, J.W., Rheaume, S.J., and Button, D.T., Water quality in the Lake Erie–Lake Saint Clair drainages, Circular 1203, U.S. Geological Survey, Reston, VA, 2000.

NJDEP GIS (New Jersey Department of Environmental Protection GIS), http://www.state.nj.us/dep/gis/; accessed July 2006.

PASDA (Pennsylvania Spatial Data Access), http://www.pasda.psu.edu/; accessed July 2006.

Peters, C.A., Robertson, D.M., Saad, D.A., Sullivan, D.J., Scudder, B.C., Fitzpatrick, F.A., Richards, K.D., Stewart, J.S., Fitzgerald, S.A., and Lenz, B.N., Water quality in the western Lake Michigan drainages, Wisconsin and Michigan, 1992–1995, Circular 1156, U.S. Geological Survey, Reston, VA, 1998.

Pielou, E.C., *After the Ice Age: The Return of Life to Glaciated North America*, University of Chicago Press, Chicago, 1991.

Plank, M.O. and Schenck, W.S., Delaware Piedmont geology, including a guide to the rocks of the Red Clay Valley, Special publication 20, Delaware Geological Survey, University of Delaware, 1998, http://www.udel.edu/dgs/Publications/pubsonline/SP20.pdf; accessed July 2006.

Plato (427–347 B.C.), *Timaeus and Critias*, H.D.P. Lee (trans.), Penguin Classics, Harmondsworth, Middlesex, 1977.

Randall, A.D., Hydrogeologic framework of stratified-drift aquifers in the glaciated northeastern United States (Regional Aquifer Analysis – Northeastern United States), Professional Paper 1415-B, U.S. Geological Survey, Reston, VA, 2001.

Richardson, H.W. and Gordon, P., U.S. population and employment trends and sprawl issues, in *Urban Sprawl in Western Europe and the United States*, Richardson, H.W. and Bae, C.-H.C. (eds.), Ashgate Publishing, Burlington, VT, 2004, pp. 217–235.

Skehan, J.W., *Roadside Geology of Massachusetts*, Mountain Press, Missoula, MT, 2001.

Steele, J., *Losing Ground: An Analysis of Recent Rates and Patterns of Development and Their Effects on Open Space in Massachusetts*, Massachusetts Audubon Society, Lincoln, MA, 1999.

Stein, S.M., McRoberts, R.E., Alig, R.J., Nelson, M.D., Theobald, D.M., Eley, M., Dechter, M., and Carr, M., Forests on the edge: housing development on America's private forests, General Technical Report PNW-GTR-636, http://www.fs.fed.us/projects/fote/reports/fote-6-9-05.pdf, USDA Forest Service, Washington, D.C., 2005.

Thomas, W.L., Jr., with the collaboration of Sauer, C.O., Bates, M., and Mumford, L. (eds.), *Man's Role in Changing the Face of the Earth*, University of Chicago Press, Chicago, 1956.

U.S. Census Bureau, http://www.census.gov; accessed July 2006.

Vitousek, P.M., Gosz, J.R., Grier, C.C., Melillo, J.M., Reiners, W.A., and Todd, R.L., Nitrate losses from disturbed ecosystems, *Science*, 204, 469–474, 1979.

Watt, M.K., A hydrological primer for New Jersey watershed management, Water Resources Investigations Report 00-4140, U.S. Geological Survey, Reston, VA, 2000.

Whitney, G.G., *From Coastal Wilderness to Fruited Plain: A History of Environmental Change in Temperate North America from 1500 to the Present*, Cambridge University Press, Cambridge, 1994.

Widmer, K., *The Geology and Geography of New Jersey*, Van Nostrand, Princeton, NJ, 1964.

Zimmerman, M.J., Grady, S.J., Todd Trench, E.C., Flanagan, S.M., and Nielsen, M.G., Water-quality assessment of the Connecticut, Housatonic, and Thames River basins study unit: analysis of available data on nutrients, suspended sediments, and pesticides, 1972–92, Water Resources Investigations Report 95-4203, U.S. Geological Survey, Reston, VA, 1996.

Zon, R., Forests and water in the light of scientific investigation, USDA Forest Service, Washington, D.C., 1927 [reprinted with revised bibliography from Appendix V of the Final Report of the National Waterways Commission 1912, Senate Document 469, 62nd Congress, 2nd Session].

2 Fundamentals of Hydrology

We forget that the water cycle and the life cycle are one.

Jacques Cousteau

2.1 INTRODUCTION

Water is essential for life in all forms and all places. Water is a prime mover of nutrients, sediment, and carbon in ecosystems. The quantity, timing, and quality of streamflow are direct and integrative reflections of watershed conditions above the point of observation. Therefore, it is not surprising that water often becomes a focal point for many environmental impact assessments and also predictions of future conditions.

It is imperative that we have an accurate understanding of hydrologic science because as Satterlund and Adams (1992) note:

> ... much of watershed management is still characterized by the overthrust of ideas beyond their time ... hence public pressures for certain types of wildland use (or non-use) continue to be based on some theories that are correct, some that are applicable in some degree, and some that are completely erroneous.

Hydrology is an interdisciplinary field that builds upon physics, chemistry, biology, and ecology and also includes elements of meteorology, plant physiology, soil science, and engineering. The integration of knowledge and concepts from the basic sciences is a prerequisite for mastery of hydrologic science that—along with civil and environmental engineering, sociology, political science, and economics—forms the foundation of watershed management.

2.1.1 ORDERLY YET VARIABLE PROCESSES

Broad seasonal patterns are evident when climatic and streamflow data are averaged over months and years. While seasonal patterns are relatively clear and predictable, spatial and temporal variations are the rule rather than the exception when we seek to describe and understand watershed systems at shorter time scales (hours, days, or weeks). Hence, in any given year, the range of variation in precipitation, solar radiation, air temperature, evaporative processes, and streamflow is often as important as the seasonal or annual totals and averages. For example, one large stormflow event may move more sediment than the total transported by all the other storms during the remainder of the year. Under other circumstances, snowmelt may release a volume of water in a single day that is equivalent to several months of rainfall later in the year. Unusually wet or dry conditions during the growing season can substantially alter vegetation and biogeochemical cycling. So, although individual hydrologic processes may be predictable and well understood, their complex interactions require careful measurements and detailed analyses.

Watershed characteristics (including soils, topography, geologic setting, vegetation, and climate) have a fundamental influence on the quantity, timing, and quality of streamflow. They are the combined source of natural variation in water yield and quality. The extent to which human activities change these characteristics is of fundamental interest and importance. As human activities become more pervasive and extreme, their "signature" or "ecological footprint" becomes more apparent. As a general rule, larger quantities of lower quality water flows more rapidly to the watershed outlet as forests are converted to other uses.

This chapter has two goals. First, to define the terms and describe the hydrologic processes that are embedded in all chapters of this book and the scientific literature upon which it is based, and second, to provide a foundation for critical thinking, deductive reasoning, professional judgment, and decision making. The chapter begins with a description of the water balance and the energy balance and includes examples of their linkage and interaction. This is followed by a sequential discussion of hydrologic processes that follows the path of water movement through ecosystems—precipitation (rain and snow), water movement and storage in soils, evaporative processes, streamflow, and open channel hydraulics. The interested reader should also be aware of the wide range of comprehensive textbooks available for further study (e.g., Brooks et al., 2003; Chang, 2003; Dingman, 1994; Dunne and Leopold, 1978; Gordon et al., 1992; Satterlund and Adams, 1992).

2.2 THE WATER BALANCE

The water balance or water budget (Equation 2.1) summarizes the relationship between the principal components of the hydrologic cycle. It is deceptively simple.

$$P - ET - Q \pm \Delta S \pm L = 0, \qquad (2.1)$$

where

P = precipitation;
ET = evapotranspiration;
Q = water yield (streamflow + groundwater recharge);
S = storage (Δ signifies "change"); and
L = leakage, in or out of the watershed.

When written in an expanded, working form (Equation 2.2), the interconnections and multiple combinations of inflow, outflow, and storage become much more evident:

$$P - (I + E + T) - (Q_{OF} + Q_{SS} + Q_{GW})$$
$$\pm \Delta(S_{soil} + S_{SWE} + S_{plants} + S_{streams} + S_{lakes} + S_{wetlands}) \pm L = 0, \qquad (2.2)$$

where

P = precipitation (rain, snow, or mixed);
I = interception;
E = evaporation;
T = transpiration;
Q_{OF} = overland flow;
Q_{SS} = shallow subsurface flow (in the root zone);
Q_{GW} = groundwater flow;
S_{soil} = soil water storage;
S_{SWE} = snow water equivalent (storage in snowpack);
S_{plants} = water stored in plant tissue;
$S_{streams}$ = water stored in streams and rivers (in transit);
S_{lakes} = water stored in lakes and ponds;
$S_{wetlands}$ = water stored in wetlands; and
L = leakage.

Whatever the magnitude and direction of change in water balance components, the law of conservation of mass reminds us that their algebraic sum must always equal zero—at all times and all places. In other words, compensatory changes maintain the water "balance." As later sections of this chapter will describe, there are multiple feedback loops that connect *ET*, *S*, and *Q*. John Muir (1911) astutely observed that "When we try to pick out anything by itself, we find it hitched to everything else in the Universe." Since water is the universal solvent and the prime mover of dissolved and suspended material, Muir's observation has important implications for biogeochemical cycling, soil erosion, and sediment transport.

It is also important to consider the temporal aspects or nature of hydrologic processes. Although the water balance suggests an orderly hierarchy or sequence of processes—this is not realistic. In nature, hydrologic processes occur simultaneously, not in series or in some consistent order. Consider the following examples as a preview of the rest of this chapter.

- During the growing season, evapotranspiration (evaporation + transpiration + interception) and gravity drainage *simultaneously* decrease soil water content (θ) on sloping sites. Because streamflow is also a function of soil water content, it decreases as well until the next rain event recharges the system.
- During the snowmelt period, the water that was stored in the snowpack flows into soil moisture storage *and* simultaneously increases the rate of subsurface flow. Under some conditions (e.g., frozen or saturated soil) snowmelt also causes overland flow. Snowmelt, subsurface flow, and overland flow combine to increase wetland water levels and streamflow. The process continues until the snowpack is depleted by melt or refreezing delays the process.
- At the end of the growing season, vegetation becomes dormant at the same time that air temperatures and evaporative demand decrease. Fall rains simultaneously recharge soil moisture and increase rates of subsurface flow, wetland water levels, and streamflow.

All three examples highlight a key theme—a change in one water balance component leads to compensatory changes in others.

2.2.1 THE WATER YEAR

In order to compare different watersheds (with different topography, vegetation, soils, or land uses) or different years (e.g., wet, dry, median) on the same site, hydrologists often shift the period of analysis from the calendar year to the "water year." A water year is any continuous 12 month period so chosen to minimize the *net* change in storage (ΔS). In many areas, wet conditions occur at relatively consistent times every year. In New England, for example, the rainy period during and after leaf fall in October and November or the snowmelt period in February and March are potentially useful start/end points for the water year. Because the accumulation and melting of snow can be highly variable from year to year, and fall rains are more consistent, selecting October or November typically yields more reliable results. Although the day-to-day (diurnal) changes in water balance components can be substantial during the 12 month period, the *net* annual change in storage can be minimized by starting and ending at consistently wet times of year ($\Delta S \rightarrow 0$). Hence the annual water balance can be simplified to three terms: $P - ET - Q \cong 0$. It follows that accurate measurements of two terms can be used to reliably estimate the third (one equation, one unknown; $P - Q \cong ET$ or $P - ET \cong Q$) for the water year. Daily data are required for tracking and tracing the movement of water *during* the year, solving Equation 2.2 an in hourly or daily time step.

2.3 THE ENERGY BALANCE

The water balance and energy balance are inextricably linked. Solar energy drives the hydrologic cycle (e.g., heating, cooling, freezing, thawing, evaporation, photosynthesis, transpiration, snowmelt, atmospheric circulation, etc.) while, at the same time, water (as water vapor and precipitation) becomes an important vehicle for energy exchange between the Earth's surface and the atmosphere (Brooks et al., 2003; Satterlund and Adams, 1992).

The energy balance equation summarizes the pathways of energy flow for any "active" surface—a plant canopy, soil surface, snowpack, etc.—wherever energy exchange or transformation takes place (Satterlund and Adams, 1992):

$$R_n + H_s + LE + G + M = 0, \tag{2.3}$$

where

R_n = net radiation (solar and terrestrial);

H_s = sensible heat exchanged with the atmosphere;

LE = latent heat exchanged with the atmosphere;

G = heat exchanged with the ground or vegetation; and

M = metabolic heat of photosynthesis or respiration.

The net change in G and M is negligible over the long run, leaving R_n, H_s, and LE as the primary components of the energy balance.

Because all energy is ultimately derived from the sun, and energy output cannot exceed input (i.e., the law of conservation of mass and energy), the energy balance can be simplified to $R_n = H_s + LE$. This simplified form shows that solar radiation input and terrestrial radiation output are counterbalanced by energy transfer via convection and latent heat exchange.

Sensible heat transfer (i.e., heat that you can feel), also called convection (H_s), is driven mainly by temperature gradients between the surface and the atmosphere. It is also affected by wind velocity and humidity. Latent heat exchange (LE) is the energy associated with changes in the physical state of water:

- Water vapor to liquid water—condensation;
- Liquid water to water vapor—evaporation;
- Liquid water to ice—solidification (freezing);
- Ice to liquid water—fusion (melting);
- Ice directly to water vapor—sublimation; and
- Water vapor to ice—sublimation (riming).

Like convection, latent heat exchange can occur in either direction—transferring energy *to* the active surface from the atmosphere or *from* the active surface to the atmosphere. Exchange is the operative word. Melting ice or snow (latent heat of fusion), evaporation (latent heat of vaporization), and sublimation all transfer energy *from* the active surface to the atmosphere since all three changes in state require heat. Ice or soil frost formation, condensation, and riming (hoarfrost) transfer energy *to* the active surface from the atmosphere since all three changes in state release heat.

2.3.1 SOLAR RADIATION

The slope, aspect, and juxtaposition of terrain features influence the intensity ("flux density," in W/m^2) of solar radiation. In addition, latitude and time of year influence sun angle and day length in predictable ways. Because the sun rises in the east, traverses the southern sky, and sets in the west, a north-facing slope may only receive indirect diffuse radiation (indirect radiation that is scattered by clouds, dust, or aerosols), while a south-facing slope is exposed to intense, direct-beam ("clear sky") solar radiation throughout most of the day. Slope steepness also influences the intensity of direct beam solar radiation. This can be visualized by holding a flashlight parallel to a tabletop and slowly raising a book or pad of paper into the light beam. The book or pad is, in effect, a planar south slope. When the surface is held at a low angle (slope) with respect to the tabletop, the light beam will illuminate a large elliptical area at a low light intensity. As the surface approaches a vertical orientation (perpendicular to the tabletop), the light beam illuminates a small, distinct circular area at a high light intensity. Since the total output of light from the flashlight is constant, the radiant energy per *unit* area varies in direct proportion to the slope (the angle formed with the tabletop). The same phenomenon occurs at the landscape scale with the sun and terrain features. In hilly or mountainous areas, the juxtaposition of terrain features (e.g., a high mountain shading adjacent hills for part of each day) introduces even more variation in radiant energy loading.

Latitude affects day length and the altitude of the sun as the Earth completes its annual orbit. As latitude (angular distance from the Equator, where day length and sun angle remain relatively constant) increases, the difference between growing season and dormant season day length also increases (Table 2.1). The maximum sun angle above the horizon (solar noon) also changes substantially with time of year. In late June the sun may be almost directly overhead, while in late December its zenith is much lower in the southern sky. The combined result of differences in day length and associated differences in sun angle (recall the simple flashlight experiment) has a substantial effect on total radiation (watts) and peak intensity (radiant flux density, W/m^2). Atmospheric conditions on any given day (i.e., clear, partly cloudy, hazy, overcast, etc.) also affect the proportion of direct beam and diffuse solar radiation.

When it reaches the active surface, solar (also called shortwave) radiation is either reflected, transmitted, or absorbed as a function of surface characteristics and material properties. For example, the reflectivity of new (clean, dry, crystalline) snow may range from 0.8 to 0.95 (80% to 95%), while pavement sealed with flat black paint may have a reflectivity as low as 0.02 (2%). While new snow transmits or absorbs only 5% to 20% of incident radiation, an adjacent area of pavement will absorb up to 98% of this energy and increases in temperature accordingly. The reflectivity of other surfaces (e.g., water, soils, forest canopy, herbaceous plants, streambeds, etc.) may range from 0.05 to 0.45. The wide variation in reflectivity means that land cover and land use—by way of vegetation and surface characteristics—can have a substantial influence on the energy balance and, as a result, the water balance via evaporative processes.

The effect of seasonal differences in solar radiation, along with the influence of terrain attributes, is also evident in the wide variation in depth and persistence of the snowpack in mountainous areas. Snow may be found in sheltered, north-facing valleys weeks, even months, after it has melted on south slopes. It has long been known that vegetative cover influences the snowmelt process (Table 2.2). Similar differences in microclimate influence growing season conditions (air temperature, soil temperature, evapotranspiration, and soil moisture depletion) as well. As a result, the quantity and timing of streamflow in headwater streams also varies significantly in relation to slope, aspect, shading, and any other influences on microclimate.

The type, age, vertical structure, height, and density of vegetation all have a substantial influence on radiant energy exchange. Consider the differences in direct and diffuse solar radiation beneath dense, mature stands of coniferous or deciduous trees over the course a year. While the leaf area,

TABLE 2.1
Variation in Day Length (hours) in Relation to Latitude and Time of Year in the Northeastern Area

Latitude	Jan	Feb	Mar	Apr	May	Jun	Jul	Aug	Sep	Oct	Nov	Dec
50° N	8.5	10.0	11.9	13.7	15.4	16.3	16.0	14.5	12.7	10.8	9.1	8.2
40° N	9.6	10.7	12.0	13.2	14.4	15.0	14.8	13.8	12.5	11.2	10.0	9.4
30° N	10.4	11.2	12.0	12.8	13.7	14.0	13.9	13.3	12.4	11.5	10.7	10.2
20° N	11.0	11.5	12.0	12.6	13.1	13.3	13.2	12.8	12.2	11.8	11.2	10.9

Adapted from Dunne and Leopold (1978).

TABLE 2.2
The Last Date with Snow Present in 1908
at the Russian Imperial Agronomic Institute

In fields, clearings, and open places	April 22
In young, open stands	April 24
In old, open stands on south slopes	April 26
In birch stands	April 29
In pine stands	May 6
In spruce stands	May 15

Zon (1927:37).

color, and density of the evergreens remain relatively constant (shading the forest floor beneath the "active" surface), the deciduous stand canopy is leafless during the dormant season, of variable density during the spring and fall transitions, and dense and layered during the 3 to 4 month growing season. The active surface—where radiant energy exchange occurs—shifts between the forest floor (under the leafless canopy during the dormant season) and the canopy (in full leaf during the growing season), affecting air and soil temperature and evaporation rates in the process. It also affects the pathway of evaporative loss—from the soil during the dormant season versus through the plants during the growing season. Recall the obvious differences in air temperature that you experience when walking along a tree-lined street with regular intervals of sun and shade on a very hot summer day. The radiant energy loading on the tree crowns and the sunny sections of the sidewalk is the same—the active surface alternates between the tree crowns (casting shade) and the sunlit sidewalk.

2.3.2 TERRESTRIAL RADIATION

Terrestrial (also called long wave) radiation is absorbed or emitted in relation to material properties, surface area, and temperature of the object or surface. The Stefan-Boltzmann equation can be used to calculate the flux of terrestrial radiation (Satterlund and Adams, 1992). It quantifies the relationship between elapsed time and physical properties (density, color, texture, etc.), surface area, and temperature of an object and the amount of terrestrial radiation that it absorbs or emits (energy *exchange*).

2.4 PRECIPITATION

Quantifying the type, timing, duration, and frequency of precipitation is the first step in anticipating and predicting the quantity, quality, and timing of streamflow. While summary statistics such as annual and monthly totals are important to assess general patterns and trends, they often do not capture the event-based relationship of water balance components. For example, 5 cm of precipitation occurring in a day-long event in mid-May, a 30 min July thunderstorm, a 2-hour November storm, a snowstorm in mid-January, or a rain-on-snow event in early April would have substantially different effects on soil water content, evapotranspiration, and streamflow response.

Three conditions must be met in order for precipitation to occur: (1) vapor pressure reaching saturation, (2) the availability of condensation nuclei (dust particles, ice crystals, or aerosols), and (3) a water droplet or ice crystal mass that exceeds the buoyant forces holding them aloft in the atmosphere. Typically the first condition is met when air is lifted and cooled in the atmosphere (approximately 1°C per 100 m or 3 to 5°F per 1000 ft). This vertical displacement may be caused by a warm front, a cold front, convection, or the orographic influence of mountains (or islands).

Frontal precipitation originates from differences in temperature, relative humidity, and density at the interface of two air masses. Warm, moist air flows up and over colder air when driven by atmospheric circulation. In many cases, high elevation cirrus clouds form at the leading edge of a warm front, followed by a progressive lowering of cloud cover (ceiling) as the front passes a given point. The distance from the leading to the trailing edge (marked by low stratus clouds) may be 150 to 500 km, depending on the temperature and humidity differences and wind speed at various levels of the atmosphere.

In contrast, the passage of a cold front drives a wedge of air underneath ambient warm, moist air. This may occur in a more compact space and with greater rapidity, contributing to more rapid lifting and cooling and greater precipitation rates. In extreme cases, a "squall line" of cumulonimbus clouds marks the interface between the air masses. Clouds may exhibit an "anvil top," as rapid lifting and cooling lofts air into the upper atmosphere (12,000 to 17,000 m) where very high winds shear the cloud tops forward of the main body. As a result of these differences in formation and structure, cold front storms tend to produce greater rainfall intensities than warm front storms.

Convective precipitation occurs as a direct result of surface heating, evapotranspiration, and the rapid vertical movement of warm, moist air. Columns of turbulent air lead to the formation of cumulus and cumulonimbus clouds and once again demonstrate the direct connections between water and energy balances. The intensity of solar radiation and the supply of water (in soils, lakes, streams, and wetlands) influence rates of surface heating, evaporation, and cloud formation. Along with the rapid passage of cold fronts, convection can lead to the formation of hail. Hailstones form when water droplets are carried rapidly aloft, gaining mass and freezing in the process. Hail may make several trips up and down through the cloud formation, each time accumulating another layer of ice. When the mass of the hailstone becomes larger than the buoyant force, or it is caught in a downdraft, it falls and strikes the Earth's surface, where it may damage property and vegetation, especially agricultural crops.

Orographic precipitation (rain or snow) occurs when an air mass flows up and over a topographic obstruction (a mountain or island). Again, the process of lifting and cooling leads to the formation of rain or snow. In many cases, the amount of precipitation is directly proportional to elevation. Over the long term, differences in the amount and type of precipitation, as well as systematic variation in the temperature regime can substantially influence the type and condition of vegetation (Kudish, 2000). Simply put, the microclimate becomes wetter and colder with increasing elevation. In the Catskill Mountains, for example, the oak–maple–white pine forests at lower elevations (up to about 450 m) transition to northern hardwoods (sugar maple–beech–yellow birch) at midelevations (about 450 to 900 m), and finally to a balsam fir, red spruce, and paper birch forest (more than 900 m) that is more like central Ontario or Quebec than the mid-Hudson River Valley. In the Adirondack Mountains, the more northerly latitude and higher elevations (up to 1580 m) support alpine vegetation at the summits.

2.4.1 SNOW

As noted in the earlier discussion of the energy balance, snow is unique among naturally occurring substances. It may range from delicate, dendritic crystals to more compact, granular forms. Snow is a very efficient reflector of solar radiation, yet nearly a "black body" (a perfect absorber and emitter) of terrestrial radiation. The ability to limit solar radiation gains while maximizing emissions of terrestrial radiation (i.e., heat loss) is a principal reason why snow can persist throughout the winter months.

The total amount of water stored in the snowpack is referred to as the snow water equivalent (SWE). Snow water content (SWC), a portion of the SWE, is the small quantity of liquid water held in the pore space of the snowpack, up to about 3% by volume. (Because most of the pore space is filled with air, snow can be an effective insulator and protective cover for many plants and animals [e.g., ruffed grouse, small mammals, forest regeneration, and herbaceous vegetation]). In areas with a persistent snowpack, a layered structure develops from the accumulated series of winter storms (Figure 2.1).

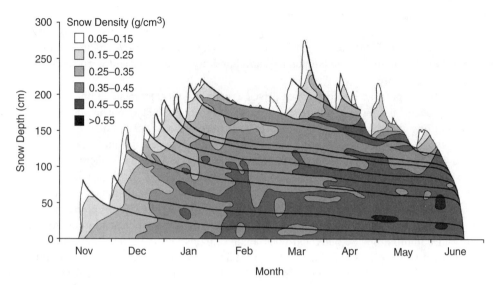

FIGURE 2.1 Variation in snow depth and density versus time at Lower Meadow, Central Sierra Snow Laboratory, California, during the winter of 1952 to 1953. (Adapted from *Snow Hydrology*, U.S. Army Corps of Engineers, 1956.)

Snow undergoes a process called metamorphism as wind, settling, and freezing and thawing gradually transforms prismatic crystals into rounded ice pellets. As the facets on snow crystals are degraded by metamorphism, the reflectivity of solar radiation decreases, energy absorption increases, and the process accelerates. Heat is transferred throughout the snowpack in response to temperature gradients until it reaches a uniform upper limit temperature of 0°C and the SWC is approximately 3%, held in thin liquid films on the remaining ice crystals. While radiant energy exchange is the dominant process affecting metamorphism and melt, latent heat exchange (condensation and solidification), convective transfer from the lower (about 100 m) portion of the atmosphere to the snowpack, conduction of energy from the soil, and advection of energy from rainfall all may have a significant influence. The darkening shades of gray in Figure 2.1 show the gradual, unidirectional increase in density resulting from the metamorphism process.

The accumulation, metamorphism, and melting of snow in many parts of the northeast is significant because of the time lag it creates between precipitation, soil moisture recharge, and streamflow. What would have been individual rainfall events at other times of the year, perhaps as many as 10 to 15, steadily contribute to the SWE over many months then are combined by the snowmelt process. In many cases, snowmelt dominates the annual hydrograph. In many public water supply systems it is needed to recharge soil moisture and groundwater, generate streamflow, and, ultimately, fill reservoirs. Hence an "open" winter without significant snowfall may forecast a summer drought and water supply shortage. In contrast, a rain-on-snow event can lead to widespread flooding. A relatively commonplace (high-frequency) maritime tropical rain storm on a deep snowpack that is at or near 0°C can be the catalyst for a low-frequency (large and damaging) flood event.

In addition to the SWE of the snowpack, the presence or absence of soil frost can have a substantial influence on the volume, timing, and rate of spring streamflow. If, for example, the snowpack forms early (i.e., November), it effectively insulates the soil surface and prevents the formation of soil frost. If, however, the soil is wetted by late fall rains then exposed to rapid and sustained decreases in temperature, the soil surface can freeze and drastically reduce infiltration capacity. This condition is more common on agricultural or developed areas than in forests, where the litter layer, organic matter, and large pores promote rapid water movement through the soil surface. If the frozen soil is covered (insulated) by subsequent snowfall, it can persist until spring. Snowmelt under these conditions can lead to a rapid and virtually complete conversion of snowmelt to streamflow.

2.5 WATER MOVEMENT AND STORAGE IN SOILS

The soil is the nexus for hydrologic processes. The characteristics and condition of a soil are the primary determinants of the pathway(s) and rate of water movement. A soil profile is described with a series of horizons and by their associated physical and hydraulic properties. In forest ecosystems, leaves and other organic material form a litter layer at the surface. Decomposition by invertebrates, microbes, and physical and chemical processes leads to the development of an organic (O) horizon (also referred to as humus and leaf mold in the early literature). The thickness of the litter layer and O horizon is closely linked with the characteristics of the organic material as well as the microclimate of the forest floor (cool and moist versus warm and dry). For example, maple or birch litter on a south slope would decompose much more rapidly than an equal amount of oak or hemlock litter (with higher lignin content and lower pH) on a north slope. For decades, forest hydrologists and soil scientists have emphasized the importance and ecological value of the litter layer and O horizon (Sartz, 1969). Even earlier, George Perkins Marsh (1864) accurately described their critical functions as follows:

> The vegetable mould, resulting from the decomposition of leaves and of wood, carpets the ground with a spongy covering which obstructs evaporation from the mineral earth below, drinks up rains and melting snows that would otherwise flow rapidly over the surface and perhaps be conveyed to the distant sea, and then gives out, by evaporation, infiltration, and percolation, the moisture thus imbibed. The roots, too, penetrate far below the superficial soil, conduct water along their surface to the lower depths to which they reach, and thus serve to drain the superior strata and remove the moisture out of the reach of evaporation.
>
> *Man and Nature, 1965:145*

   Beneath the O horizon is the A, the uppermost mineral horizon. It typically ranges in thickness from about 10 to 30 cm. It is the zone of mixing (by small mammals and the decomposition mechanisms described earlier) of mineral soil from below and organic matter from above. Referred to as "topsoil" in common parlance, the A horizon often is the most fertile part of the profile, as evinced by the high concentration of rooting. In some parts of the northeastern United States, principally the glacial tills, a thin (5 to 10 cm), light gray layer—the E horizon—forms when humic acids and other chemical reactions mobilize and leach iron and dark-colored soil colloids. Beneath the A (or A and E) horizon is the zone of accumulation—the B horizon. Clay particles, other colloids, and iron oxides fill some of the soil pores and typically impart a rusty or reddish brown color to this layer. The influence of this abrupt change in soil properties (i.e., increased density and decreased porosity and permeability) on water movement and storage will be discussed in more detail below. The C horizon, or subsoil, is the portion of the profile that is least affected by soil forming processes. In general, it has a higher bulk density and a lower concentration of rooting and organic matter. In the parts of the northeast with thin (less than 2 m) glacial tills, the C horizon may lie directly on bedrock. In other areas with deeper soils it may transition to the D horizon—the parent material. This boundary may be indistinct, but it can be functionally defined by the maximum rooting depth (approximately 2 to 3 m) of mature trees.

   When forests are converted to agricultural, residential, or commercial uses, changes in the soil profile can be considerable. Tillage of agricultural soils mixes and homogenizes the O and A horizon, forming a plow layer or AP horizon. In pastures with high animal densities, the combination of grazing, soil compaction, and mixing, the addition and transformation of nutrients in manure, and accelerated decomposition of organic matter can rapidly deplete the O horizon and alter the favorable properties of the A horizon. In developed areas, excavation and earthmoving operations can dramatically alter the native soil. Even if the topsoil (litter layer, O, and A horizons) is stockpiled during construction and spread back on the completed site, the mixing and compaction caused by heavy equipment substantially alters soil physical and hydraulic properties. The largest pores are the least resistant to soil compaction.

The legacy effects of land use should also be considered. The hydrologic differences between the soils in a virgin forest (never harvested or altered), primary forest (cut, perhaps repeatedly, but never converted to agricultural use), and secondary forest (established on abandoned agricultural land) can influence or help to explain differences in vertical rooting distribution, nutrient cycling, soil frost, and streamflow response.

2.5.1 Soil Physical and Hydraulic Properties

The quantification of soil physical and hydraulic properties begins with the determination of bulk density (oven dry weight ÷ undisturbed sample volume) and soil texture (the relative proportions of gravel [coarse fraction], sand, silt, and clay particles). Bulk density is used to calculate porosity—the total void space in a soil. The porosity of a soil indicates its ability to store water. There are 11 textural classes designated by the U.S. Department of Agriculture (USDA) Natural Resources Conservation Service (ranging from homogeneous sand, silt, or clay soils, to a relatively uniform mixture, loam, to other admixtures such as silty clay loam, loamy sand, etc.). Sand particle diameters range from 0.05 to 2.0 mm, with silts from 0.002 to 0.05 mm, and clays less than 0.002 mm. The bulk density and porosity of soils varies over a very limited range with respect to texture (density from about 1.0 to 1.6 g/cm³; porosity from about 40% to 60% by volume). In contrast, the median particle size and corresponding pore diameter as well as the number of particles and the internal surface area per unit volume vary by orders of magnitude (Table 2.3). The combined influence of pore diameter and particle surface area in relation to water movement through soil is discussed below.

Soil structure, a qualitative description of the shape and arrangement of soil aggregates, also provides helpful circumstantial evidence about water movement and storage. Soil structure may be granular, platy, columnar, blocky, massive, etc.—each indicating relative rates and directional differences in water movement. A uniform granular structure allows water to flow in any direction, whereas a columnar structure promotes vertical flow and a platy structure leads to horizontal or lateral flow. The combination of porosity, texture (particle size distribution), and structure indicate the size of soil pores, the surface area of soil particles, conduciveness to plant growth, and, taken together, the water holding capacity and permeability of the soil profile or horizon.

2.5.2 Soil Water Potential (Energy)

Traditionally water storage in soils has been described in three states or conditions: gravitational water (between saturation and field capacity), plant available water (between field capacity and

---

TABLE 2.3
A Comparison of Physical Properties for Various Soil Textures

Soil properties	Soil texture			
	Sand	Sandy loam	Loam	Silty clay loam
Bulk density (g/cm³)	1.6	1.5	1.45	1.39
Porosity (%)	40	44	45	48
Mean particle diameter (mm)	2.0	0.67	0.42	0.12
Approximate number of particles/cm³	125	3,320	13,500	579,000
Total surface area (cm²/cm³)	16	47	75	262
Permeability (m/day)	15.2	3.0	0.6	0.15
Matric potential at 50% of saturation (bars)	0.2	0.6	0.8	7.5

wilting point), and hygroscopic water (a very thin film of water tightly held on soil particles under dry conditions). Field capacity is variously defined as the water content (1) when gravity drainage stops, (2) 2 to 3 days after a soaking rain, or, most accurately, (3) when the rate of water movement becomes negligibly small (e.g., about 2 mm/day). It is typically measured at 0.33 bars (337 cm H_2O) in the laboratory using a device called a pressure plate. Undisturbed soil cores are saturated, then subjected to successively higher pressures (0.1, 0.33, 1, 3, 5, and 15 bars) to simulate the effect of increasing suction (negative pressure). The water released from the sample at each pressure level is accurately measured to determine the volume of water remaining in the soil sample (alternatively, soil water content is measured directly; [wet weight – oven dry weight]/sample volume). The wilting point is defined as the water content when plants can no longer extract water from the soil. It is measured at 15 bars (15,300 cm H_2O) in the laboratory. In the field, however, wilting point varies in relation to vegetation type and is indicative of physiological differences. For example, the wilting point of aspens or cottonwood (*Populus* spp.) of –13 bars contrasts with the –23 to –25 bars value for many oaks (*Quercus* spp.). This explains, in part, their relative success on wet sites and dry sites, respectively.

In general terms, the concepts of field capacity and wilting point are helpful in understanding the interplay between water storage, the rate of water movement, and water availability for evaporative processes. Yet in order to accurately describe key phenomena, a more comprehensive, physically based description of the system is needed. It is important to understand the mechanisms across the continuous spectrum from saturated, unsaturated, and air-dry conditions. In order to quantify and interpret the movement and storage of water it is necessary to quantify its energy status at different points in the soil profile. "Potential" is short for potential energy—or energy due to position. (Because rates of water movement in soils are comparatively slow, kinetic energy—energy due to motion—is negligibly small.) Differences in soil water potential (energy) generates a gradient that induces water movement from ("wet") areas of high energy to ("dry") areas of lower energy. When soil water potential becomes uniform, the hydraulic gradient approaches zero. As the driving force for water movement decreases, flow slows then ceases. Inputs of rain or snowmelt, or outputs via evaporation or water use by plants lead to localized changes in water content and soil water potential. Any difference or change in soil water potential once again generates a hydraulic gradient and induces water movement.

Soil water potential (ψ_w), is the sum of the gravitational potential (ψ_z), pressure potential (ψ_p), matric potential (ψ_m), thermal potential (ψ_t), and solute or osmotic potential (ψ_s). Thermal and solute potential are typically small unless permafrost, extreme surface heating, fertilization, or other conditions that produce substantial temperature or chemical concentration gradients are present. In most other cases, thermal and solute potential are too small to measurably contribute to soil water movement. As a result, soil water potential is approximated by the hydraulic potential (ψ_h), the sum of the gravitational potential and matric potential or the gravitational potential and pressure potential. Gravity is always present, so gravitational potential is always a contributor to hydraulic potential and hydraulic gradients. The pressure potential is only operative when the soil is saturated. This is a relatively rare condition in forests, but may become more common when forests are converted to other uses (e.g., agricultural or residential use). At all other times when the soil is unsaturated, matric potential or matric suction is generated. Matric potential or suction is subject to the most variation in relation to water content (θ); it is caused by capillary and adsorptive forces.

Capillary suction in a soil is related to the pore size distribution (matric suction is inversely proportional to pore diameter) which, in turn, is related to soil texture (particle size distribution), structure, and land use history. The adsorption of water to soil particles is directly related to the total surface area of the soil particles (Table 2.3). Hence fine textured soils (silts, clays, silt loams, clay loams, etc.) with small particles, small pores, and large surface areas exert more matric suction per unit volume than coarse textured soils (sands and sandy loams). Since soils are unsaturated most of the time, matric potential or suction is usually an important, if not dominant, determinant of the direction and rate of water movement.

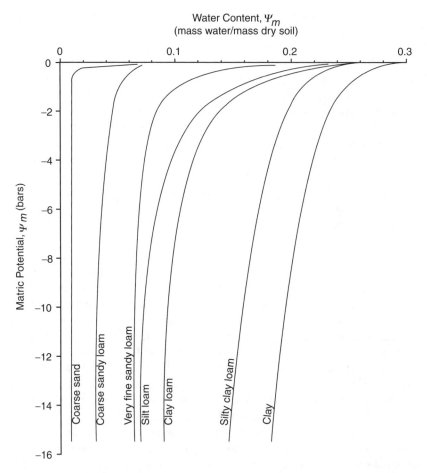

FIGURE 2.2 Soil water characteristics curves—matric potential (suction) versus water content—for a range of soil textures. (From Hanks and Ashcroft [1980], redrawn with kind permission, Springer Science and Business Media; originally from Taylor and Ashcroft, *Physical Edaphology*, W.H. Freeman, San Francisco, 1972.)

The relationship between soil water content (θ) and matric potential (suction) is described by the soil water characteristic curve (also called the soil water retention curve or soil water release curve) (Hanks and Ashcroft, 1980; Hillel, 1982) (Figure 2.2). The data series (water content at 0.1, 0.33, 1, 3, 5, and 15 bars) gathered with the pressure plate apparatus described earlier is used to derive the curve(s). Figure 2.2 clearly shows the expected variation with soil texture—particle and pore size distribution. Coarse textured soils with large pores drain rapidly, while fine textured soils offer much greater resistance at a given suction and retain more water. Furthermore, there is a systematic and orderly progression of water retention properties. Each textural class has a "characteristic" curve; every soil series or horizon has a characteristic curve that represents its unique features.

2.5.3 PERMEABILITY

The flux of water through soil (or any porous media) is represented by Darcy's law (flux equals the volume per unit time moving through a unit area). Like other governing equations from physics and chemistry—Fourier's law for heat flux, Ohm's law for electricity, and Fick's law for chemical diffusion—it is comprised of a gradient (driving force) and a material property that describes the flux or flow of water, heat, or electricity. In Darcy's law (Equation 2.4), the driving force is the difference in soil water potential between two points—the hydraulic gradient. The material property is saturated hydraulic conductivity, also called permeability. It is a function of the size, shape, and arrangement of soil pores. (Permeability and porosity should not be used interchangeably because one refers to flow and the other to storage.)

$$q = K_s(\Delta\psi_h/L),\tag{2.4}$$

where

q = flux of water [(cm^3/hr)/cm^2];

K_s = saturated hydraulic conductivity [(cm^3/hr)/cm^2];

$\Delta\psi_h/L$ = hydraulic gradient; difference in hydraulic potential/length of the flow path [cm/cm]; and

$\psi_h = \psi_m + \psi_z$ = matric potential + gravitational potential.

Consider the analogy to electrical conductivity. The amount of electricity that a wire can transmit or conduct is a function of the material from which it is manufactured (i.e., copper, aluminum, brass, steel, etc.) and its diameter. As a result, a large diameter pure copper wire will conduct more electricity than a small diameter wire manufactured from an inferior alloy when connected to the same power source (driving force). Similarly the large (about 1 mm) interconnected pores of coarse sand will conduct more water than a clay in which the largest pores are 0.001 mm in diameter. In other words, the frictional resistance to water flow in clay may be two or more orders of magnitude greater than coarse sand when subjected to the same hydraulic gradient (driving force). Referring again to Table 2.3, the difference in the number of particles and collective surface area helps to explain the difference in permeability for the sand and the clay, about 15 and 0.15 m/day, respectively. Less obvious site-specific differences in soil texture and structure (particle and pore size distribution) help to explain the observed variation between different soils.

The saturated hydraulic conductivity of a soil sample or soil horizon can be determined in the laboratory or the field (Hillel, 1982). In the laboratory, an undisturbed soil core is enclosed in a column between two porous ceramic plates. The sample is saturated, then the driving force is held constant and the volume of water that passes through the sample per unit time is carefully measured. In the field, adaptation of this constant-head permeameter is used for *in situ* measurements. Unless a soil is unusually homogeneous, both techniques can yield highly variable results. This is caused by small-scale differences in pore structure. For example, a worm hole or an old root channel may be present in one area and absent in another. These macropores (also called soil pipes or preferential flow paths) can short-circuit the small pores in the soil matrix and substantially increase effective permeability. Because water finds the path of least resistance, a single 1 cm macropore may transmit more water than thousands of 1 μm pores in the surrounding soil matrix. In other settings, a minor difference in organic matter or the proportion of rock fragments may have a major effect on saturated hydraulic conductivity. Again, the legacy effects of land use may be evident in some areas. The inherent variability of soil physical and hydraulic properties does not, however, negate their usefulness. It is a reminder that the relative differences—rather than the absolute differences—between soil horizons, soil types, and watersheds may be the most useful means of discerning and predicting patterns and trends at the landscape scale.

Because soils are unsaturated most of the time (snowmelt, hurricanes, other unusually large storms may lead to saturated conditions), it is important to consider how water moves under unsaturated conditions. Water is least tightly held in the center of the largest soil pores. As noted earlier, the largest pores are also the most efficient conductors of water. Therefore, when drainage and evapotranspiration reduce the soil water content, the first increment of water comes from the center of the largest pores. Air flows into this void space and hydraulic conductivity is reduced. The pore structure and internal surface area of a soil remain constant (with the exception of shrink/swell clays). Therefore the capillary and adsorptive forces generated by soil pores and particle surfaces affect smaller and smaller quantities of water. Instead of flowing through the center of the largest interconnected pores, water must flow across the surface of soil particles—a longer and much less efficient flow path. This produces exponential decreases in unsaturated hydraulic conductivity for small, incremental decreases in soil water content.

2.6 EVAPOTRANSPIRATION

Evapotranspiration is the sum of interception (rain or snow that is evaporated off the plant canopy), evaporation directly from the soil, open water, or wetlands, and transpiration (water use by plants). It is typically the second largest component of the water balance, accounting for up to three-fourths of annual precipitation. The net effect or ecological signature of evapotranspiration is plainly evident in the annual pattern of streamflow. During the portions of the dormant season that are unaffected by snow accumulation and melt, evapotranspiration is minimal and streamflow is a muted reflection of rainfall. During the growing season, the general pattern of streamflow is a mirror image of evapotranspiration—such is the dominant biophysical influence of the process (Figure 2.3). In effect, water must escape the evapotranspiration process before it can contribute to growing season streamflow or groundwater recharge.

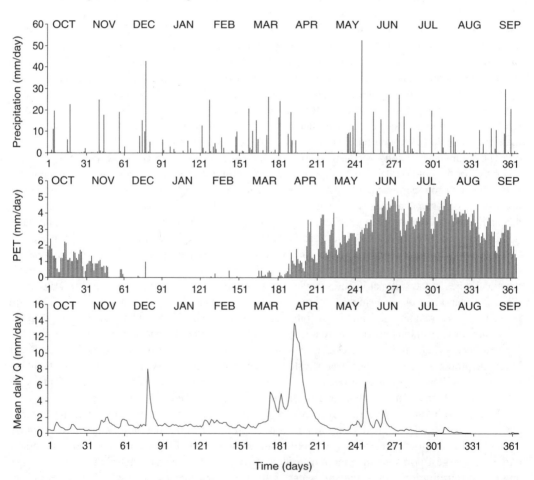

FIGURE 2.3 Precipitation, potential evapotranspiration, and mean daily streamflow (mm/day; note difference in y-axis scales), October 1, 2000 to September 30, 2001. Precipitation and air temperature data from the Orange, Massachusetts (airport), National Oceanic and Atmospheric Administration (NOAA) meteorological station; streamflow data from the U.S. Geological Survey gauging station on the East Branch of the Swift River, Hardwick, Massachusetts. Potential evapotranspiration was calculated with the Thornthwaite method.

Potential evapotranspiration is defined as the amount of water that could enter the atmosphere under prevailing conditions (wind velocity, air temperature, vapor pressure) when the supply of water is not limiting. That is not to say that potential evapotranspiration is constant. It may be very low on a cool, wet day, intermediate on a warm, humid, breezy day, or very high on a hot, dry, windy day. Potential evapotranspiration can be estimated using air temperature data (e.g., the Thornthwaite, Hamon, or Blaney-Criddle equations) or more complex and physically representative energy budget methods (Penman-Montieth) (Brooks et al., 2003).

Actual evapotranspiration shows the effects of limitations on the supply of water. It is, by definition, less than or equal to potential evapotranspiration. The rate of actual evapotranspiration reflects the combined influence of soil water retention characteristics and the active biological regulation of the transpiration process by plants. As soil water content decreases, the remainder is more tightly held and, as a result, it becomes more difficult for plants to extract. This leads to changes in plant water use to avoid desiccation and wilting. Actual evapotranspiration can be estimated from a time series of soil water content measurements or as the water balance residual with reliable measurements of annual precipitation and water yield ($AET \cong P - Q$). A more detailed description of each of the components of evapotranspiration appears below.

2.6.1 INTERCEPTION

Interception is influenced by the type, density, and vertical structure of the plant canopy. Naturally, tree species such as pine, spruce, and fir and evergreen shrubs such as mountain laurel, holly, and rhododendron may intercept rain or snow throughout the year. The finely divided foliage and layered canopy architecture of conifers presents a large, complex surface for interception. Thin films of water form on and between the needles or leaves and are then subject to evaporation. This may account for 20% to 30% of annual precipitation. In contrast, the interception capacity of deciduous trees and herbaceous plants varies widely through the year. During the dormant season, interception by leafless vegetation may be negligibly small. During the growing season, the total leaf area (interception capacity) of the forest canopy, midstory, understory, and herbaceous vegetation may exceed that of a dense, even-aged conifer stand. The length of the growing season (approximated by the number of frost-free days) and the frequency, intensity, and duration of rain events also affect total interception loss. Because interception capacity is finite, small, frequently spaced rain events lead to greater cumulative interception loss than a few large storms over an equivalent time period.

In addition to precipitation that passes through openings in the canopy or drips off branches and foliage (throughfall), water may reach the soil surface via stemflow. The sum of throughfall and stemflow is referred to as net precipitation. Again, differences in canopy architecture and branching habit strongly influence this component. Consider the vase-shaped crown of American elm (*Ulmus americana*) versus the pendant branches of Norway spruce (*Picea abies*) in relation to interception, throughfall, and stemflow.

2.6.2 EVAPORATION

Evaporation is a passive physical process that, depending on site conditions and the time of year, may comprise a large or small component of evapotranspiration. In a mature upland conifer forest, interception and transpiration may far exceed evaporation. The forest floor is shaded by evergreen canopy and covered with a thick litter layer of needles and other woody debris. In contrast, evaporation from a sparsely vegetated forested wetland in the Mid-Atlantic region may far exceed annual interception and transpiration. Open water or saturated conditions near the surface provide an abundant supply of water in a relatively moderate climate. In the case of a tilled farm field during the dormant season, a small amount of water may be intercepted by crop residues, while transpiration has been effectively eliminated. Clearly, in tilled fields, evaporation from the soil is the principal component of evapotranspiration.

Recalling the earlier discussion of the energy balance and the location of the "active surfaces," it is clear that radiant energy loading, temperature and vapor pressure gradients, and wind speed are key determinants of the amount of evaporation. In other words, evaporation is directly proportional to exposure of the site and the availability of water. Water in wetlands, vernal pools, shallow depressions, and streams is either present (available) for evaporation or absent—slowing or stopping the process until the next rain event refills the system. In soils, the availability of water is directly influenced by surface conditions and physical and hydraulic properties. Because evaporation occurs at or near the soil surface, the zone of drying extends downward into the profile. The hydraulic gradient (driving force for water flow) increases as the difference between the soil water content at the surface and the water content deeper in the profile increases. However, recall from the earlier discussion that unsaturated hydraulic conductivity decreases exponentially with incremental decreases in soil water content. These countervailing forces (represented in Darcy's law) regulate the rate of water movement to the active surface and therefore the rate of evaporation from the soil. The next rain event reverses the hydraulic gradient. The rapid increase in water content at the soil surface is subject to matric suction in the zone of drying produced by evaporation (and/or transpiration) as well as the force of gravity. This causes a "wetting front" to move through the soil profile and replenish soil water content in the process. As conditions change after the storm event, evaporation begins again and the zone of drying begins to form and extend downward from the soil surface.

2.6.3 TRANSPIRATION

Transpiration is the primary component of evapotranspiration in forest ecosystems. Far from being passive wicks that transmit water from the soil to the atmosphere, trees, shrubs, and many herbaceous plants actively regulate the movement of water through the soil–plant–atmosphere continuum (Brooks et al., 2003). Solar energy and wind combine to generate a demand for water vapor at and around the canopy surface—lowering water potential to as little as –30 or –40 bars. This, in turn, generates a leaf water potential of –20 or –30 bars and a xylem potential (the conductive vessels in the stem and roots) of –10 or –20 bars, until suction is exerted at the interface between the roots and the soil. This very powerful energy (suction) gradient induces water movement through the soil-plant-atmosphere continuum. As the zone of drying extends around the root surfaces, water moves from the surrounding soil toward the plant. When the soil water potential reaches –15 bars in the vicinity of the roots, the rate of supply cannot keep pace with the evaporative demand at the canopy surface and wilting ensues.

Water vapor exits the leaves through pore-like structures called stomates (white, gray, or light green spots that are visible without magnification on the underside of many evergreen needles). In addition to water vapor, carbon dioxide flows in through the stomates, where it is absorbed and used as a basic constituent for photosynthesis. Oxygen, a product of photosynthesis, flows out through the stomates along with the water vapor. A waxy layer called the cuticle substantially limits water loss through the rest of the leaf surface. It is the stomates that actively regulate transpiration losses—closing when leaf water potential falls below a physiologically sustainable level and opening when the water content and turgor pressure of the leaf are restored by inflow from the soil (via the xylem). The equilibrium may also be restored if the evaporative demand abruptly decreases (e.g., hot, dry, windy conditions are replaced by cool, moist air as a cold front passes the site).

Field measurements of soil water content in forest ecosystems cannot clearly separate the influences of interception, transpiration, and evaporation. Their combined effect (along with drainage of the soil water that contributes to streamflow and groundwater recharge) is most clearly shown by successive measurements during dry weather. Along with the differences in interception noted earlier, there can be substantial differences in soil water depletion by deep-rooted woody (perennial) vegetation versus shallow-rooted annual plants. At any given time, soil water content in the profile will be higher (and available storage lower) for shallow-rooted annual vegetation. Hence an equal amount of precipitation will generate a larger water yield.

The effects of land use on streamflow and water quality usually unfold as follows. Land use affects vegetation type, structure, and biomass. The character and condition of the vegetation has a primary influence on evapotranspiration and, as a direct result, the spatial and temporal patterns of soil water content in a watershed. Soil water content affects the soil water potential and hydraulic conductivity. The energy status of soil water affects the volume and rate of overland flow and subsurface flow, and ultimately the quantity and timing of streamflow and groundwater recharge. In addition, the character and condition of vegetation influences soil properties such as organic matter content, microbial activity, macropore abundance, porosity, permeability, and biogeochemistry — amplifying its direct and indirect effects upon the water balance. The direction of the change is a key consideration. As a general rule, the loss of forest vegetation produces unfavorable changes in the quantity, quality, and timing of streamflow; the restoration of forest vegetation produces favorable changes.

2.7 STREAMFLOW

The streamflow response of a watershed depends on several interrelated characteristics and processes. Soil properties, vegetative cover, land use, and seasonal variations in climate combine to influence the amount of water in storage (antecedent conditions) and, as a result, the speed with which streamflow is generated. The condition of the soil surface (i.e., infiltration capacity) and available storage in the soil profile are especially important. As noted earlier, the amount and intensity (rate) of rainfall or snowmelt influence the volume and timing of streamflow. It is, however, the seasonal variation of evapotranspiration that accounts for the substantial differences in growing season versus dormant season streamflow response to a rain event of a given size.

2.7.1 THE HYDROGRAPH

When scientists and watershed managers use the term *hydrograph*, they may be referring to changes in the rate of streamflow with respect to time for a storm event *or* for the entire water year. The components of a storm event or snowmelt hydrograph include (1) the rising limb, (2) the peak discharge, and (3) the recession or falling limb (Figure 2.4). *Baseflow* refers to the rate of streamflow before a rain or snowmelt event causes an increase. *Stormflow* is the area under the curve (hydrograph) after a graphical or mathematical process to "separate" baseflow and stormflow has been completed. The *time to peak* or *time of concentration* refers to the elapsed time between the centroid of the rain or snowmelt and the peak discharge. It reflects watershed characteristics such as land slope, channel slope, and hydraulic roughness, drainage density, wetland area, and other influences on the travel time of stormwater through the system. A hydrograph records the passage of a wave past a fixed observation point. Visualize the changes in stream stage (water level) when a very large thunderstorm occurs during the summer low flow period. Initially there is little discernable response as interception and infiltration delay the progress of water toward the stream. After this initial delay, water reaching the stream channel causes rapid increases in stage (the rising limb or leading edge of the wave). When rainfall intensity reaches its maximum value, stormwater is collected, combined, and conveyed to the observation point (e.g., stream gauging station). After the wave crest (peak discharge) passes, a long period of gradually decreasing flow (relative to the rising limb) occurs — this is the recession limb. The "trailing edge" of the wave marks the transition back to baseflow (sometimes called "dry weather flow"). Thus a stormflow hydrograph (discharge versus time) is a reflection of the wave as it passes the observation or measurement point.

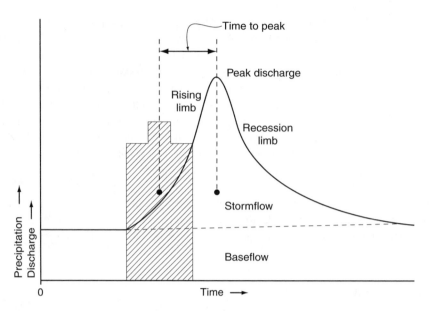

FIGURE 2.4 Definition sketch for a streamflow hydrograph (discharge versus time). The cross-hatched histogram represents the rainfall event that generated the hydrograph (stormflow event).

2.7.2 PATHWAYS OF FLOW

Three general pathways of flow have been identified. They are Hortonian overland flow, saturation overland flow, and subsurface flow (either shallow or deep). Hortonian overland flow occurs when the rainfall intensity or snowmelt rate exceeds the infiltration capacity of soil. (It is named in honor of Robert Horton, whose pioneering work on the infiltration process in the 1930s formed the basis of subsequent prediction and modeling methods.) Some water can (and usually does) flow through the soil surface, but any "rainfall excess" flows downslope along the path of least resistance, usually the path of steepest descent. This form of overland flow is most common on sites that (1) lack the protection of an organic litter layer, (2) are compacted, (3) have been eroded by earlier events, or (4) have fine textured, low permeability soils. This pathway of flow is not limited to low permeability soils, however. A continuous or extensive ice layer in any soil can cause Hortonian overland flow until it is thawed and the infiltration capacity of the soil is restored.

Saturation overland flow occurs when available storage rather than infiltration capacity is limiting. This pathway of flow is more common during the dormant season (when cumulative evapotranspiration is limited), in gently sloping terrain, or in areas of the watershed where subsurface flow converges at the base of a concave slope. It is less common in steep terrain with convex or planar landforms as well as during the growing season when evapotranspiration increases available storage. As a result, saturation overland flow is largely comprised of "old" water that has already entered the soil, contributed to the thickness or areal extent of the saturated zone, then reemerged wherever the saturated zone intersects the soil surface. Naturally any rain or snowmelt ("new" water) on the saturated area of the watershed immediately contributes to overland flow. The likelihood of saturation overland flow decreases as slope steepness increases. Because the rate of subsurface flow is directly proportional to slope steepness (hydraulic gradient), water moves through the soil mantle instead of contributing to the thickness of the saturated zone and rising to the surface.

Both pathways of overland flow are important determinants of the quantity, quality, and timing of streamflow. Unlike subsurface flow, which is filtered through the soil profile, overland flow may detach, lift, and carry soil particles, organic matter, nutrients, and pathogens from the soil surface. Other characteristics (slope, landform, rainfall or snowmelt volume and rate) being equal, overland flow is also the most rapid pathway. As such, overland flow minimizes the residence time and opportunity for biological, physical, and chemical transformation of potential pollutants within a watershed.

If overland flow, soil erosion, and nutrient export occur repeatedly, the ability of the plant community to recover from and resist this chronic stress steadily declines. Soil erosion reduces site fertility by removing clay and silt particles and organic matter while leaving the heavier, less fertile sand particles behind. In severe cases, a soil crust or erosion pavement forms on the surface. This surface condition further limits infiltration capacity and generates overland flow and additional erosion even though the soil permeability 5 or 10 cm below the surface is very high. Lower fertility and water holding capacity in turn limit nutrient assimilation, organic matter production, and rooting—all of which are needed to favorably influence the physical and hydraulic properties of soil. The cycle (overland flow → soil erosion → nutrient depletion → reduced site productivity) is bound to continue until deliberate and persistent restoration efforts reduce the frequency and impact of overland flow and allow the site to recover.

If neither infiltration nor storage are limiting then rain or snowmelt will enter the soil profile and move laterally as subsurface flow. Whether subsurface flow is shallow (within the root zone) or deep (in the groundwater flow system beneath the root zone) is a function of the hydrogeologic characteristics of the site and the genesis and morphology of the soil profile. In the shallow, stony glacial tills in the uplands of New England, infiltration capacity is rarely, if ever, exceeded by rainfall intensity. Water enters the soil and flows vertically through the soil horizons until it encounters bedrock or the shallow saturated zone, whichever comes first. It then flows laterally through the soil profile at a rate that is directly proportional to the soil's saturated hydraulic conductivity and slope (hydraulic gradient). In contrast, in some parts of the northern Lake States, water flows through the soil surface and the O and A horizons until an abrupt change in hydraulic conductivity causes the B horizon to be an impeding layer. A transient saturated zone in the A horizon leads to lateral flow, while at the same time, some of the water flows vertically into the B and C horizons. In both parts of the region, steep slopes and highly permeable soils combine to produce rates of shallow subsurface flow that can generate stormflow (in hours or days), especially during the dormant season.

In the deep (unglaciated) residual soils of the Mid-Atlantic region there is nothing to impede the vertical flow of water if the infiltration capacity is greater than rainfall intensity. (The exception may be a dense, low permeability "plow pan" at the base of the AP horizon—the legacy of earlier agricultural use.) Any water that is not detained in the root zone will reach the saturated zone (the "water table" in common parlance). The direction and rate of groundwater flow in this unconfined aquifer is, like any porous media, governed by its saturated hydraulic conductivity and the hydraulic gradient (the slope of the water table surface; usually a muted reflection of surface topography). Because of the time required to reach the aquifer through the unsaturated soil, and the time required to reach the stream channel when driven by a very low hydraulic gradient, water from any given rain event may emerge as baseflow weeks or months later.

2.7.3 Variable Source Area Concept

Until John Hewlett proposed the variable source area concept in the late-1950s there was no universally applicable description of streamflow generation (summarized in Hewlett and Troendle [1975]). Hortonian overland flow was supposed to be the cause of all stormflow even though it was not observed in the field during rainfall events on many forested watersheds. Robert Horton's experimental work in arid areas correctly identified infiltration capacity (the rate of water movement through the soil surface) as the limiting condition and primary pathway for conversion of rainfall into streamflow. As noted in earlier sections of this chapter, in temperate and boreal forests, the intensity of rainfall or the rate of snowmelt rarely exceeds the infiltration capacity of forest soils. The litter layer protects the soil surface from raindrop splash. This organic matter also favorably influences soil structure and hydraulic properties. Field measurements and observations during rain events at the Coweeta Hydrologic Laboratory near Asheville, North Carolina, clearly demonstrated that other mechanisms and pathways of flow needed to be identified and explained—namely saturation overland flow and shallow subsurface flow. Accurately describing the pathways of flow was necessary but not sufficient to explain the watershed-scale streamflow response. When Hewlett developed the variable (or saturated) source area concept and it was validated by other researchers (most notably Dunne and Black [1970]), differences in climate and watershed characteristics no longer confounded the explanation of the spatial and temporal patterns *and* processes of streamflow generation. It also serves as a clear and dynamic example of the compensatory changes and countervailing forces that characterize the water balance equation.

The saturated source area varies on two time scales: (1) seasonally and (2) during rain and snowmelt events. The second is superimposed upon the first. As the forest ecosystem advances into the dormant season, evapotranspiration accounts for a progressively smaller portion of precipitation until it becomes negligibly small. As a result, the saturated source area of the watershed expands in proportion to the additional water in soils, wetlands, and streams. The channel network also expands and conveys more streamflow. During this period, intermittent streams flow continuously and ephemeral streams increase in length (extending upslope) and discharge. Because the saturated source area and stream channel network are more extensive (expanding), any "new water" from subsequent rain or snowmelt events has less distance to travel and therefore requires less time before it contributes to streamflow. In addition, rates of subsurface flow can increase exponentially in relation to small increases in soil water content. (Recall the relation of unsaturated hydraulic conductivity to soil water content discussed earlier.) As a result, both the volume and the rate (peak discharge) of streamflow increase during the dormant season rain and snowmelt events. In colder climates, the formation of snowpack and ice in wetlands, lakes, ponds, and stream channels limits the availability of water. As a result, the variable source area contracts and streamflow is gradually reduced. In fact, the minimum annual streamflow in many watersheds occurs in February or March rather than August or September as is usually supposed. Snowmelt causes a rapid reversal as the saturated source area expands and typically reaches its maximum annual extent.

During the growing season the opposite set of conditions occurs. Evapotranspiration accounts for a progressively larger proportion of precipitation. At the beginning of the growing season, ephemeral streams dry up and flow in intermittent streams becomes erratic—depending solely on rain events. The saturated source area contracts and rates of subsurface flow decrease exponentially in relation to soil water content. This leads to increased time (opportunity) for evapotranspiration to withdraw water from the soil. This in turn leads to increased available storage for subsequent rainfall events. The difference in the recession limb slope of stormflow hydrographs for the growing versus the dormant season shows the direct and immediate effect of evapotranspiration on the availability of water, the size and configuration of the variable source area, and streamflow volume and rate. Calculating a traditional metric (hydrologic response ratio = depth of streamflow/depth of precipitation) from the precomputer era of hydrologic science illustrates the cumulative effect of evapotranspiration (or lack thereof) on the size of the saturated source area and consequent volume of streamflow throughout the water year (Table 2.4).

The difference in dormant season and growing season antecedent conditions forecasts the size and rapidity of the stormflow response throughout the year. Simply put, the chance of rain landing (or snowmelt originating) on or near the saturated source area and rapidly contributing to streamflow during the dormant season is substantially higher than during the growing season. Hence a storm of the same size in December versus August will produce very different streamflow responses. The saturated source area continues to expand as the rain progresses—producing the rising limb of the hydrograph. The peak discharge of the hydrograph corresponds to the maximum rainfall intensity (or snowmelt rate), the maximum extent of the saturated source area, and the routing time required for water to reach the measurement point. When the rain ends, the saturated source area begins to contract, leading to the recession limb of the hydrograph.

TABLE 2.4
Monthly Precipitation, Streamflow, and Hydrologic Response Ratio for East Branch Swift River, Central Massachusetts, October 2000 to September 2001

Month	Precipitation (mm)	Streamflow (mm)	Q/P (%)
October	65.5	19.5	30
November	68.4	31.5	46
December	89.8	53.7	60
January	30.7	30.7	100
February	65.2	34.3	53
March	138.5	61.0	44
April	41.2	169.1	410
May	63.6	24.9	39
June	138.8	46.6	34
July	95.3	9.4	10
August	43.8	5.4	12
September	104.2	0.7	1

Daily data are presented in Figure 2.4.

The dynamic nature of the saturated source area through the seasons and during storm events has important implications for watershed managers and regulators. It reminds us that riparian areas—the interface between upland and aquatic ecosystems—should not be treated as static entities protected by a narrow "one-size-fits-all" buffer strip when it is clear that biophysical boundaries are subject to systematic, sometimes very rapid change. The structure and function of riparian areas are discussed in detail in Chapter 4.

2.8   STREAM CHANNEL FORM AND FUNCTION

Once water reaches a stream channel, its movement is affected by a number of interrelated channel characteristics. Flowing water and variations in streamflow in turn affect the form and function of stream channels. The flow of water in open channels can be described mathematically in many ways. The most basic method is the continuity equation:

$$Q = AV,$$

where Q is discharge (length^3/time or volume/time), A is the cross-sectional area of flow (length^2), and V is the mean velocity (length/time). The velocity of flowing water in open channels varies in two directions: (1) vertically with respect to distance from the stream bottom or bed, and (2) horizontally with respect to distance from the streambanks (Figure 2.5). The mean velocity occurs at approximately $0.6 \times$ flow depth, relative to the water surface. In a symmetrical channel, the maximum velocity of flow occurs at the point that is most distant from the frictional resistance imparted by the streambed and banks.

In addition to the continuity equation, perhaps the most common and useful method of describing open channel flow is Manning's (1890) equation (Equation 2.5 or 2.6) (Figure 2.6). Manning was an Irish engineer who quantified velocity and flow in streams and canals in relation to their shape, slope, and hydraulic "roughness" (Henderson, 1966). Manning's equation is now used for a wide range of applications, including the estimation of peak discharge from high water marks and channel survey data, flood wave routing through stream networks, and design and restoration projects.

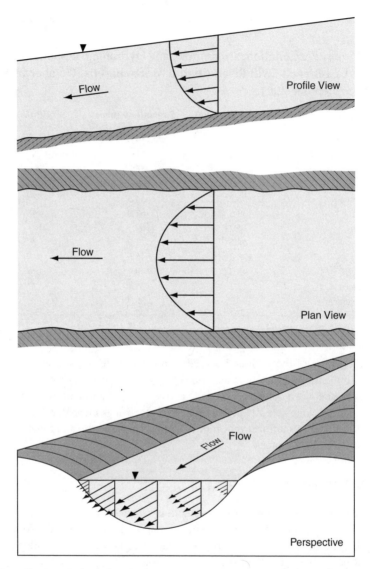

FIGURE 2.5 Generalized vertical (profile view) and horizontal (plan view) velocity variation of water flowing in an open channel.

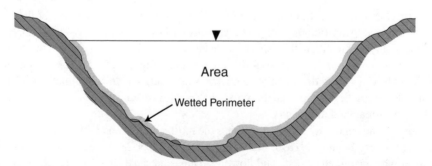

FIGURE 2.6 Definition sketch for the cross-sectional area and wetted perimeter of an open channel. The solid triangle marks the water surface.

$$V = C/n \cdot R_\mathrm{h}^{2/3} \cdot s_\mathrm{o}^{1/2}. \tag{2.5}$$

Substituting Q/A for V yields

$$Q = C/n \cdot A \cdot R_\mathrm{h} \cdot s_\mathrm{o}^{1/2}, \tag{2.6}$$

where

  C = conversion constant; 1.0 for SI units and 1.486 for FSS units;
  V = mean velocity (ft/sec or m/sec);
  Q = discharge (ft³/sec or m³/sec);
  A = cross-sectional area (ft² or m²);
  n = Manning's (hydraulic) roughness coefficient;
  R_h = hydraulic radius = area/wetted perimeter (ft or m); and
  s_o = bed slope (ft/ft or m/m).

Note that V and Q are inversely proportional to hydraulic roughness and directly proportional to hydraulic radius (hydraulic mean depth) and slope. The relation of V and Q to slope and roughness (n) is self-evident. The relation to hydraulic radius requires more explanation. Consider the form and function of a stream channel that is a perfect semicircle in cross section. Because a circle is the geometric shape with the minimum ratio of perimeter to area, a semicircular channel has the minimum contact (frictional resistance) between water and the substrate. In contrast, a broad shallow channel of the same cross-sectional area will have a much smaller hydraulic radius (mean depth). As a result of the increased contact between water and the substrate, greater total frictional resistance will reduce flow velocity and discharge. The hydraulic radius term is, in effect, a shape factor.

Hydraulic roughness (n) is the fitted parameter or empirical coefficient in Manning's equation. Using a current velocity meter for *in situ* measurements, accurately surveying the stream channel cross section to determine the area and hydraulic radius, measuring channel slope, and specifying the units (metric or English) reduces Equation 2.5 to one unknown. In addition to narrative descriptions, some reference manuals contain photographs of typical stream reaches. (A reach is a segment or section of a stream with relatively uniform characteristics. It can be defined in the field or based on a detailed channel survey, cross sections, and a longitudinal profile.) This enables the analyst to make reasonable estimates of velocity and discharge for a variety of purposes, such as peak discharge from high water marks and the evaluation of channels and hydraulic structures at road-stream crossings.

Manning's equation is also a valuable conceptual model for understanding open channel hydraulics and the dynamic equilibrium concept. Consider the hydraulic consequences—changes in velocity or discharge—of changing any one parameter while holding the others constant. Increasing slope naturally increases velocity and discharge. Decreasing hydraulic roughness also increases velocity and discharge. Dredging a channel and increasing flow depth increases the hydraulic radius and, as a result, velocity and discharge. Now consider the potential of making all three changes simultaneously (increasing slope, decreasing roughness, and increasing depth by narrowing the channel) as often occurs in urban and suburban areas on velocity and discharge. Also consider the corresponding increase in the force of the flowing water, its ability to lift and carry pollutants, and the physical stress imposed on aquatic organisms.

2.8.1 Energy Status of Flowing Water

The continuity equation, $Q = AV$, shows that an infinite combination of areas (width and mean depth) and velocities can produce the same discharge (e.g., $Q = 10$ m³/sec = 10 m² × 1 m/sec, or 1 m² × 10 m/sec, or 5 m² × 2 m/sec, or 2 m² × 5 m/sec → ∞). In common parlance, "still waters run deep," or vice

PLAN VIEW

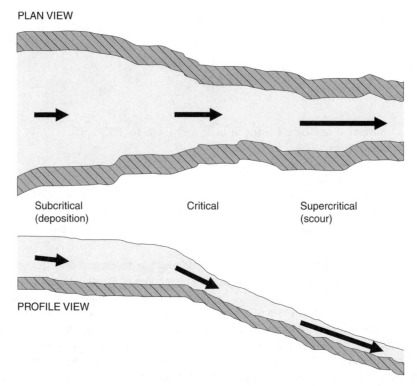

Subcritical   Critical   Supercritical
(deposition)   (scour)

PROFILE VIEW

FIGURE 2.7 Variation in the energy status of flowing water in relation to width, depth, cross-sectional area, and channel slope. The arrows are velocity vectors for subcritical (tranquil), critical, and supercritical (shooting) flow.

versa. There is, however, one particular combination of area (flow depth) and velocity that defines the boundary between inertial (due to motion) and gravitational (due to position) energy. When inertial forces and gravitational forces are exactly in balance, this defines *critical flow* (with a corresponding critical depth, critical velocity, and critical slope). Its primary importance is the demarcation of the two alternate energy statuses: (1) *subcritical* or tranquil flow and (2) *supercritical* or shooting flow. Consider a stream reach where tranquil flow reaches a transition to a steeper section of the streambed, accelerates, and becomes turbulent enough to produce "white water" (Figure 2.7). In this example, subcritical (tranquil) flow passes through critical depth and velocity at the slope transition and becomes supercritical (shooting) flow. The arrows in Figure 2.7 are velocity vectors; they indicate direction and magnitude. The transition from a *mild* (subcritical) to a *steep* (supercritical) slope leads to simultaneous decreases in width, depth, and area as a result of the increase in velocity (at a constant discharge).

2.8.2 Dynamic Equilibrium

Stream channels exist in a state of *dynamic equilibrium.* This is not an oxymoron, but an accurate scientific description based on the traditional definition of these two terms in physics and chemistry. Dynamic refers to objects in motion; equilibrium refers to a state of balance. A brief review of the laws of motion (Sir Isaac Newton [1642–1727]) in relation to flowing water helps to define and describe the dynamic equilibrium concept. Simply paraphrased:

1. An object at rest tends to stay at rest; an object in motion tends to stay in motion.
2. Force equals mass times acceleration ($F = ma$).
3. For every action there is an equal and opposite reaction.

It follows, for example, that removing streamside trees leads to channel scour and bank erosion until the dynamic equilibrium is reestablished. The sediment generated by bank erosion is transported downstream and deposited when flow velocity decreases in subcritical reaches. The deposition of sediment reduces cross-sectional area, which in turn changes hydraulic radius and flow velocity. The complex balance between the hydrologic regime, sediment loading, woody debris inputs, riparian vegetation, geologic setting, characteristics of the bed and bank material, and legacy effects (e.g., log drives in the 1800s) is described in sum by the dynamic equilibrium concept.

The same principle of compensatory change discussed at the beginning of this chapter in relation to the water balance also applies to the dynamic equilibrium concept—the balance of ecological and hydraulic forces—in stream and rivers. Later chapters present detailed examples from the literature of changes in land use that lead directly to changes in the quantity (mass) and timing (rate or acceleration) of stormflow. Increasing the mass and acceleration of stormflow must also increase the force of flowing water (quantified as shear stress and stream power [Gordon et al., 1992; Henderson, 1966]). The "equal and opposite reaction" to this change may be an increase in cross-sectional area, bed slope, or hydraulic roughness (as fine material such as sand and gravel is eroded, leaving larger, more massive cobbles and boulders behind). Woody debris that was relatively stable under predevelopment conditions may be swept downstream. Similarly streamside trees and other vegetation may be undermined and toppled into the channel. The loss of root support further destabilizes the banks, while the temporary increase in hydraulic roughness may cause overbank flow, braiding, and other diversions from the original channel. In theory, the dynamic equilibrium of a stream or river can be reestablished in relation to the new flow regime. In practice, the encroachment of development on the channel and its floodplain, channel alteration and armoring (grading, shaping, smoothing, adding riprap, etc.), and continued alteration of the water balance may make the attainment of a new force balance functionally impossible.

2.8.3 BANKFULL DISCHARGE

Stream and river channels evolve over millennia in relation to regional climate, geologic setting, soils, topography, vegetation, and the natural disturbance regime of their watershed. Unusual circumstances (e.g., a rain-on-snow event on a deep snowpack over frozen soils, a hurricane that generates 25 cm of rain in 24 hours, severe wildfires that temporarily denude the watershed, etc.) produce unusual hydrologic and hydraulic results. These low frequency events are important, yet by definition they are relatively rare (i.e., probability of occurrence of 1% or 0.5% in any given year). Until recently, the importance of relatively common, high frequency events—bankfull discharge—was underestimated or at least overshadowed by the dramatic effects of extreme events. A brief overview of hydrologic frequency analysis provides helpful background; detailed procedures can be obtained from most hydrology textbooks (e.g., Brooks et al., 2003).

The process of frequency analysis (empirically estimating the likelihood or probability of the occurrence of floods, droughts, or other hydrometeorological phenomena) begins with the compilation of a reliable long-term data series from sources such as the U.S. Geological Survey Water Resources Division or the National Climatic Data Center. In most cases, an annual maximum series is extracted from the daily records and sorted in rank order (from largest to smallest) without regard for the year of occurrence. (With rare exceptions, annual peak discharge is a statistically independent event. However, when a single major storm event produced successive peak discharges on December 31 and January 1, one would be excluded from the annual maximum data series. The larger of the two discharges is used to represent the flood event.) A statistical method (e.g., log-Pearson type III, L-moments, etc.) is used to derive a flood frequency curve—the best fit model for the log-transformed annual peak discharge versus probability of occurrence. The frequency curve ranges from annual peak discharges during very dry years that, having stayed well within the stream channel, would not even be considered "flood" events by nonhydrologists, to very large (low probability or frequency) floods that are remembered for generations. While the large, low frequency floods inundate the floodplain, damage property, and endanger people, the "mean annual flood" (probability of occurrence is 50% in any given year, or alternatively, a flood event that, on average, is equaled or exceeded every other year) is also important with respect to channel-forming processes. As noted above, streams and rivers evolve over long time periods based on watershed conditions and the consequent hydrologic regime. Bankfull discharge occurs, on average, every other year ("mean annual flood") and is an important, high frequency driver for channel structure, function, and dynamic equilibrium. It follows therefore that when land use changes alter the frequency of bankfull discharge (let alone overbank flow), the new hydrologic regime necessarily alters channel form and equilibrium. In other words, because the conversion of forests to other land covers and uses increases the volume and rate of stormflow, a peak discharge that was equaled or exceeded every other year may now occur many times in a single year.

2.9   SUMMARY OF FUNDAMENTAL PRINCIPLES

- The water balance and energy balance are expressions of the law of conservation of mass and energy. They are inextricably linked. Therefore, changing vegetative cover (micro-climate, evapotranspiration, snowmelt, etc.) or land use (storage capacity, flow routing, etc.) generates compensatory changes in the water yield (Q) and quality.
- The relation of the soil surface conditions and infiltration capacity to rainfall intensity (or snowmelt rate) is a key determinant of flow path (overland, subsurface, or both), soil erosion, and nonpoint source pollutant transport.
- The soil is the nexus for hydrologic processes and biogeochemical cycling. Therefore, changing the rate and volume of water movement into and through the soil alters the rate and efficiency of nutrient cycling and plant uptake.
- Forest soils are unsaturated most of the time. As a result, overland flow is rare and rates of subsurface flow are comparatively slow. The combination of high infiltration capacity and unsaturated subsurface flow is a primary reason why forests sustain the flow of streams during dry weather.
- Evapotranspiration (interception + evaporation + transpiration) is directly related to vegetation type (e.g., deep-rooted woody plants versus shallow-rooted annual plants, evergreen versus deciduous trees, native herbaceous vegetation versus agricultural crops, etc.). After precipitation, evapotranspiration is the largest term (variable) in the water balance equation and a direct reflection of land cover and land use.
- Volume, flow path, and flow rate are interrelated. When rainfall intensity or snowmelt rate exceeds the infiltration capacity *or* storage capacity of a soil, overland flow is the obligate result. When infiltration capacity exceeds the rainfall or snowmelt rate *and* storage capacity exceeds total rain or snowmelt volume, subsurface flow through the soil (or aquifer) is the result. The streamflow response of any watershed, as well as the relative proportions of stormflow and baseflow, is strongly influenced by the volume, pathway, and rate of water movement.
- The variable source area concept explains the systematic variation in streamflow for seasons (growing and dormant) and storm events. It describes the combined effects of climate (precipitation and evapotranspiration), topography, soils, and geologic character-istics on the expansion and contraction of the saturated source area for streamflow. It is also a reminder that "one-size-fits-all" riparian area management strategies may be insufficient (too narrow) in some areas and excessive (too wide) in others.
- The dynamic equilibrium concept describes the shifting balance between the major component parts or characteristics of streams and rivers: flow regime, sediment load, bed and bank material, woody debris, riparian vegetation, recent alteration by people, and the legacy effects of earlier uses and abuses. When a deliberate or inadvertent alteration

of one component creates an imbalance with other components it will generate change (recall the laws of motion).

• Water is the universal solvent and a prime mover of sediment, nutrients, and organic matter. Water is essential for life.

REFERENCES

Brooks, K.N., Folliott, P.F., Gregersen, H.M., and DeBano, L.F., *Hydrology and the Management of Watersheds*, 3rd ed., Iowa State University Press, Ames, 2003.

Chang, M., *Forest Hydrology: An Introduction to Water and Forests*, CRC Press, Boca Raton, FL, 2003.

Cousteau, Jacques, http://www.americanrivers.org/site/PageServer?pagename=AMR_content_09eb; accessed July 2006.

Dingman, S.L., *Physical Hydrology*, Macmillan, New York, 1994.

Dunne, T. and Black, R.D., Partial area contributions to storm runoff in a small New England watershed, *Water Resour. Res.*, 6, 1296–1311, 1970.

Dunne, T. and Leopold, L.B., *Water in Environmental Planning*, W.H. Freeman, New York, 1978.

Gordon, N.D., McMahon, T.A., and Findlayson, B.L., *Stream Hydrology: An Introduction for Ecologists*, John Wiley & Sons, New York, 1992.

Hanks, R.J. and Ashcroft, G.L., *Applied Soil Physics*, Springer-Verlag, New York, 1980.

Henderson, F.M., *Open Channel Flow*, Macmillan, New York, 1966.

Hewlett, J.D. and Troendle, C.A., Nonpoint and diffuse water resources: a variable source problem, in *Proceedings of Watershed Management*, American Society of Civil Engineers, New York, 1975, pp. 21–46.

Hillel, D., *Introduction to Soil Physics*, Academic Press, New York, 1982.

Kudish, M., *The Catskill Forest: A History*, Purple Mountain Press, Fleischmanns, NY, 2000.

Marsh, G.P., *Man and Nature, or Physical Geography as Modified by Human Action*, Charles Scribner, New York, 1864 [reprinted with annotations, Lowenthal, D., (ed.), Harvard University Press, Cambridge, MA, 1965].

Muir, J., *My First Summer in the Sierra*, Houghton Mifflin, Boston, 1911 [reprinted by Sierra Club Books, San Francisco, 1988].

Sartz, R.S., Effects of watershed cover on overland flow from a major storm in southwestern Wisconsin, Research Note NC-82, U.S. Department of Agriculture, Forest Service, St. Paul, MN, 1969.

Satterlund, D.R. and Adams, P.W., *Wildland Watershed Management*, 2nd ed., John Wiley & Sons, New York, 1992.

Taylor, S.A. and Ashcroft, G.L., *Physical Edaphology. The Physics of Irrigated and Nonirrigated Soils*, W.H. Freeman, San Francisco, 1972.

U.S. Army Corps of Engineers, Summary Report of the Snow Investigations: Snow Hydrology, North Pacific Division, U.S. Army Corps of Engineers, Portland, OR, 1956.

Zon, R., Forests and water in the light of scientific investigation, USDA Forest Service, Washington, D.C., 1927 [reprinted with revised bibliography from Appendix V of the Final Report of the National Waterways Commission 1912, Senate Document 469, 62nd Congress, 2nd Session].

3 Nutrients, Pesticides, and Metals

X had marked time in the limestone ledge since the Paleozoic seas covered the land. Time to an atom locked in a rock, does not pass. The break came when a bur-oak root nosed down a crack and began prying and sucking. In the flash of a century the rock decayed, and X was pulled out and up into the world of living things. He helped build a flower, which became an acorn, which fattened a deer, which fed an Indian, all in a single year.

Aldo Leopold, 1949, *A Sand County Almanac*

3.1 INTRODUCTION

Nutrients, pesticides, and metals are three major categories of chemical compounds that can pollute streams in forested, agricultural, urban, and mixed land use watersheds. In this chapter we examine the chemical interactions and physical processes that influence the movement of these pollutants from the land to waterways, regardless of local land use. Atmospheric deposition of pollutants and eutrophication are also defined and discussed in this chapter. Atmospheric deposition is a regional phenomenon that is partly or completely independent of local land cover or use. Eutrophic conditions, particularly in coastal estuaries, are the result of nutrient loading from large areas, including loading from atmospheric deposition, and often cannot be attributed to a particular local land use.

3.2 NUTRIENT CYCLING AND CONSERVATION

Living organisms require mineral nutrients. Absent human disturbance, nutrients enter forest and grassland ecosystems as atmospheric gases, airborne particles, and from the weathering of bedrock and soil minerals. Nutrients are taken up by living vegetation, stored in plant tissues, and released to the soil when plants die and decay, when animal tissue decomposes after the death of an organism, or in the feces of animals that have consumed plant material. These nutrients are then available for new plant growth. As long as site-appropriate vegetation is in place, nutrients are efficiently recycled through the ecosystem (Likens et al., 1967, 1977). Nutrients that are required in relatively large quantities by plants are called macronutrients. They include nitrogen, phosphorus, potassium, calcium, magnesium, and sulfur (Kimmerer, 1990). The nutrients that most directly affect water quality are nitrogen and phosphorus.

3.2.1 NITROGEN

Nitrogen combines with carbon, hydrogen, and oxygen to form the amino acids and proteins in the living cells of plants and animals. In general, the growth of terrestrial vegetation is limited by the amount of available nitrogen (Fenn et al., 1998; Vitousek et al., 2002).

3.2.1.1 The Nitrogen Cycle (Figure 3.1A,B)

3.2.1.1.1 Nitrogen Fixation

Approximately 80% of the atmosphere is composed of nitrogen gas, but nitrogen in this form (N_2) is unavailable to plants. Nitrogen is converted to biologically available compounds (ammonium

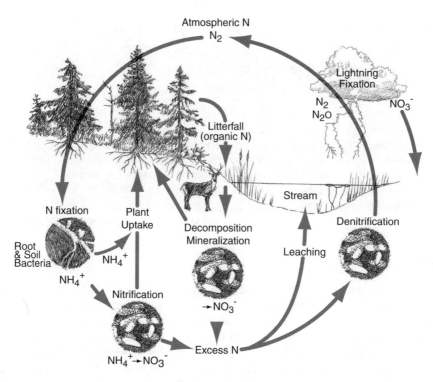

FIGURE 3.1A The nitrogen cycle in an undisturbed forest ecosystem. Soil- and root-associated organisms convert atmospheric nitrogen to ammonium (NH_4^+) (nitrogen fixation). Fixed nitrogen is stored in soil microbial tissue and in plants as NH_4^+ and nitrate (NO_3). Following decomposition and mineralization of organic matter, biologically available forms of nitrogen are again available for plant uptake. Nitrogen is returned to the atmosphere through denitrification (primarily in wetlands). Leaching losses to streams are small. (Original drawing by Ethan Nedeau.)

and nitrate) in the atmosphere by lightning, and in terrestrial ecosystems by free-living soil bacteria and lichens that inhabit the bark of trees and decaying wood. This process is called nitrogen fixation. Symbiotic nitrogen-fixing bacteria, such as *Rhizobium* or *Frankia*, that live in association with the roots of some plant species (e.g., legumes and alders) also process nitrogen. Many genera of cyanobacteria (blue-green algae) fix nitrogen in aquatic environments. In the process of nitrogen fixation, atmospheric nitrogen (N_2) is first reduced to ammonium ions (NH_4^+).

$$\text{(nitrogenase)*}$$
$$N_2 + 10H^+ \rightarrow 2NH_4^+ + H_2 \tag{3.1}$$

*Nitrogenase is an enzyme provided by biological organisms that facilitates this reaction. Nitrogen in this form may be taken up directly by plants or consumed by bacteria and "immobilized" in organic tissue (see the section on mineralization and immobilization).

3.2.1.1.2 Nitrification

Bacteria chemically transform ammonium. *Nitrosomas* bacteria oxidize NH_4^+, converting it to nitrite (NO_2). Another bacterial genus, *Nitrobacter*, then consumes and processes nitrite, converting it to the nitrate anion (NO_3). This entire process (the conversion of ammonium to nitrate) is called nitrification (Raven et al., 1999).

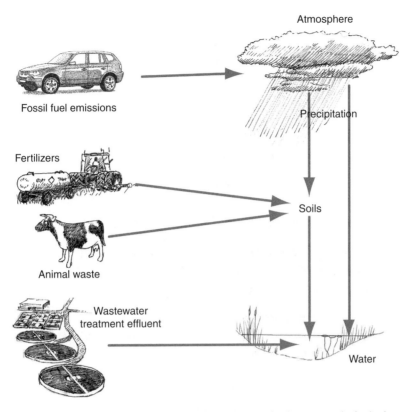

FIGURE 3.1B Human activity supplies additional fixed nitrogen to mixed-use watersheds. At the same time, the loss of forest vegetation and wetlands reduces watershed nitrogen storage and nitrogen loss through denitrification. The proportion and amount of nitrogen leaching to streams increases. (Original drawing by Ethan Nedeau.)

Step 1 (Nitrosomas bacteria)

$$NH_4^+ + 1.5O_2 \rightarrow NO_2^- + 2H^+ + H_2O^* \tag{3.2}$$

*The two hydrogen ions produced in the oxidation of NH_4^+ can decrease soil pH (i.e., increase soil acidity) (see Box 3.1).

Box 3.1   pH: A Measure of Soil Acidity

In pure water, a small number (approximately one in 10 million) H_2O molecules dissociates to form H^+ and OH ions. At 25°C the product of these concentrations is 10 to 14 mol/L. Since the number of H^+ ions must equal the number of OH ions, the concentration of H^+ ions is 7 to 10 mol/L. The pH scale is a measure of the acidity of a solution—the negative logarithm of hydrogen ion (H^+) concentration. In a neutral solution, the concentration of H^+ ions and OH ions are equal (7 to 10 mol/L) and the pH is 7. As the concentration of H^+ ions increases, the pH decreases and the resulting solution becomes more acidic. When the pH is greater than 7, the concentration of OH ions is greater than that of the H^+ ions and the solution is alkaline (Figure 3.2). Soil pH varies regionally based on the chemical composition of the soil parent material, the type of vegetation, and patterns of atmospheric deposition (Brady and Weil, 2002).

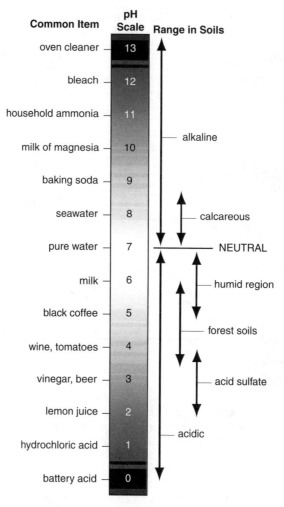

FIGURE 3.2 The pH scale and forest soil equivalents. (Brady, N.C. and Weil, R.R., *The Nature and Properties of Soils*, 13th ed. © 2002. Electronically reproduced by permission of Pearson Education, Inc., Upper Saddle River, NJ.)

Step 2 (Nitrobacter bacteria)

$$NO_2^- + H_2O \rightarrow H_2O \cdot NO_2^- \rightarrow NO_3^- + 2H \qquad (3.3)$$

(Barnes et al., 1998:561).

3.2.1.1.3 Mineralization and Immobilization

Bacteria also act to convert organic nitrogen compounds in dead plant tissue to ammonium and nitrate during decomposition, a process known as "mineralization" (Brady and Weil, 2002).

$$
\begin{array}{c}
\text{-----------------mineralization----------------->} \\
\begin{array}{ccc}
+2H_2O & +O_2 & +1/2O_2 \\
\end{array} \\
R\text{--}NH_2 \rightarrow\rightarrow OH^- + R\text{--}OH + NH_4^+ \rightarrow\rightarrow 4H^+ + \text{energy} + NO_2^- \rightarrow\rightarrow \text{energy} + NO_3^- \qquad (3.4) \\
\begin{array}{ccc}
-2H_2O & -O_2 & -1/2O_2 \\
\end{array} \\
\text{<-----------------immobilization----------------} \\
\text{(where R represents an organic group)}
\end{array}
$$

Carbon also is consumed and released as carbon dioxide (CO_2) during the decomposition process. Immobilization (the reverse of mineralization) is the incorporation of nitrate into organic tissue during bacterial consumption.

3.2.1.1.4 Denitrification

Nitrogen is returned to the atmosphere through denitrification. Various genera of bacteria, including *Agrobacterium* and *Pseudomonas*, living in intermittently anaerobic environments (these include areas with rising and falling saturated zones such as wetlands, riparian zones, and tidal mud flats) are able to use NO_2^- or NO_3^- as a substitute for oxygen in metabolic processes. Organic material in these environments supplies carbon compounds that act as electron donors in a series of chemical reactions that constitute bacterial denitrification. In summary, in the process of denitrification, specialized bacteria consume carbon compounds, nitrates, and nitrites and release atmospheric gases N_2 and N_2O and CO_2.

$$\text{(reductase)*}$$
$$5CH_2O + 4H^+ + 4NO_3^- \rightarrow 2N_2 + 5CO_2 + 7H_2O* \tag{3.5}$$

*Reductase is an enzyme provided by biological organisms that facilitates this reaction.
The reduction of NO_3^- to $2N_2$ requires several intermediate steps (Barnes et al., 1998:564; Raven et al., 1999; Stevenson, 1986).

At any given time, nitrogen in the forest floor and in soil and water exists in different organic and inorganic forms. Organic nitrogen includes nitrogen bound in organic compounds, such as leaf litter and microbial tissue, and dissolved organic nitrogen (DON). Inorganic or mineralized nitrogen is found as ammonium and nitrate and is available for biological uptake. The relative amount of nitrogen in each of these soil nitrogen pools depends to a large extent on factors that affect the rate of biological activity which determines the rates of plant uptake and microbial decomposition, mineralization, nitrification, and denitrification. Seasonal variations in soil water content and temperature produce seasonal patterns in the amount of mineralized nitrogen in the soil. The nitrate (NO_3^-) concentration in the soil is of particular concern because it is biologically available, more soluble than ammonium (NH_4^+), and more likely to leach to streams. Nitrate tends to accumulate in the soil during the dormant season when biological demand is low. Ammonium and nitrate are taken up by plants and microbes during the growing season. The rate of microbial decomposition is reduced in soils as either acidity (low pH) or alkalinity (high pH) increases (Brady and Weil, 2002; Creed and Band, 1998a,b).

3.2.2 Phosphorus

High-energy phosphate bonds in adenosine triphosphate (ATP) constitute the primary means of energy transfer in plants. Deoxyribonucleic acid (DNA), the primary genetic material in plants and animals, and ribonucleic acid (RNA), essential to protein synthesis, also contain phosphorus (Brady and Weil, 2002; Kozlowski and Pallardy, 1997).

3.2.2.1 The Phosphorus Cycle (Figure 3.3A,B)

Phosphorus (P) is released through the weathering of phosphorus-bearing rocks. It is taken up by plants where it may be stored long term in woody stems, returned to the soil through litterfall, or consumed by animals. Decomposing bacteria break down the phosphorus-containing organic compounds in animal waste and in dead plant and animal tissue, releasing inorganic phosphorus; this is again made available for plant uptake, starting the cycle anew. At any given time, phosphorus exists in the soil in various organic and inorganic forms. Only a small portion of the total phosphorus in soils is in soluble form and available for plant uptake. Inorganic phosphate anions

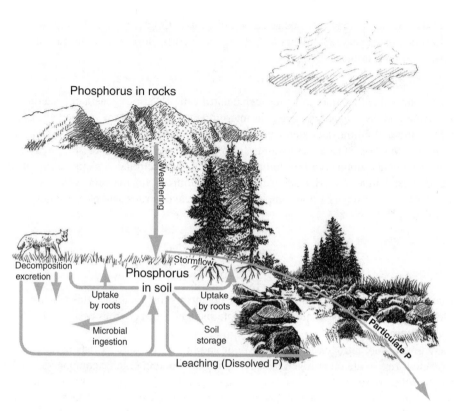

FIGURE 3.3A The phosphorus cycle in an undisturbed forest ecosystem. Phosphorus is released from rocks through weathering and erosion. Phosphorus in the form of phosphates is taken up by soil microbes and plants. Phosphates also combine with other minerals and are held in the soil. Following decomposition of organic matter, phosphorus is again available for plant uptake. Leaching losses to streams are very small and occur primarily with sediment loss during stormflow. (Original drawing by Ethan Nedeau.)

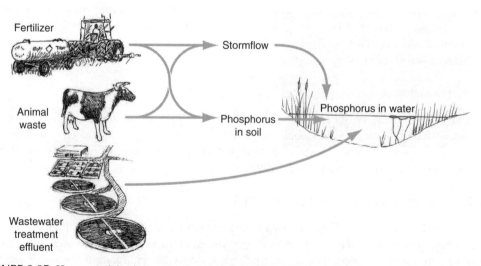

FIGURE 3.3B Human activity increases phosphorus loading. The loss of forest vegetation reduces plant uptake and increases sediment loss through erosion. The amount of phosphorus lost to streams with sediment and in dissolved form increases. (Original drawing by Ethan Nedeau.)

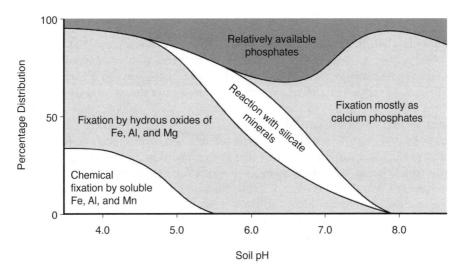

FIGURE 3.4 Reaction of phosphate with soil minerals. (Brady, N.C. and Weil, R.R., *Nature and Properties of Soils*, 13th ed. © 2002. Electronically reproduced by permission of Pearson Education, Inc., Upper Saddle River, NJ.)

(anions containing a phosphate group [PO_4]) that are potentially available to plants tend to form compounds with mineral cations. In an acidic environment (pH ≤ 5.0), phosphate anions will combine with iron (Fe^{3+}) or aluminum (Al^{3+}) ions if those elements are available. At pH levels of 7.0 or higher, phosphates form compounds with calcium (Ca^{2+}) (Brady and Weil, 2002; Edwards and Withers, 1998; Sharpley et al., 1981). Neutral soils, soils with a pH between 6 and 7, contain the highest levels of dissolved, plant-available phosphate, usually in the form of $H_2PO_4^{2-}$ (Figure 3.4). Phosphate anions can also adhere to and form compounds with silicate clay minerals. Phosphorus in newly precipitated mineral compounds may still be somewhat available to growing plants and for transport in the dissolved form because the phosphate ions initially adhere to the surface of the soil particles and can be displaced fairly easily. As the compounds age, the phosphate ions migrate to the interior of the particle and the compound becomes increasingly stable (Brady and Weil, 2002; Froelich, 1988). Larger amounts of biologically available phosphorus compounds are present in soils high in organic matter because humic and organic acids form compounds with iron and aluminum as well, reducing the number of these cations available to fix phosphates (Brady and Weil, 2002; Stevenson, 1986).

The phosphorus bound in most organic compounds is unavailable for plant uptake until the organic matter decomposes. Soil bacteria release inorganic phosphates (mineralization) during decomposition of organic material. Mineral phosphate released by decomposition enters the soil solution, where it may (1) be consumed by bacteria, and thus stored again in microbial tissue (immobilization); (2) form compounds with minerals, calcium, aluminum, or iron; or (3) be taken up by plants (Stevenson, 1986).

3.2.3 PROPERTIES OF NUTRIENT COMPOUNDS AND PATHWAYS OF FLOW

Various nitrogen and phosphorus compounds are lost from the soil through different hydrologic pathways. The pathway of transport depends on the chemical characteristics of a particular compound, the climate, and the properties and condition of the soil. Soluble compounds tend to enter the soil and are transported (leached) via subsurface flow. Compounds that adsorb to soil particles are typically transported with overland flow, with much of the transport occurring during stormflow events. In undisturbed forested and prairie ecosystems, nutrients are conserved and recycled, there is very little overland flow to carry sediment, and leaching losses are minimal. In agricultural and urban systems, however, there are imbalances between the supply and consumption of nitrogen and phosphorus, and excess loading of nitrogen and phosphorus can have severe environmental consequences.

Biologically available nitrogen is transported to streams as ammonium (NH_4^+), nitrate (NO_3^-), and dissolved organic nitrogen (DON). Nitrate is more soluble than ammonium, and abundant in agricultural and urban environments. Dissolved nitrate enters the soil and is carried by subsurface flow to streams. Stream water nitrate concentrations are highest during periods of baseflow when groundwater inputs are the primary source of streamflow (Watt, 2000). Soluble nitrate can also contaminate groundwater (Persky, 1986).

When soils with high levels of phosphorus are exposed to precipitation, phosphorus that is adsorbed to eroding soil particles is transported in overland flow. Stream water concentrations of phosphorus tend to be highest during stormflow events when the greatest quantities of sediment are carried to streams (Watt, 2000) (Figure 3.5). Phosphorus can also be transported in dissolved form as soluble reactive phosphorus (SRP), which is more readily available to plants than sediment-adsorbed phosphorus. When phosphorus loading is greater than the amount required by plants, excess phosphorus is stored in the soil. When sorption points on soil particles are saturated, more of the phosphorus remains in solution and is transported as soluble reactive phosphorus by subsurface flow (McDowell and Sharpley, 2001). These topics are discussed in more detail in Chapters 7, 8, and 9.

3.2.4 CATION EXCHANGE

Other nutrients are present in soil and soil water as mineral cations, positive ions such as calcium (Ca^{2+}), magnesium (Mg^{2+}), potassium (K^+), and sodium (Na^+) dissolved in the soil solution or adsorbed to clay (soil micelles), and humus particles. Ca^{2+}, Mg^{2+}, K^+, and Na^+ ions are essential plant nutrients. They are released to the soil solution from weathered bedrock and till minerals. Additional inputs come from airborne dust particles. Other cations in solution in the soil include ammonium (NH_4^+), aluminum (Al^{3+}) from mineral weathering, and hydrogen (H^+).

Soil micelles are very small (less than 0.001 mm) colloidal particles of clay suspended in soil solution. In general, the surfaces of soil micelles have a net negative charge, although this may vary with soil pH. Cations in soil solution are attracted to and held by the negative charges on the surface of the soil micelles. These negative charges are called cation exchange sites. The cation exchange capacity (CEC) of a soil, defined as the number of cations that can be adsorbed by a kilogram of soil, is an indication of the number of negative charges on soil particles and a measure of the capacity of a soil to retain mineral cation nutrients. Cations move back and forth between the soil solution and cation exchange sites. Whether a particular type of cation (H^+ or Ca^{2+}, for example) is held at a cation exchange site or released to the soil solution depends upon the strength of the bond and the concentration of that type of cation in the soil relative to the concentration of other cations. The list below shows soil cations in order of decreasing bond strength:

$$Al^{3+} > H^+ > Ca^{2+} > Mg^{2+} > K^+ = NH_4^+ > Na^+ \text{ (Barnes et al., 1998).}$$

An example of a cation exchange chemical reaction is shown below:

$$\boxed{\text{Micelle}} - Ca^{2+} + 2H^+ \leftrightarrow \overset{H^+}{\underset{H^+}{\boxed{\text{Micelle}}}} + Ca^{2+} \qquad (3.6)$$
(in soil solution)   (in soil solution)

Note that the charges must balance, one Ca^{2+} is replaced by two H^+, and that the chemical reaction can go in either direction.

Each micelle may have thousands of cation exchange sites. Even though H^+ ions form a stronger bond with the soil micelle, more Ca^{2+} ions can replace H^+ ions at cation exchange sites if there are

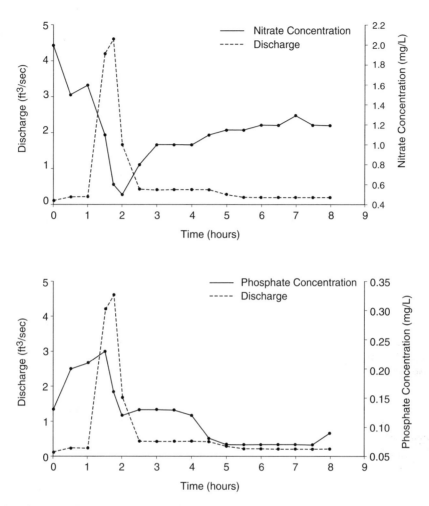

FIGURE 3.5 Concentrations of two nutrients, nitrate and phosphate, in streamwater during a typical storm. Nitrate and phosphate, because of their physical properties, act differently during storms. Nitrate dissolves easily in water, but phosphate tends to cling to sediments. (Top) The concentration of nitrate in the stream is greater than 2 mg/L before the storm begins. Most of the nitrate of the nitrate in the stream is derived from groundwater discharge rather than from overland flow. Therefore, as overland flow, corresponding to the peak discharge, reaches the stream, nitrate that was in the stream before the storm is diluted. As the storm passes and discharge decreases, the nitrate concentration begins to rise again. (Bottom) Phosphates react differently. Before the storm, the phosphate concentration is about 0.13 mg/L. The phosphate concentration peaks when the overland flow reaches the stream. Because the overland flow washes phosphate-containing sediments into the stream, the peaks in discharge and concentration coincide. As the discharge decreases, the phosphate concentration also decreases (Watt, 2000). (Courtesy of the U.S. Geological Survey.)

more of them in the soil solution. Likewise, an excessive input of H^+ ions will result in the displacement and release of Ca^{2+} ions. The continual release of a proportion of nutrient cations is required so that they can be available for plant uptake. However, excessive loss of nutrients from cation exchange sites may result in the net loss of cation nutrients, since cations that are not absorbed by plants are leached to streams. Substitution of H^+ ions for NH_4^+ ions on cation exchange sites can increase the supply of NH_4^+. This increases the rate of nitrification and the supply of NO_3^- ions in soil solution (Barnes et al., 1998; Brady and Weil, 2002).

Negatively charged cation exchange sites are also located on humus particles, but these sites are only available or active when the pH of the soil is relatively high (generally greater than 5.5).

The CEC of a soil depends on the proportion and type of clay in the soil and the amount of organic matter. The CEC varies regionally depending on the predominant regional soil types. Clay-rich alfisols in the Midwest have relatively high CECs, while the CECs of New England inceptisols are relatively low (Barnes et al., 1998; Brady and Weil, 2002; Holmgren et al., 1993).

3.2.5 Nutrient Conservation

3.2.5.1 Forests

In an undisturbed forested landscape, nutrient loss is variable, but generally small relative to agricultural or urban settings. Forest vegetation provides a variety of mechanisms that serve to retain and store nutrients in the forest ecosystem. This begins with throughflow. Many nutrients enter the forest as atmospheric deposition with precipitation. Natural rainwater is slightly acidic (pH approximately 5.6). This is due to the reaction of water with carbon dioxide in the atmosphere, resulting in the formation of carbonic acid (Brady and Weil, 2002):

$$CO_2 + H_2O \rightarrow H_2CO_3 \rightarrow\rightarrow HCO_3^- + H^+. \tag{3.7}$$

As precipitation comes in contact with the surface of leaves, branches, and stems, it loses hydrogen ions, thus becoming less acidic, and accumulates cations. These nutrient cations are carried into the soil solution where they are available for uptake and assimilation by plant roots. Helvey and Kunkle (1986), working at the Fernow Experimental Forest in West Virginia, found that precipitation that passed through the forest canopy (throughfall) contained 38% fewer hydrogen ions. Throughfall also accumulated calcium ions as it came in contact with the leaves and branches of trees. At this site, slightly greater quantities of calcium, nitrate, and sulfate entered the system with precipitation then left the watershed in streamflow. Stream pH values averaged 5.8 during the 2 years of the study, values that were typical of the 15 year period of record. Thus the forest canopy protected both soil and stream water from further acidification. Studies at the Hubbard Brook Experimental Forest in the White Mountains of New Hampshire (Likens et al., 1977) also showed that throughfall and stemflow were enriched while passing through the forest canopy. In this area, 88% of the hydrogen ions in precipitation were adsorbed while passing through the forest canopy. There were 10 times more calcium ions in throughfall reaching the forest floor after passing through the tree canopy (1.59 mg/L versus 0.16 mg/L). Magnesium was 15 times higher (0.45 versus 0.03 mg/L), phosphorus 18 times higher (0.15 versus 0.0026 mg/L), and potassium 91 times higher (6.37 versus 0.07 mg/L).

Trees and other plants act quickly to absorb available nutrients and store them in living tissue. Nitrification rates are low because NH_4^+ resulting from decomposition is quickly taken up by plants and microbes (Bormann and Likens, 1994). Often these nutrients are recycled within the living plant. For example, nitrogen is withdrawn from leaves in the fall, stored in plant tissue during the dormant season, and used to fuel new growth the following spring (Kozlowski and Pallardy, 1997). Much of the remaining on-site nutrient capital is consumed and immobilized by soil bacteria and fungi that thrive in forest soils. The overall loss of mineralized nutrients (nitrates, phosphates, and nutrient cations) to leaching and streamflow is very small (Hobbie and Likens, 1973; Likens et al., 1967; Meyer and Likens, 1979).

Box 3.2   Changes in Nutrient Cycling Following Complete Removal of Forest Vegetation

An experiment in the 1960s on Watershed 2 of the Hubbard Brook Experimental Forest in the White Mountains of New Hampshire investigated the effect of removing trees from the forest ecosystem. The results of this study showed clearly the disruption in ecosystem processes that can occur when the forest vegetation is removed from a site for an extended time period

(Bormann and Likens, 1994; Likens et al., 1970). This ecological experiment does not reflect the effects of conventional timber harvesting on the nutrient capital of forested sites. Other studies have shown that nutrient loss following timber harvesting is quite variable. In some instances, losses of nitrate and calcium may be minimal or nonexistent (Binkley and Brown, 1993; Lynch and Corbett, 1994; Verry, 1972) (see Chapter 6).

At Hubbard Brook, all the vegetation (trees and shrubs) were cut and left on the site—the soil was not disturbed. The watershed was treated with herbicide for 3 years to eliminate regrowth (Likens et al., 1970). The removal of the tree canopy resulted in higher soil temperature and increased net precipitation and soil moisture. This increased the rate of decomposition during the first year following treatment. The absence of vegetative uptake meant that more nitrogen in the form of ammonium (NH_4^+) was available to soil bacteria. These two factors—more decomposition and more available ammonium—resulted in increased nitrification (the conversion of ammonium to nitrate) by soil bacteria. Nitrification increased 7-, 11-, and 9-fold during the 3 years of herbicide treatments following forest removal. During the 3 year herbicide period, mean annual stream water nitrate concentrations ranged from 38.4 to 58.9 mg/L, compared to concentrations near 1 mg/L before the experimental treatment.

Bacteria release two hydrogen ions (H^+) in the process of converting the NH_4^+ cation to NO_3, thus increased nitrification can lower the pH of the soil solution. In addition, precipitation reaching the ground on deforested sites was more acidic because the forest canopy that had acted to remove and store hydrogen ions was removed. This flush of hydrogen ions into the soil caused the displacement of Ca^{2+}, Mg^{2+}, K^+, and Na^+ from cation exchange sites. In the absence of vegetative uptake, these nutrient cations were then carried with subsurface flow and flushed into streams (Bormann and Likens, 1994; Bormann et al., 1974). High concentrations of hydrogen ions can also lead to the mobilization of aluminum cations (Al^{3+}).

Stream water chemistry returned to near pretreatment conditions within 3 to 5 years following the cessation of herbicide treatments and the regrowth of vegetation. Nitrate concentrations were reduced to 0.1 mg/L, far below pretreatment levels, as young, vigorously growing trees filled the site.

3.2.5.2 Grasslands

Grasslands develop in regions where the annual precipitation (250 to 800 mm/year) is less than required for tree growth, but greater than found in desert regions. Plants that thrive in grasslands have basal meristems, primary growth tissue near the ground, where it is not damaged by fire, and extensive dense root systems that store nutrients and support annual regrowth of aboveground stems, leaves, and flowers. Prior to European settlement, large areas of Illinois, Minnesota, Missouri, parts of Wisconsin, and most of Iowa were prairie grasslands. Today less than 1% of the native prairie remains in Minnesota, Missouri, Iowa, and Illinois (Whitney, 1994). Before European settlement, regional climate, topography, and native vegetation led to high levels of nutrient storage in prairie soils. Grasses protect the soils from the force of raindrops and infiltration rates are high. High levels of calcium and magnesium compounds accumulate in soils because rainfall is insufficient to carry these substances to streams. Grasses absorb large quantities of nutrients that are then returned to the soil and reabsorbed as stems, leaves, flowers, and roots die, decay, and regrow each year (Smith, 1992). The soils that develop in these ecosystems (mollisols) are characterized by high levels of calcium-rich organic matter and the highest CECs in the United States (Brady and Weil, 2002).

As in other systems, phosphorus in grasslands tends to be immobile and often occurs in forms that are unavailable to plants (Seastedt, 1988). Phosphorus is transferred among neighboring plant species in grassland communities, thus retaining the nutrient in biologically active form (Walter et al., 1996).

3.2.5.3 Variability in Nutrient Processing

There are differences in nutrient processing even among forested ecosystems that appear to be relatively similar. In this section, using nitrogen loss from forested ecosystems as an example, we discuss possible sources of local variability. In addition to annual and seasonal variation resulting from climate, these include intrinsic environmental characteristics of particular watersheds such as soil type, topography, species composition, stand age, and land use history, and extrinsic factors such as nitrogen loading from atmospheric deposition.

3.2.5.3.1 Topography

Creed and Band (1998a,b) examined nitrogen export from 13 small watersheds within the Turkey Lakes watershed in central Ontario, Canada. The watersheds are remarkably similar except for variations in topography. All the watersheds are located within an uneven-aged old-growth (120 to 180 years) forest dominated by sugar maple (*Acer saccharum*) with small stands of yellow birch (*Betula alleghaniensis*). Both the concentrations of nitrate in discharge waters and the seasonal pattern of peak nitrate concentrations (10-year averages) varied among the different watersheds. Peak nitrate concentrations occurred during spring snowmelt in all the watersheds, ranging from approximately 0.8 to 1.5 mg/L. Eight of the 13 watersheds showed additional peaks in nitrate concentration at other times during the year. Creed and Band hypothesized that variations in nitrate export were the result of topographic patterns that controlled the expansion and contraction of variable source areas within the watershed. Nitrate builds up in the soil during cold and dry periods and is flushed to surface waters with subsurface flow when a rising saturated zone intersects nitrate-rich material at or near the forest floor. This increase in the thickness of the saturated zone accelerates the movement of water and accumulated nutrients to streams (ephemeral, intermittent, and perennial). The proportion of watershed area that came in contact with the saturated source area was influenced by the spatial arrangement, form (concave, planar, convex), and steepness of the hills and valleys within that particular watershed.

3.2.5.3.2 Geology and Soils

In general, soils with higher CECs are more likely to hold ammonium ions (NH_4^+) on soil micelles. This limits the supply of NH_4^+ ions that would be transformed to soluble NO_3^- through nitrification. The CEC is highest in silt loam and silty clay soils, and in soils with large amounts of humus or organic matter (Barnes et al., 1998).

A study by Hornbeck et al. (1997) examined differences among nitrate concentrations in streams from three small upland forested watersheds (Cone Pond, Hubbard Brook, and Sleepers River) located within an 80 km radius in northern New Hampshire and Vermont. Despite the proximity and general similarity of these sites, there were substantial differences in the amount of nitrate lost to stream water. No nitrate loss was detected at the Cone Pond site, losses of 0.3 to 0.6 kg/ha/yr were measured at Hubbard Brook, and losses of 1.1 to 1.6 kg/ha/yr were measured at Sleepers River. (Please note that nutrient losses in stream water may be described as concentrations [mg/L] as in the Hubbard Brook Watershed 2 study [Box 3.2] or as loads [kg/ha/yr]. The difference between these two measurements is explained in Box 3.3.) The authors speculate that the higher losses at the Sleepers River site may be due to soil differences (higher soil pH, total nitrogen pool, available NH_4^+, and carbon:nitrogen ratios) that favor nitrification. Both the bedrock and till at the Sleepers River site were composed of calcium and magnesium carbonates; bedrock and surficial geology at the other two sites resulted in acidic soils. The authors were puzzled by the unexpected absence of nitrate in streamflow at the Cone Pond site. Precipitation and soil water measurements at this site showed that nitrate in precipitation was immobilized and stored in forest floor material and not released to streams and that little nitrification took place. They speculate that this condition may have been the legacy effect of an intense fire that occurred on the site in 1820.

Soil characteristics also influence denitrification rates. Groffman and Tiedje (1989a,b) found that soil texture and drainage class were the primary landscape factors governing rates of denitrification. In this study, which examined multiple sites in Michigan, the estimated annual nitrogen loss through denitrification ranged from less than 1 kg N/ha/yr in a well-drained sand soil to more than 40 kg N/ha/yr in a poorly drained clay loam soil.

Box 3.3   Concentration (mg/L) versus Loading ([kg/ha]/yr)

Most water quality data are measured and reported as a concentration (e.g., mg/L, or mass/volume). This is consistent with the purpose of many water quality monitoring programs—to assess (1) compliance with environmental regulations, (2) suitability for human consumption or contact, or (3) proximity to ecological or physiological thresholds. There are at least two shortcomings of limiting water quality measurements and subsequent analyses to concentration data. First, the relative ecological impact at the landscape scale may be insignificant. In other words, a high concentration of a water quality constituent in a very small volume of water may have little or no effect downstream when mixed with larger volumes of water. Therefore, when management actions are designed in relation to concentration data, there is a risk that time and resources will be expended on interventions that have little or no effect at the landscape scale. Second, a comparatively low concentration in a stream or river with a substantial discharge may nominally comply with environmental regulations, but deliver a large mass of undesirable material to a receiving water (e.g., reservoir, lake, or wetland). This is the complement of the first case. By simultaneously measuring the discharge of a stream or aquifer (volume/time) as the multiplier for concentration (mass/volume), loading (mass/time) can be calculated. For example, conversion of a forest to another land use typically increases the generation of nutrients and sediment *and* the volume of stormflow. Hence the dilution of nutrient and sediment concentrations in larger volumes of stormwater masks the adverse impact of the land use change and the need for prevention, mitigation, or restoration. The variation of streamflow in relation to seasons and storm events (often by orders of magnitude) underscores the need to design representative sampling strategies. Systematic measurements of water quality constituents and discharge can be used to construct a budget or mass balance for a watershed (outflow or export) or receiving water (inflow or loading to a reservoir). This is the rationale of the total maximum daily load (TMDL) concept and watershed management plans based on proportional contributions to pollutant loading versus compliance programs based solely on concentration (National Research Council, 2001).

3.2.5.3.3   Forest Floor Carbon:Nitrogen Ratio

The amount of carbon relative to the amount of nitrogen (C:N ratio) of the forest floor material is very important in determining the amount of mineral nitrogen in soil solution (Gunderson et al., 1998). Soil microbes need 1 g of nitrogen for every 24 g of carbon they consume to produce new microbial tissue. The C:N ratio of microbial tissue is approximately 8:1 and the remaining carbon is consumed and released as CO_2 during microbial respiration, which provides the energy for metabolic processes. When the C:N ratio is high, microbes consume most of the available NH_4^+ and NO_3^-. It is thus immobilized in microbial tissue and the pool of dissolved mineralized nitrogen is very small.

If the C:N ratio is smaller, pools of dissolved, mineralized, biologically available nitrogen, NH_4^+, and NO_3^- remain in soil solution. The rate of nitrification—the conversion of ammonium to soluble nitrate—increases, leading to the leaching of nitrate to forest streams. Yoh (2001) found very little nitrogen present as NO_3^- in the soil solution at soil C:N ratios greater than 20. Based on data from forests in Europe and Scandinavia, Gunderson et al. (1998) recommend that the following guidelines might be used to predict the likelihood of nitrate leaching in conifer stands: C:N > 30 = low risk; 25 < C:N < 30 = moderate risk; C:N < 25 = high risk.

3.2.5.3.4 Species Composition

In a review of studies from the Northeast, Binkley (2001; Binkley et al., 2004) found that nitrate concentrations in streams in hardwood forest stands were greater (mean 0.46 mg/L) than concentrations in conifer stands (mean 0.15 mg/L).

A group of researchers at the Institute of Ecosystem Studies, Millbrook, New York, working primarily in the Catskill Mountains, found higher rates of mineralization and nitrification and higher concentrations of soil solution nitrogen in sugar maple stands than in stands where beech (*Fagus grandifolia*) or red oak (*Quercus rubra*) trees were dominant in the overstory (Lovett and Rueth, 1999; Lovett et al., 2002). Lovett et al. (2002) hypothesize that the C:N ratio in sugar maple stands is low due to the characteristics of sugar maple leaf litter (low C:N ratio, high nitrogen content). An increased supply of ammonium resulting from the decomposition of these leaves undergoes nitrification, producing more dissolved nitrate in the soil and an increase in nitrogen leaching. Peterjohn et al. (1999), working in the Fernow Experimental Forest in West Virginia also found greater extractable pools of nitrate in sugar maple dominated forests. In a study in the Allegheny Forest of western Pennsylvania, Lewis and Likens (2000) found that stream nitrate concentrations were 2.6 to 7.0 times lower in red oak stands than in forest stands dominated by other hardwoods—sugar maple, American beech, yellow birch, black cherry (*Prunus serotina*), and white ash (*Fraxinus americana*). Lewis and Likens concluded that rather than controlling the rate of nitrate loss to streams, the red oaks "merely occur" on soils that had low nitrate losses. This is consistent with other studies that find that red oak seedlings are shaded out by other faster growing hardwoods such as red and sugar maple on more fertile sites (Lorimer, 1992).

Higher concentrations of nitrogen are found in forest streams when nitrogen fixing species are present in riparian areas. Hurd et al. (2001) examined the nitrogen contribution of nitrogen fixing speckled alders (*Alnus incana* spp. *rugosa*) in forested watersheds in the Adirondacks. They estimated that 85% to 100% of the foliar nitrogen content of alder leaves came from nitrogen fixation and that alder leaves contributed 43 kg/ha of fixed nitrogen in the year of the study (1997). Hurd et al. concluded that speckled alder in the wetlands of northern New York "added substantial amounts of N to alder dominated wetlands in the Adirondack Mountains" and may increase nitrogen loading in streams as well.

3.2.5.3.5 Seasonal Variation and Groundwater Contributions

In general, stream water nitrate concentrations are highest during spring snowmelt, when nitrate stored in the soil during the winter dormant season is flushed out of the soil. Nitrate concentrations decrease during the growing season when vegetative demand is high. These patterns may be modified by nitrate that reaches deeper groundwater flow paths.

Burns et al. (1998) found that the nitrate concentrations in water from some perennial springs in the Neversink River watershed in the Catskill Mountains were higher than those found in nearby shallow groundwater nearby during the summer months. These springs were connected to and supplied by older (6 to 22 months), deeper groundwater that had accumulated nitrate in previous dormant seasons.

3.2.5.3.6 Stand Age

Studies at Hubbard Brook Experimental Forest in New Hampshire and Coweeta Hydrologic Laboratory in North Carolina have clearly shown that nitrogen leaching is relatively low in rapidly growing, early successional or aggrading forest stands due to high rates of vegetative uptake (Boring et al., 1988; Bormann and Likens, 1979; Pardo et al., 1995). Vitousek and Reiners (1975) (see also Gorham et al., 1979) developed the "nutrient retention hypothesis," linking nutrient export to the age or successional stage of a forest stand. Following a stand-initiating disturbance, the forest regrows and biomass accumulates in living plant tissue, dead wood, and in forest floor material. As a result, carbon and nutrients are stored in the forest ecosystem. As long as biomass is accumulating, a large percentage of the annual nutrient input to a site will be stored on-site in living tissue. Nutrient export is equal to nutrient import minus storage. As a stand matures, stored biomass approaches a maximum. The amount of additional nutrients retained each year is reduced and nutrient export increases. Vitousek and Reiners compared stream water concentrations of nitrate and several other mineral nutrients from mature (logged more than 100 years ago) and approximately 25-year-old stands at Hubbard Brook to test this hypothesis. They found that nitrate concentrations in streams from the older watersheds were significantly higher than concentrations in streams from the 25-year-old stands.

Recently additional studies have been completed, adding new information regarding the effect of stand age on nitrogen leaching. Hedin et al. (1995) investigated nitrogen processing in old-growth conifer forests in Chile. Atmospheric pollution complicates studies of nitrogen processing in the northern hemisphere, and this study provided an opportunity to investigate old-growth temperate forests in a relatively pollution-free, undisturbed area. Thirty-one forested watersheds were included in this study. The findings supported Vitousek's nutrient retention hypothesis. Very low levels of nutrients were taken up by these mature forests, and for the most part the amounts exported in stream water reflected the atmospheric input from marine aerosols — sea salt blown in from the Pacific Ocean. In other words, there was no net storage of new nutrient input. Nitrogen processing was slightly different than that of other mineral nutrients. While stream water nitrate concentrations were several orders of magnitude lower than mean concentrations for North American streams, nitrogen was being leached from these old-growth forests, predominantly in the form of DON derived from accumulated organic matter in the forest floor. In general, nitrogen in streams in temperate forests of the northeastern United States occurs as 45% nitrate, 10% ammonium, and 45% DON, with mean concentrations of 0.5 mg/L nitrate, 0.09 mg/L ammonium, and 0.32 mg/L DON (Binkley, 2001). In Chile, DON constituted 95% of the nitrogen in stream water, with a mean concentration of 153 µg/L (0.153 mg/L); ammonium was 4.8% (7.4 µg/L), and nitrate only 0.2% (0.37 µg/L). If DON is included in the nitrogen loss analysis, these forests might be described as leaking nitrogen. Hedin et al. (1995) suggest that "the relative abundance of NO_3^- vs. NH_4^+ may have changed as a function of N loading" in temperate forests of the Northern Hemisphere and that the proportion of NO_3^- in stream water from old-growth stands may be a sensitive indicator of the influence of atmospheric pollution.

Mitchell et al. (1992) compared biogeochemical cycling of nitrogen and sulfur in a 300-year-old stand in the Turkey Lakes watershed in Ontario, Canada, and a 100-year-old stand in the Huntington Forest in the Adirondack Mountains of New York. They found much higher levels of nitrate leaching (1300 versus 18 Mol_c/ha/yr) in the Turkey Lakes stand despite reduced nitrogen input at the Ontario site. Both stands were dominated by sugar maple, however, there was a higher proportion of mature beech trees in the overstory of the Huntington Forest. Mitchell et al. attributed the higher rate of nitrate leaching at the Turkey Lakes site to stand age, but acknowledged that higher proportions of slowly decomposing beach leaves in forest litter and a higher soil C:N ratio may have increased nitrate retention at Huntington Forest.

A study in upper Michigan (Fisk et al., 2002) compared nitrogen storage and cycling in three 80-year-old midsuccessional forest stands and in three uneven-aged old-growth stands in which the oldest trees were 200 to 300 years old. In this case, while there were statistically significant differences among stands in nitrogen export measured in kilograms of nitrogen per hectare per year, the differences were not related to stand age. Nitrate was the primary form of nitrogen exported from two of the old-growth and two of the midsuccessional stands, while DON was the primary form of nitrogen exported from the remaining stands. It appeared that intrinsic site factors were more important than stand age in determining the magnitude of nitrogen loss (note the Creed and Band [1998a,b] study of site topography, discussed previously).

3.2.5.3.7 Land Use History

There is some indication that past land use and forest management practices may influence nitrogen export on currently forested sites. A study conducted in southern Germany (Feger, 1992) compared nitrate export from two stands of Norway spruce subject to similar levels of atmospheric nitrogen deposition and containing similar soil nitrogen reserves but differing substantially in management history. Historical forest management practices had included litter raking and removal and grazing at one of the stands, but not at the other. Nitrogen fertilizer (700 kg/ha of $[NH_4]_2SO_4$) was added to both plots. The raked and grazed site retained nearly all of the added nitrogen, while the nitrogen additions on the other site led to increased rates of nitrification and aluminum mobilization and leaching of nonacid cations and nitrate. Latty et al. (2004) reported that logging and burning had reduced the carbon and nitrogen pools in managed compared to old-growth stands; however, this did not result in differences in the C:N ratio or in net nitrification rates.

Studies of nitrogen addition experiments at Harvard Forest, Petersham, Massachusetts (Aber et al., 1998; Magill et al., 1997, 2000), and in Maine (Magill et al., 1996) (see Box 3.4) attributed stand differences in nitrogen export to land use history. A Harvard Forest red pine (*Pinus resinosa*) stand that was plowed and used for pasture until 1926 exported far more nitrogen than a hardwood stand that had been clearcut once following the 1938 hurricane but never plowed. In this case, it seemed possible that agricultural land use had resulted in increased soil nitrogen levels, shown in higher initial rates of nitrogen mineralization in the soil of the pine stand. Thus it was possible that the effect of prior land use was still evident 70 years after reforestation (Aber et al., 1998).

Box 3.4   Nitrogen Addition Experiments

In order to predict the effects of atmospheric deposition of excess nitrogen over long periods of time, researchers are currently involved in a variety of nitrogen addition experiments (Magill et al., 1997). In these experiments, nitrogen is artificially added to experimental stands for a number of years while chemical changes in the forest canopy, forest floor, and forest soils are measured. The purpose of these experiments is to test the ability of different forest ecosystems to retain nitrogen, to determine the vulnerability of these ecosystems to nitrogen saturation, and to measure nitrogen leaching to streams that results when nitrogen saturation occurs.

Harvard Forest, Massachusetts

The Harvard Forest experiment was begun in 1988. Ammonium nitrate (NH_4NO_3) was applied to two stands at four rates: control, low nitrogen (5 g $N/m^2/yr$), high nitrogen (15 g $N/m^2/yr$), and low nitrogen plus sulfur. One stand was a 70-year-old red pine plantation planted on agricultural pasture land in 1926. The other stand was a 50-year-old hardwood forest dominated by black and red oak with black birch, red maple, and American beech; it had been clearcut once, following the 1938 hurricane, but never plowed. Soils and topography were similar in the two stands. Nitrogen retention was quite high in both stands. In the low nitrogen treatments, 97% to 100% of the added nitrogen was retained in both stands. In the high nitrogen treatments, the hardwood stand retained 96% of the added nitrogen, while the pine stand retained 85%. The pine stand showed a significant increase in nitrate leaching in 1989 after the first year of nitrogen additions. In the hardwood stand, a significant increase in nitrogen leaching did not occur until 1995. Wood production declined in the pine high nitrogen plot (Aber et al., 1998; Magill et al., 1997, 2000).

Bear Brook Watershed, Maine

This is a paired watershed experiment begun in 1987. The topography, soils, and bedrock of both watersheds are similar. Tree species composition is similar as well—the upper areas of both watersheds are 80- to 120-year-old red spruce and balsam fir, while the lower portions of the watersheds consist of 45- to 55-year-old mixed northern hardwoods, American beech, sugar maple, red maple, yellow and white birch, and eastern hemlock. Patterns of nutrient fluxes were similar prior to the initiation of experimental treatments in 1989. Since that time the West Bear watershed has received 25.2 kg N/ha/yr in addition to the 8.4 kg N/ha/yr from atmospheric deposition that falls on both watersheds. There were significant differences in nitrogen mineralization between the watersheds and between forest types within the watersheds. Net nitrogen mineralization was higher in the forest floor of hardwood areas compared to softwood areas. Nitrification rates did not differ significantly by forest type. In 2001, after 12 years of nitrogen additions, the experimental West Bear watershed had significantly higher rates of nitrogen mineralization and nitrification and lost 20% of nitrogen input (7.3 kg N/ha/yr) in stream export, while the untreated East Bear watershed lost 4% (0.30 kg N/ha/yr) (Jefts et al., 2004).

The Great Lakes Region

Two related studies (Pregitzer et al., 2004; Zak et al., 2004) tested the vulnerability of northern hardwood stands to chronic nitrogen additions. These studies began in 1987 and involved control and experimental nitrogen addition plots in four forest sites located along a 500 km gradient in Upper and northern Lower Michigan. The four sites are all second-growth hardwood stands, approximately 90 years old, dominated by sugar maple. Soils are similar throughout the region. There is some variation in average temperature, nitrogen deposition, and inherent soil nitrogen availability. Three 30 m × 30 m control plots were established at each site in 1987; nitrogen additions (3 g $NO_3/m^2/yr$) were begun on comparable experimental plots established in 1993 and results were reported after an 8 year study period. Highly significant increases occurred in the concentrations of dissolved organic carbon (DOC), NO_3, and DON in the soil solution in the experimental stands and an average of 2.2 g N/m^2 (72% of the added NO_3) leached from the soils in the last 2 years of the study period.

It is important to recognize that nitrogen processing and export in a particular forest stand (as well as the processes and export of other nutrients) may be influenced by many of the factors discussed above to varying degrees. In general, we would expect higher rates of nitrogen export in mature hardwood forests (especially those with high proportions of sugar maple) with large amounts of nitrogen stored in soil organic matter, high rates of nitrification due to favorable conditions (moderate pH, temperature, etc.) for biological activity (Fenn et al., 1998), and watershed characteristics (soil and topography) that lead to more extensive saturated source areas (Creed and Band, 1998a,b).

3.3 PESTICIDES

The U.S. Geological Survey (1999) found that "streams and groundwater in basins with significant agricultural or urban development, or with a mix of these land uses, almost always contain complex mixtures of nutrients and pesticides." As with nutrients, the mode of transport depends on the chemical characteristics of the pesticide. More soluble pesticides tend to be transported in dissolved form with subsurface flow and may be more likely to contaminate groundwater. Other pesticides adhere to sediments and move with eroded soil particles. In general, pesticides that have a solubility in water of less than 1 mg/L and a soil half-life of more than 30 days have the potential to accumulate in sediment and aquatic biota (U.S. Geological Survey, 2000). Pesticides that are retained in the soil degrade into other compounds; however, these secondary metabolites may not be entirely harmless (Larson et al., 1997). Pesticide transport tends to be highest in the first storm event following application, and concentration varies with stream size (dilution effects are lessened in smaller streams) (Fenelon, 1998; Gaynor et al., 1995; Mullaney and Zimmerman, 1997). Thus aquatic biota may be subjected to acute concentrations that are well above those measured during periods of baseflow.

The use of organochlorine pesticides such as DDT, chlordane, and dieldrin was discontinued in the 1970s. These compounds have low solubility in water, high solubility in fat, and are resistant to degradation. They tend to adsorb to sediments and organic material. DDT, chlordane, and dieldrin are still frequently detected in streambed sediments. Levels of organochlorine pesticides in fish tissue have declined considerably over the last 20 years, but concentrations of DDT in fish tissues at the most contaminated sites may still pose a risk to human health (Myers et al., 2000; Wall et al., 1998). Pesticides currently in use tend to have greater solubility and shorter half-lives than the organochlorine pesticides used in the past (Larson et al., 1997; U.S. Geological Survey, 2000).

More soluble pesticide compounds can be found both in dissolved form and adsorbed to particles. The proportion of the pesticide compound that sorbs to particles depends on its chemical characteristics, the concentration of the pesticide in the soil solution, the soil type, soil pH, the organic matter content of the soil, and temperature (Larson et al., 1997). Pesticide breakdown rates in both soil and surface waters may be enhanced by exposure to sunlight. The rate of photodecomposition is greater in salt water than in fresh water and greater for some compounds than others (Lin et al., 1999).

A study in northern Missouri (Blanchard and Lerch, 2000) concluded that agricultural herbicides, including atrazine and cyanazine, were detected at the highest concentrations in streams in areas with low permeability soils. In these areas, herbicides were carried to streams by overland flow. When soil infiltration rates were greater, moderately soluble compounds such as atrazine were retained and degraded in the soil and very little was leached with subsurface flow.

Leu et al. (2004a,b) examined herbicide loss to surface water from an agricultural region in Switzerland with moderate slopes (less than 10%) by applying identical mixtures of herbicides (atrazine, dimethenamid, and metolachlor) on the same day to different fields. The major herbicide loss occurred during the first two storm events following herbicide application. As in the Blanchard and Lerch (2000) study, herbicide loss was greatest in areas where soil permeability and infiltration rates were low, resulting in overland flow that carried herbicides directly to streams or to tile drains.

A study in Iowa (Squillace and Thurman, 1992) also examined the transport of herbicides from cropland. This study found that while atrazine adsorbed to sediments in the fields, the compound began to dissolve when the sediment concentration was less than 50,000 mg/L. Atrazine was transported in association with soil particles in overland flow, dissolved in transport during intense rainstorms, or dissolved after reaching the stream channel. The proportion of dissolved versus sediment-adsorbed atrazine increased as the concentration of sediment in the water decreased. Ninety-four percent of the annual load of atrazine in the river was transported with overland flow and 6% entered through groundwater transport (leaching). Between 1.5% and 5% of the atrazine applied to crops during the year was transported to stream channels.

Dissolved pesticide compounds can volatilize and be distributed over broad areas through atmospheric deposition (Larson et al., 1997). A study in Canada (Donald et al., 2001) found that there was no difference in the concentrations of herbicides in water from wetlands in wildlife areas, farms with no pesticide use, conventional farms with moderate pesticide use, and no-till farms with high pesticide use. Agricultural herbicides were lost to the atmosphere by volatilization from soil and through plant evapotranspiration. The herbicides were then redeposited locally or became entrained in local convective clouds and were redistributed by rainfall in a relatively homogeneous mixture over the broader landscape (Donald et al., 2001). Table 3.1A and Table 3.1B list several commonly used herbicides and insecticides, with half-lives, solubility, and the associated probability of groundwater contamination.

TABLE 3.1A
Properties of Commonly Used Pesticides: Herbicides

Name	Toxicity to mammals and fish	Mean half-life	Solubility	Transport (sediment adhesion)	Means of degradation	Groundwater contamination risk
Alachlor	Slight to moderate (use restricted; carcinogenic in laboratory animals)	8 days	High: 240 mg/L at 25°C	Mostly dissolves in water, low to moderate soil adhesion	Soil microbes	Little (breaks down rapidly in water)
Atrazine	Slight to moderate	60 to 100 days (can persist more than 1 year in cold or dry conditions)	Moderate: 28 mg/L at 20°C	Moves with sediments and dissolves in water	Soil microbes	High due to persistence
Cyanazine	Moderate	2 to 4 weeks in air-dried sandy clay loam; 7 to 10 weeks in sandy loam; 9 weeks in fresh sandy clay soil	Low	Can be transported in overland flow and sediment and can leach through soil to groundwater	Soil microbes	High
Simazine	Slight (sheep and cattle especially sensitive)	28 to 149 days	Low: 5 mg/L at 20°C	Adsorbs to clay and muck; poorly binds to other soils	Soil microbes	Moderate
Prometon	Slightly to moderately toxic to fish; slightly toxic to aquatic invertebrates	139 to 2227 days	High	Adsorbs to soils with high organic or clay content	Soil microbes	High

EXTOXNET, http://extoxnet.orst.edu/pips/ghindex.html (July 2006).

TABLE 3.1B
Properties of Commonly Used Pesticides: Insecticides

Name	Toxicity to nontargeted insects, mammals, and fish	Mean half-life	Solubility in water	Transport	Means of degradation	Groundwater contamination risk
Diazinon	Highly toxic to birds, fish, and bees	14 to 28 days	High: 40 mg/L at 20°C	Moderately adsorbed to soil	Bacterial enzymes	Seldom migrates below the top 0.5 inch of soil, but has been found in wells in California
Endosulfan	High	50 days	Low: 0.32 mg/L at 22°C	Adsorbs to soil particles (with sediment)	Soil microbes	Low
Carbaryl	Moderately toxic to aquatic organisms (LC_{50} of 1.3 mg/L in rainbow trout); lethal to bees and other nontargeted insects	7 to 14 days in sandy soil; 14 to 28 days in clay soil; persists longer in low pH water	High: 40 mg/L at 30°C	Bound by organic matter (in overland flow)	Sunlight and bacteria	Detected in groundwater in three cases in California
Chlorpyrifos	Moderately to highly toxic to birds, fish, aquatic invertebrates, wildlife, and bees	60 to 120 days; can range from 2 weeks to more than 1 year	Low: 2 mg/L at 25°C	Adsorbs strongly to soil particles (with sediment)	Ultraviolet light, chemical hydrolysis, and soil microbes	Low

EXTOXNET, http://extoxnet.orst.edu/pips/ghindex.html (July 2006).

3.4 METALS

Pollution by metals is widespread. Metals from industrial processes, military operations, and mining enter the environment through atmospheric deposition and improper disposal of waste materials (Breault and Harris, 1997; Chen and Cutright, 2003; Long et al., 2000; Rheaume et al., 2001). Atmospheric deposition of lead from leaded gasoline has resulted in the worldwide dispersal of lead aerosols. Metals are found in agricultural pesticides, fertilizers, and manure (from animal feed) (Alloway, 1995). Lead-arsenic pesticides were widely applied on apple, blueberry, and potato crops in New England from about 1900 to 1950 and streams and sediments near these sites still contain high contaminant concentrations (Robinson and Ayuso, 2004). The application of composted sludge as fertilizer may release associated heavy metals into the environment through leaching (Leita and De Nobili, 1991; Sawhney et al., 1994). Traces of arsenic, mercury, lead, cadmium, chromium, copper, nickel, and zinc are routinely found in sediments in streams in the Northeast (Lindsey et al., 1998; Myers et al., 2000; Wall et al., 1998).

These elements vary in toxicity from arsenic and cadmium, which are very toxic to humans, animals, fish, and birds, to copper and zinc, which are essential nutrients for plants and animals, but can be toxic when absorbed or ingested in excessive amounts (Brady and Weil, 2002). Metallic ions frequently enter stream systems adsorbed to fine-grained sediments (Long et al., 2000). Metallic ions can also enter streams dissolved in stormflow and be deposited in channel sediments. From stream sediments, some metals (mercury in particular) may be absorbed by plants and enter the food chain, bioaccumulating in fish tissue and presenting health hazards to humans who consume the fish (Haitzer et al., 2003; Rheaume et al., 2001). Metals form many different types of compounds in soils and the transport and bioavailability of these compounds will vary depending upon their chemical properties, the pH of the soil solution, and the nature and abundance of various other soil components.

We will examine the chemistry and transport of zinc compounds as one example of metal transport from land to streams. As noted earlier, zinc is an essential nutrient, a component of various plant and animal enzymes involved in protein and carbohydrate metabolism. Excess zinc levels are principally toxic to plants, but may also endanger the health of fish (Brady and Weil, 2002; Kiekens, 1990). Zinc is commonly found in stream sediments in the Northeast, although concentrations are usually below probable toxic effect levels. Breault and Harris (1997) found zinc in every sample from 43 sampling sites throughout New England. All concentration levels were higher than the estimated background concentration of zinc in the Earth's crust, implying anthropogenic sources.

Zinc is released naturally through the weathering of bedrock and soil minerals. The zinc cation (Zn^{2+}) competes with calcium cations (Ca^{2+}) to adsorb to cation exchange sites on soil micelles. A certain portion of zinc ions in soil solution become incorporated into the crystalline structure of the soil particles. This occurs more frequently in neutral or alkaline soils (pH $\geq$ 7) and this portion of Zn^{2+} is then permanently fixed in the soil. Zn^{2+} forms complexes with chloride (ZnCl), phosphate ($ZnHPO_4$), nitrate ($Zn[NO_3]_2$), and sulfate ($ZnSO_4$), and with water to form zinc hydroxide (ZnOH). Zinc ions also form complexes with organic compounds. These are reversible reactions and the Zn^{2+} ions may be released into the soil solution if the compounds dissolve. Under minimally disturbed conditions, the zinc concentration in the soil solution is very low (Kiekens, 1990). The solubility of zinc compounds depends on soil pH and is highest in acid soils. Leita and De Nobili (1991) recorded a 10-fold decrease in water-extractable zinc when pH increased from 6 to 9. At high pH, zinc hydroxides are too insoluble to supply the ions needed for plant growth and plants may suffer from zinc deficiencies.

Human activities have increased the amount of zinc released to soils, water, and the atmosphere. Galvanizing is a process in which metals are coated with zinc to prevent rust. Water draining off galvanized roofs and through galvanized pipes carries dissolved zinc into stream channels (Good, 1993; Rheaume et al., 2001). Discarded batteries and old tires contain zinc compounds that can dissolve and release zinc in stormflows (Brady and Weil, 2002). Zinc is used extensively in industrial processes and may enter streams with wastewater and as leachate from landfills. Zinc can also enter streams and other water bodies directly with atmospheric deposition (Holmgren et al., 1993). Zinc phosphide is a restricted use pesticide used to kill rodents (EXTOXNET, 1996). Another source of zinc in streams is overland flow and erosion from fields where municipal sewage sludge has been applied as fertilizer (Alloway, 1995; Berti and Jacobs, 1996; Brady and Weil, 2002; Leita and De Nobili, 1991; Richards et al., 1998). Concern about heavy metal contamination from sludge applications led to efforts to remove the sometimes valuable metals from wastewater before releasing it to wastewater treatment plants; the metal content is now lower than it was in the past (Brady and Weil, 2002). Controversy continues concerning the mobility and pathways of metal ions in soil fertilized with sludge (Richards et al., 1998). Careful monitoring and treatment to maintain neutral pH on sludge application sites appear to be one way of limiting the mobility and bioavailability of zinc.

3.5   ATMOSPHERIC DEPOSITION

In addition to loading directly from the land surface to streams, pollutants are carried as gases and particles by air currents and distributed over vast regions of North America and the entire world. Changes in atmospheric chemistry resulting from pollutant emissions have adversely affected water quality in the Northeast and may be implicated in declines in forest health and productivity. While this discussion focuses on acid deposition resulting from sulfur and nitrogen emissions, atmospheric processes are also responsible for the widespread transport of many other pollutants, including pesticides (Cooter and Hutzell, 2002; Donald et al., 2001) and metals such as mercury and lead (Alloway, 1995; Landis et al., 2002).

During the 20th century, the increasing use of fossil fuel for power generation and transportation in growing urban and industrial societies resulted in pollutant emissions that lowered the pH of precipitation across much of the northern hemisphere (acid rain). Nitrous oxides (NO_x), primarily released from vehicle exhaust, and sulfates (SO_4^{2-}), from coal burning power plants and metal refining processes, react with water to form nitric acid (HNO_3) and sulfuric acid (H_2SO_4) (Brady and Weil, 2002). These are strong acids that dissociate easily to release H^+ ions into water in soils and streams. The pH of rain in affected areas commonly fell to between 4.0 and 4.5 (Brady and Weil, 2002; Likens et al., 1996). Regions most severely affected are located downwind of industrial centers, and have naturally acidic soils with limited sources of nutrient cations.

Large sections of the northeastern United States, northern European and the Scandinavian countries, central Russia, and eastern China have been adversely affected by acid deposition (Stoddard et al., 1999). In the eastern United States, agricultural and industrial activities (especially power generation by electrical utilities), particularly in the Ohio River Valley, produce pollutants that are carried by air currents to New England and the Mid-Atlantic region. Forty-one percent of the sulfur dioxide emissions in the United States comes from seven states in the Ohio River Valley (Illinois, Indiana, Kentucky, Ohio, Pennsylvania, Tennessee, and West Virginia). In addition, much of the ammonium emitted to the atmosphere is related to agricultural activity in the Ohio River Valley and other midwestern states (EPA cited in Driscoll et al., 2001; NADP, 2004).

Acid precipitation is thought to have begun in the early 1950s, resulting in an increase in soil acidity in many areas in the northeastern United States. When the concentration of H^+ ions in the soil is relatively high, nutrient cations such as Ca^{2+} and Mg^{2+} are replaced by H^+ ions on cation exchange sites and enter the soil solution, where they are leached into streams and lost from forested ecosystems (Lawrence et al., 1995). Likens et al. (1996) estimate that stream water losses of Ca^{2+} measured at Hubbard Brook Experimental Forest were 2.2 times greater than inputs from precipitation and weathering from 1955 to 1975 and 2.7 times greater from 1976 to 1993. This was a result both of increased net soil release of Ca^{2+} and decreased Ca^{2+} input in precipitation.

Increased concentrations of H^+ and sulfate (SO_4) in soil water also facilitate the release of toxic aluminum ions. Aluminum, after oxygen and silica, is the third most abundant element in the Earth's crust. It is insoluble in alkaline or neutral water. Excess hydrogen ions from acid precipitation can break down the crystalline structure of soil particles releasing some Al^{3+} ions from their central position and allowing them to adhere to external cation exchange sites, often replacing nutrient cations. When the ratio of calcium to aluminum in the soil is less than 1, Al^{3+} is more readily absorbed by plant roots, inhibiting the absorption of Ca^{2+} and potentially compromising the health of the plant. The reduction in vegetative Ca^{2+} uptake also increases the concentration of Ca^{2+} in the soil solution and the amount of Ca^{2+} leaching to streams (Driscoll et al., 2001; Lawrence and Huntington, 1999). Exchangeable Al^{3+} ions move in and out of the soil solution, either held at cation exchange sites or released as dissolved aluminum. Al^{3+} is involved in hydrolysis reactions, splitting water molecules and releasing H^+, thus lowering soil pH and increasing soil acidity (Ridley et al., 1997):

$$Al^{3+} + nH_2O \leftrightarrow Al(OH)_n^{3-n} + nH^+. \tag{3.8}$$

Because Al^{3+} can increase soil acidity, it is called an acid cation. Free aluminum ions in the soil solution and stream water are toxic. In fact, the primary initial danger from acidification may be the increase in biologically available aluminum (Baldigo and Lawrence, 2001; Barnes et al., 1998; Stevenson, 1986). In addition to damaging plant life, Al^{3+} is toxic to soil microbes and aquatic biota. Dissolved aluminum directly damages fish gills. While some fish species are more acid tolerant than others, populations generally decline when pH falls below 5.5 (Baldigo and Lawrence, 2001).

Increased nitrogen loading from atmospheric deposition affects forest health and nutrient cycling. Although tree growth in forests is usually limited by nitrogen availability, years of excess nitrogen loading can create a condition of "nitrogen saturation" in which existing vegetation is unable to take up the available nitrogen and excess nitrogen is released to soil water and leaching increases (Magill et al., 1996; Paerl et al., 2002). Even though nitrogen is an essential nutrient, too much nitrogen results in reduced tree growth and a decline in forest health due to imbalances between nitrogen and other mineral nutrients (e.g., calcium, magnesium, and potassium) (Aber et al., 1989; Kelty et al., 2004; Magill et al., 1997). Excess nitrate from vehicle exhaust in urban areas and along travel corridors increases nitrogen deposition in populated areas, adding local sources to regional nitrogen deposition (Freedman, 1989). A study conducted by the U.S. Geological Survey concluded that watersheds in Connecticut received between 850 and 1500 $kg/yr/km^2$ of nitrate in atmospheric deposition, with the variation primarily due to proximity to urban centers (Zimmerman et al., 1996). The cumulative impact of nitrogen leaching from upland watershed land use and atmospheric deposition causes water quality problems and overfertilization of aquatic ecosystems. This is a particular problem in East Coast estuaries (Anderson et al., 2002).

The Clean Air Act, passed in 1970 (with subsequent amendments in 1990) in the United States, and similar regulations in northern European countries have resulted in substantial reductions in sulfate emissions from power plants and other industrial point sources. Sulfate emissions have decreased 38% from their highest estimated levels (28.8 Tg/yr in 1973) (Driscoll et al., 2003, after U.S. Environmental Protection Agency, 2000). Nitrous oxide emissions (primarily from nonpoint automobile exhaust) remain relatively unchanged (Aber et al., 1998; Likens et al., 1996; Stoddard et al., 1999) (Figure 3.6).

Likens et al. (1996) observed that the reduction in atmospheric sulfate had not produced a corresponding recovery in pH levels and acid neutralizing capacity in forest soils and streams in the Hubbard Brook Experimental Forest in New Hampshire. They attributed this lag in recovery to the earlier depletion of nonacid cations (Ca^{2+} and Mg^{2+}). Inputs from weathering of bedrock minerals and atmospheric input were insufficient to compensate for the loss. In an ironic twist, it has been determined that calcium inputs from the atmosphere have diminished due to pollution regulations that have reduced atmospheric dust (Clow and Mast, 1999; Driscoll et al., 2001). Likens et al. (1996) predicted that "even with major reductions in emissions predicted from the 1990 amendments to the Clean Air Act, it is unlikely that the acid-base status of stream water will return to pre-Industrial Revolution levels in the foreseeable future."

Stoddard et al. (1999) analyzed trends in water quality data from 205 lakes and streams in eight regions of North America and Europe between 1980 and 1995. They found that, in general, surface

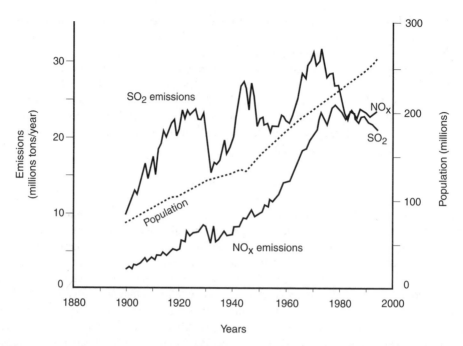

FIGURE 3.6 Trends in population, sulfur dioxide (SO_2), and nitrogen oxide (NO_x) emissions in the United States for 1900 to 1995 (Clow and Mast, 1999). (Courtesy of the U.S. Geological Survey.)

water sulfate concentrations had decreased and that the decline was more rapid in the 1990s than in the 1980s. In areas of Europe where sulfate levels had decreased, pH increased, indicating a reduction in acidity and the possible beginning of recovery. In the Adirondack and Catskill Mountain regions of New York and south-central Ontario, there were large decreases in sulfate (SO_4^{2-}) and nitrate (NO_3^-) anions, but no associated increase in surface water pH. This was attributed to the depletion of nonacid cations in watershed soils. Palmer and Driscoll (2002) examined long-term trends in soil solution (1984 to 1998) and stream water (1982 to 1998) chemistry at an undisturbed watershed at Hubbard Brook and found significant declines in aluminum concentrations in soils at higher elevations (730 m and above). Stream water concentrations of aluminum also declined significantly. Palmer and Driscoll believe that this is the result of continued decreases in the SO_4^{2-} concentration in the soil. They predicted "concentrations of Al_i [inorganic aluminum]...will fall to below the toxic threshold for fish within about 10 years if current rates of decline are maintained." They noted that "fish survival probably responds to peak conditions (stormflow) rather than to average conditions (baseflow). Aquatic biota may still be at risk from intermittently high concentrations of Al_i in stream water" (Palmer and Driscoll, 2002).

The Adirondack Long-Term Monitoring Program has been recording water chemistry data in lakes in primarily forested regions of the Adirondack Mountains since 1982. Driscoll et al. (2003) analyzed trends in water chemistry from 52 lakes in this program and results of precipitation sampling in the same area conducted under the National Atmospheric Deposition Program (NADP). They found that there had been "marked decreases" in the concentrations of H^+ and SO_4^- in precipitation since the late 1970s. At the same time, almost all the lakes in the region had shown declines in SO_4^- and NO_3^- even though NO_3^- had not declined in precipitation. The apparent increase in watershed/lake nitrogen retention may be associated with climate fluctuations, increased atmospheric CO_2, improved forest health, or some combination of conditions that increased nitrogen uptake and assimilation. There were corresponding trends showing increases in lake water pH and acid neutralizing capacity. The concentration of inorganic aluminum was declining, while the concentration of less toxic aluminum organic compounds, leached from soils, appeared to be increasing. Still, 16 of the 52 lakes in the study showed mean concentrations of inorganic aluminum that exceeded the toxic thresholds for aquatic organisms. The authors conclude that, given current pH levels and aluminum concentrations, recovery may require several decades, assuming that recent declines in acid deposition continue.

3.6 THE RESULTS OF NUTRIENT POLLUTION: ACCELERATED EUTROPHICATION

In undisturbed environments, eutrophication is a natural process during which lakes progress from a condition of low productivity to higher productivity over time. Sediment and organic matter from upland watersheds enter deep oligotrophic (low productivity) lakes. As lakes fill in and organic matter increases, conditions for plant growth improve and the rate of biomass production increases (Wetzel, 1983) (Figure 3.7). Excess supplies of phosphorus and nitrogen from human activity can accelerate eutrophication, a condition that currently threatens lakes and estuaries throughout the Northeast and the world. The three primary sources of these nutrients are fertilizers, wastewater, and atmospheric deposition (Bowen and Valiela, 2001; Jaworski et al., 1997; Valiela et al., 1997).

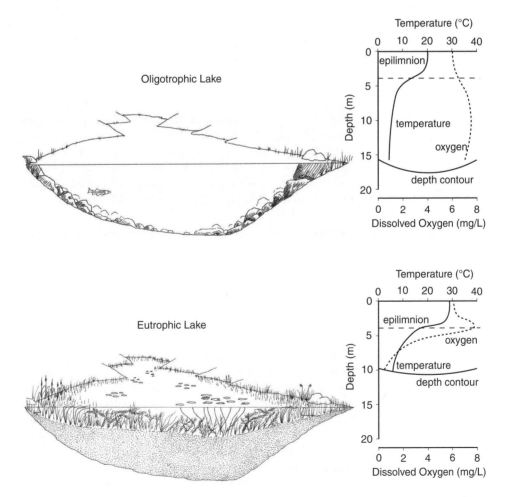

FIGURE 3.7 Oligotrophic versus eutrophic lakes. The oligotrophic lake is well supplied with oxygen at all depths. In the eutrophic lake, the decomposition of organic matter consumes oxygen and reduces the oxygen supply in bottom waters and sediments (modified from Smith, 1992).

TABLE 3.2
A Classification System for the Status of Trophic Status of Lakes

Annual mean concentration	Oligotrophic	Mesotrophic	Eutrophic
Total phosphorus (mg/L)	0.008	0.027	0.084
Total nitrogen (mg/L)	0.661	0.753	1.875
Chlorophyll a (mg/L)	0.0017	0.0047	0.014
Secchi transparency depth (m)	9.9	4.2	2.45

Modified from Wetzel (1983).

Eutrophication is defined by Nixon (1995) as an "increase in the rate of supply of organic matter to an ecosystem." This increase in organic matter is the result of the accelerated growth of plants and algae (i.e., an increase in productivity). Table 3.2 shows the relationship between phosphorus, nitrogen, chlorophyll (a measure of phytoplankton biomass), and water transparency in lakes classified as oligotrophic (low productivity), mesotrophic (intermediate productivity), eutrophic, and hypereutrophic.

Algal and aquatic plant tissue is composed of phosphorus, nitrogen, and carbon in approximate ratios of 1 P:7 N:40 C per 100 dry weight or 500 wet weight (Wetzel, 1983). Phosphorus is the limiting nutrient in most freshwater systems (Correll, 1998). In other words, freshwater streams, rivers, and lakes are, for the most part, adequately supplied with nitrogen through biological processes, while phosphorus is normally in short supply. In coastal ecosystems, the environment favors the loss of nitrogen through denitrification, so the supply of nitrogen may be limited relative to the amount of available phosphorus. Thus, in salt marshes and estuaries, excessive inputs of nitrogen are more likely to initiate accelerated eutrophication. There are notable exceptions: freshwater systems may be nitrogen limited in some situations and the same body of water may switch from nitrogen to phosphorus or from phosphorus to nitrogen limitation as conditions change. The relative quantities of plant-available nitrogen and phosphorus may also depend upon the amount and availability of associated micronutrients, such as iron, the organic matter content of the soil, and the health and abundance of microbial populations that process nitrogen and phosphorus compounds as they decompose organic matter (Harris, 1999; Howarth et al., 1988a).

Box 3.5   Harmful Algal Blooms

Harmful or toxic algal blooms (red and brown tides) are naturally occurring phenomena that cause discolored water and poisonous shellfish (Anderson et al., 2002; Falconer, 1993). Shellfish assimilate the toxins produced by these algae. Whales, porpoises, seabirds, and humans that consume the contaminated fish may succumb to paralytic, diarrhetic, neurotoxic, or amnesic shellfish poisoning. While eutrophication is not the only cause of these toxic algal blooms, it appears that overfertilization of the oceans can be correlated with an increase in the number of toxic algal blooms that occur each year (National Research Council, 2000) (Figure 3.8). *Pfiesteria*, a dinoflagellate, is a single-cell organism with multiple life stages that occurs naturally in coastal waters. Excess phosphorus is known to stimulate the growth of *Pfiesteria* populations during at least one stage in the *Pfiesteria* life cycle. *Pfiesteria* release a toxin that paralyzes respiratory systems in fish. As the fish die and decay, the organisms attack and feed on fish flesh, producing large bleeding sores. *Pfiesteria* outbreaks in the Albemarle and Pamlico estuaries of North Carolina (beginning in 1991) and in Chesapeake Bay (1997) have been associated with the deaths of millions of fish and shellfish, and with human health problems among fishermen and laboratory workers investigating the organism (Burkholder and Glasgow, 2001; Grattan et al., 2001; Magnien, 2001; National Research Council; 2000, Vitousek et al., 1997; Vogelbein et al., 2002).

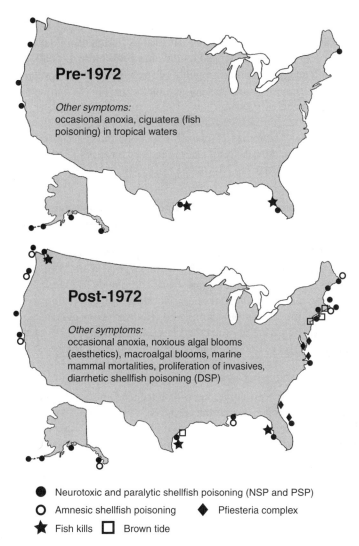

FIGURE 3.8 Harmful algal blooms have increased in recent years with increases in nutrient pollution of coastal estuaries (Anderson, 1995). (Courtesy of the U.S. Environmental Protection Agency.)

Schindler (1974, 1977) experimentally fertilized several small freshwater lakes in northwestern Ontario with nitrogen, phosphorus, and carbon to test the algal response. He described one of the experimental lakes as turning into a "teeming green soup within weeks after the nutrient additions began." Adding 0.4 g of phosphorus, 5.2 g of nitrogen, and 5.5 g of carbon per square meter caused a substantial increase in chlorophyll a concentrations in the lake, from less than 0.03 mg/L to more than 0.110 mg/L. After 2 years of treatment, the phosphorus additions were discontinued and only nitrogen and carbon were added. The lake quickly returned to its pre-eutrophication condition. The link between excess phosphorus and algal growth can also be seen in cases where a reduction in phosphorus loading has led to decreases in algal biomass. Two well-known examples are Lake Washington (Edmondson, 1970) in Seattle, and Lake Erie (Makarewicz, 1993). In Lake Washington, diversion of sewage effluent caused the phosphate concentration to decrease by 72% between 1963 and 1969, while the nitrogen concentration decreased by only 20%. During this same time period, the condition of the lake improved markedly. Phytoplankton abundance decreased and transparency (Secchi disk) increased from 1 m to 2.8 m. In Lake Erie, improved sewage treatment and a partial ban on phosphate detergents in some portions of the watershed reduced phosphorus loading to the lake from about 14,000 to less than 2,500 tonnes/yr between 1972 and 1982 (Dolan, 1993). At the same time, algal biomass dropped 52% to 89% in various basins of the lake (Makarewicz, 1993).

Nitrogen limitation in marine coastal ecosystems has been more difficult to establish; however, recent experiments and studies linking nitrogen increases with eutrophication support this theory. Researchers at the University of Rhode Island attempted to replicate Schindler's lake experiments in enclosed marine ecosystems (Oviatt et al., 1995). In enclosures where nitrogen was added, maximum chlorophyll a concentrations reached 0.070 and 0.114 mg/L^{-1}, a three- to fivefold increase over control concentrations. Additions of phosphorus alone did not cause increased productivity. In Waquoit Bay (Cape Cod, Massachusetts), researchers have also linked increases in nitrogen loading with eutrophication (Box 3.6).

Box 3.6   Eutrophication in a Coastal Marine Ecosystem: Waquoit Bay, Massachusetts

The Waquoit Bay National Estuarine Research Reserve (http://www.waquoitbayreserve.org/) is located on the south shore of Cape Cod, Massachusetts. It includes approximately 3000 acres of open water, barrier beaches, salt marshes, and uplands. Between 1938 and 1990 there was a substantial increase in residential and urban development within the watershed, while areas of natural vegetation, agricultural areas, and cranberry bogs decreased. In 1938 naturally vegetated areas comprised 84% of watershed land. In 1990 only 68% of the watershed was covered with natural vegetation. The number of housing units increased from 250 to more than 4000.

As development progressed, wastewater from septic systems and fertilizers used on lawns and golf courses became the primary source of nitrogen inputs to Waquoit Bay (Table 3.3). The path of wastewater nitrogen can be traced through the aquatic food web by measuring the proportion of the nitrogen isotope 15N in the groundwater, the water column, and in aquatic biota (McClelland and Valiela, 1998). Both producers and consumers in the Waquoit Bay estuary had higher concentrations of 15N, a characteristic signal of wastewater, than those found in similar organisms in a pristine environment. Phytoplankton and algal biomass were greatest in those areas with the greatest nitrogen loading (Bowen and Valiela, 2001).

TABLE 3.3
Nitrogen in the Waquoit Bay Estuary, 1938 and 1990

Waquoit Bay estuary, Massachusetts	1938	1990
Homes	250	4,000
Nitrogen entering the watershed	96,600 kg N/yr	161,700 kg N/yr
Nitrogen input to the bay	10,900 kg N/yr	24,300 kg N/yr
Sources of nitrogen loading to the bay		
Atmospheric deposition	8,400 kg N/yr (77%)	9,100 kg N/yr (38%)
Wastewater (septic systems)	700 kg N/yr (7 %)	10,500 kg N/yr (43%)
Fertilizer	1,700 kg N/yr (16%)	4,700 kg N/yr (19%)

Bowen and Valiela (2001), Valiela and Bowen (2002).

As the biomass of phytoplankton and algae increased, eelgrass, a seagrass that tends to be limited more by light than by nutrient availability, declined. Estimates from aerial photographs found that eelgrass meadow area decreased by 90% as nitrogen loads increased from

15 to 30 kg·N/ha/yr. The loss of seagrasses can have devastating effects on the ecosystem. Many species require this habitat and some of these are commercially important shellfish and finfish. The annual harvest of bay scallops in Waquoit Bay decreased from more than 100,000 L/yr in the early 1960s to less than 100 L/yr in the late 1990s (Bowen and Valiela, 2001; Valiela and Bowen, 2002).

3.6.1 THE EFFECTS OF EUTROPHICATION

Eutrophication initiates a chain of events that disrupt ecosystem processes. These phenomena may be mitigated or exacerbated by seasonal and annual variations in climate and streamflow, and by local physical factors such as the depth and the hydrologic residence time or flushing rate of particular water bodies (Smith, 2003).

3.6.1.1 Changes in Species Composition in Algal Communities

Downing et al. (2001) performed regression analyses on 269 observations from 99 lakes around the world and concluded that cyanobacteria dominance of the algal community increased as the concentration of total phosphorus (TP) in lake waters increased. Cyanobacteria constituted a minimal fraction of phytoplankton biomass in low nutrient oligotrophic lakes, but averaged 60% of phytoplankton biomass when TP concentrations rose above 0.08 to 0.09 mg/L. Blooms of cyanobacteria form dense mats of scum and also produce toxins that adversely affect the health of humans and domestic animals (Downing et al., 2001). Cyanobacteria are capable of fixing atmospheric nitrogen, increasing the supply of that nutrient (Howarth et al., 1988b).

3.6.1.2 Oxygen Depletion

Oxygen depletion in eutrophic waters is the result of the decay of excess organic matter combined with naturally occurring seasonal patterns of thermal stratification in lakes, and saline and thermal stratification in estuaries. In the winter, many temperate lakes are covered with ice, which is both colder at 0°C and less dense than the water below. Water is at maximum density at 4°C. As the temperatures increase in the spring, the ice melts and for a brief period (days to a few weeks, depending on the size of the lake and the surrounding topography and vegetation) water at all depths in the lake is at the same temperature (4°C) and density. Winds blowing across the lake surface create circulation patterns and a general mixing of water at different levels—spring turnover. As temperatures continue to increase and the water absorbs more solar energy, surface water is heated more rapidly than water at greater depths; circulation slows and then stops. At this point the lake waters become stratified into three layers: the epilimnion or surface water layer, the metalimnion (a layer where the thermal gradient is quite steep), and the hypolimnion or bottom water layer. This condition persists until autumn, when surface waters cool and the lake once again approaches a uniform temperature, resulting in fall turnover.

In estuaries, stratification may be caused by variations in salinity, with a layer of fresh water over denser, more saline bottom waters (Wetzel, 1983). Warming temperatures and freshwater inputs in the spring tend to increase stratification in the spring and through the summer months. Cooler temperatures and late summer storms increase mixing and reduce stratification in the fall. In the Chesapeake Bay, saline stratification is controlled by inputs of fresh water from the Susquehanna River during March and April. During wet years, more fresh water enters the Chesapeake Bay, prolonging and expanding areas of stratification (Cronin and Vann, 2003; Officer et al., 1984).

Each summer, large amounts of organic matter consisting of dead algal biomass settles through the water column of eutrophic lakes and estuaries. Decomposing bacteria break down the organic matter into simpler organic compounds. Oxygen is consumed in this process through microbial

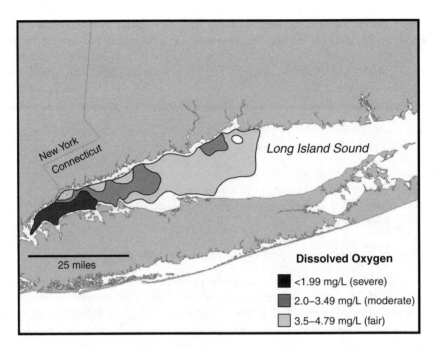

FIGURE 3.9 Dissolved oxygen concentrations in Long Island Sound, August 5–8, 2002 (Connecticut Department of Environmental Protection).

respiration. While levels of dissolved oxygen in surface waters may be high, water near the bed of the lake or estuary can become hypoxic (less than 2 mg O_2/L) or anoxic (0 mg O_2/L) due to seasonal stratification (Cooper and Brush, 1993; Officer et al., 1984; Wetzel, 1983).

Officer et al. (1984) analyzed the conditions that created hypoxic and anoxic conditions (often referred to as dead zones) in the central portion of the Chesapeake Bay that lasted from May until September. During these months, dissolved oxygen concentrations in the lower waters decreased to zero. The 2003 State of the Bay Report published by the Chesapeake Bay Foundation indicates that there has been little improvement in environmental conditions in the bay since the time of the 1984 study. In Long Island Sound, improvements in sewage treatment plants since 1990 have substantially reduced nitrogen inputs. Sound Health 2003 (Long Island Sound Study, 2003) found that hypoxic conditions still occurred in up to two-thirds of the sound during summer months (Figure 3.9)

Studies of sediment cores (Cooper, 1995; Cooper and Brush, 1993) showed that eutrophication and anoxia in the Chesapeake Bay have increased since European settlement of the watershed in the 17th century. Hypoxia and anoxia in bottom waters have been linked with the decline and loss of shellfish and finfish in Chesapeake Bay (Officer et al., 1984) and elsewhere. Howell and Simpson (1994) found that both the abundance and diversity of desirable fish species have decreased markedly with decreases in bottom dissolved oxygen, associated with nutrient loading in Long Island Sound, New York. Algal activity also can cause diurnal shifts in oxygen availability and pH in streams. These changes can increase the conversion of ammonium ions (NH_4^+) to toxic un-ionized ammonia (NH_3), endangering fish and other aquatic life (Sharpley et al., 1994).

3.6.1.3 Loss of Habitat

The floating mats of algae that form in eutrophic systems intercept sunlight and can reduce the survival of aquatic vascular plants (macrophytes) that provide a continuing supply of oxygen through photosynthesis (Staver and Brinsfield, 2001). In coastal ecosystems, algal blooms created by excess nutrients shade out seagrasses that provide critical spawning and rearing habitat for a variety of finfish and shellfish species (Box 3.6).

3.6.1.4 Internal Processing of Excess Nutrients

Some evidence suggests that restoration of eutrophic lakes may be difficult because added phosphorus will continue to be recycled internally. Anoxic conditions in the sediments of eutrophic lakes can favor the dissolution of sediment-associated phosphorus. In an anaerobic environment, ferric iron phosphate ($FePO_4$) is reduced to soluble ferrous iron phosphate ($Fe_3(PO_4)_2$) that dissolves, releasing phosphate ions into the water column. This dissolved phosphate is then available to nourish the growth of more algae, which dies and adds to the decaying organic matter at the lake bottom (Golterman, 1972; Sharpley et al. 1994). Studies in shallow eutrophic lakes in Florida have shown that much of the phosphorus required for algae growth is supplied by phosphorus released from bottom sediments (Sharpley et al., 1994).

Schindler (1974) disputed this notion in his original paper on phosphorus limitation. In his experiments, little or no phosphorus was released from bottom sediments, despite several months of anoxic conditions, and the lake recovered quickly, indicated by a drop in chlorophyll a concentrations once phosphorus additions were stopped. A recent experiment (Müller and Stadelmann, 2004) attempted to promote phosphorus immobilization by artificially aerating the bottom sediments of a eutrophic lake. While oxygenation did increase habitat for fish, it did not appear to increase phosphorus retention in lake bed sediments. Gächter and Müller (2003) argue that the release of phosphorus does not depend solely on the dissolved oxygen concentration in bottom sediments, but also on the availability of iron and the supply of sulfate that is reduced to sulfide in anaerobic conditions and can also combine with iron to form iron sulfide (FeS), decreasing the supply of iron ions that would combine with phosphate.

A study evaluating the effects of pollution controls in Lake Erie (Charlton et al., 1993) found that phosphorus generation from bottom sediments was quite variable over time. In 1970, prior to pollution control measures, the amount of phosphorus released by bottom sediments during anoxic periods was equivalent to the external loading of phosphorus to the central basin of the lake. Phosphorus regeneration from bottom sediments was lower in subsequent years, but in 1990 the amounts released rose again to the levels found in 1970, despite significant improvement in surface water quality. The authors concluded that phosphorus regeneration was most likely a sporadic process that is controlled by a number of variables and may occur naturally in some instances.

3.7 SUMMARY

3.7.1 NUTRIENTS

3.7.1.1 Nitrogen Cycle

Atmospheric nitrogen is transformed into biologically available nitrogen compounds by nitrogen-fixing bacteria. It is taken up by plants and animals and returned to the soil through the decay of organic matter. At any one time, nitrogen can be found in various forms in the soil (in bacterial tissue, in organic matter, and in soil water) and in vegetation. Nitrogen is returned to the atmosphere through denitrification.

The key pathways in the nitrogen cycle are

- Nitrogen fixation: the conversion of atmospheric nitrogen (N_2) to ammonium (NH_4^+) through bacterial action.
- Nitrification: the conversion of ammonium to nitrate (NO_3^-).
- Immobilization or the incorporation of ammonium and nitrate into organic tissue.
- Decomposition and mineralization: the conversion of organic nitrogen compounds to ammonium and nitrate.
- Denitrification: the conversion of nitrate to atmospheric nitrogen gases (N_2 and N_2O).

3.7.1.2 Phosphorus Cycle

Phosphorus is released through the weathering of phosphorus-bearing rocks. Dissolved phosphates are taken up by plants and animals and returned to the soil when organic matter from plant and animal tissue and animal waste decays. Phosphates are easily adsorbed to minerals (iron, aluminum, and calcium) and to silicate clay minerals.

3.7.1.3 Properties of Nutrient Compounds and Pathways of Flow

When nutrients are available in excess of plant requirements, nutrient loading to streams increases. This occurs when trees are removed, reducing the vegetative demand, and especially in agricultural and urban systems, where nutrients are added as fertilizer and wastewater. Nitrogen in the form of nitrate is soluble and is transported in subsurface flow. Phosphorus is less soluble and more likely to be transported in overland flow adsorbed to sediment particles. When concentrations are very high, the transport of dissolved phosphorus will increase in subsurface flow.

3.7.1.4 Cation Exchange

Soil micelles have negatively charged surfaces. Positively charged particles (cations) in the soil are attracted and held to the thousands of negatively charged cation exchange sites on each soil micelle. Cations of one type may be replaced (exchanged) by other cations with a stronger positive charge or a greater concentration in soil solution. Many essential nutrients such as calcium (Ca^{2+}) and magnesium (Mg^{2+}) are cations involved in cation exchange interactions. The CEC is defined as the number of cations that can be adsorbed by a kilogram of soil. It is a measure of the capacity of a soil to retain mineral cation nutrients and can vary substantially with soil type.

3.7.1.5 Nutrient Conservation in Forests and Grasslands

Forest foliage accumulates and holds acid cations (H^+) from precipitation. Nitrogen, phosphorus, and nutrient cations are held in living vegetation and in the organic litter layer on the forest floor. Nutrients released through the decomposition of organic matter are quickly reabsorbed by bacteria, trees, and other plants so nutrient loss is minimal. Grassland plants develop extensive root systems. Root systems and aboveground foliage protects the soil from erosion and nutrient loss. Nutrients are held in the soils and in plant material and recycled. Rainfall is lower (250 to 800 mm) than in forested ecosystems and insufficient for the leaching of nutrients to streams in undisturbed ecosystems. As a result, soils that develop in grasslands are rich in mineral nutrients (calcium and magnesium) and organic matter.

3.7.1.6 Variability in Nutrient Processing

The spatial and temporal patterns of nutrient processing are inherently variable, even in relatively undisturbed forested systems. Factors that influence processing and export to streams include climatic variation, topography, geology, soil type and carbon content, tree species composition, stand age, and land use history.

3.7.2 PESTICIDES

Pesticides differ in solubility and persistence (measured as half-life). Soluble pesticides may be found in stream water. Insoluble pesticides may be found adsorbed to sediments in streambeds. Organochlorine compounds such as DDT are the most persistent and insoluble pesticides. Although banned in the 1970s, they are still widely detected in streambeds and fish tissue. Recently developed pesticides are generally more soluble and have shorter half-lives. Not surprisingly, the greatest loss of agricultural herbicides to stream water occurs during first rain events following herbicide application. Loss of moderately soluble herbicides is greatest in low permeability soils that are more likely to generate overland flow. Pesticides can also be carried as atmospheric deposition. The widespread use of pesticides has led to their detection in lakes, streams, and rivers throughout the northeastern United States.

3.7.3 METALS

Contamination of stream sediments by metals is common. While trace concentrations of metals occur naturally, industrial activity and other forms of land use have introduced excess quantities of metals into the environment. Metallic ions may travel in dissolved form in stormflow. They also may adsorb to sediment particles and be carried with sediment. The solubility of some metallic compounds is pH dependent. Zinc compounds, for example, are more soluble in acid soils. Toxic metals are also transported in atmospheric deposition.

3.7.4 ATMOSPHERIC DEPOSITION

Emissions from power plants and vehicles release large quantities of sulfates and nitrous oxides (NO_x) into the air. These emissions react with water in the atmosphere to form sulfuric acid (H_2SO_4) and nitric acid (HNO_3). Acidic precipitation may fall on forests and streams far from the original source of the emissions. This has caused the acidification of lakes and streams and the leaching of nutrient cations (Ca^{2+} and Mg^{2+}) from forest soils. Acidification also results in the release of toxic aluminum. Atmospheric deposition of nitrates can increase nitrate leaching from forested regions as well as nitrogen loading to receiving waters. Nitrogen from atmospheric deposition contributes to eutrophication, especially in coastal estuaries. While sulfate emissions (primarily from power plants) have decreased since the 1970s, nitrogen emissions (from automobile exhaust) have not. Although there have been some measured improvements in water quality during the 1990s, it may take decades for acidified lake ecosystems to recover.

3.7.5 ACCELERATED EUTROPHICATION

Eutrophication is an "increase in the rate of supply of organic matter to an ecosystem" caused by fertilization from excess nitrogen and phosphorus. Phosphorus is typically the limiting nutrient in freshwater systems, while nitrogen tends to be the limiting nutrient in brackish tidal estuaries. The ecological consequences of eutrophication are summarized below.

3.7.5.1 Algal Species Change

Fertilization with nitrogen and phosphorus stimulates rapid growth in algal populations. Cyanobacteria (blue-green algae) constitute a greater proportion of the algal community in eutrophic systems. Cyanobacteria form dense mats or algal blooms on the surface of the water. They are capable of fixing nitrogen, thus exacerbating the problem.

3.7.5.2 Oxygen Depletion

Excess algae die and settle through the water column to lake and sea beds and decompose. Oxygen is consumed in the process of decomposition. Thermal and saline stratification of waters during the summer months results in anoxic conditions in deeper waters (the hypolimnion) and bed sediments. Seasonal anoxia has led to declines in finfish and shellfish populations in East Coast estuaries such as Long Island Sound and Chesapeake Bay.

3.7.5.3 Habitat Loss

Algal blooms shade out seagrasses that provide critical spawning and rearing habitat for finfish and shellfish.

REFERENCES

Aber, J.D., McDowell, W., Nadelhoffer, K., Magill, A., Bernston, G., Kamakea, M., McNulty, S., Currie, W., Rustad, L., and Fernandez, I., Nitrogen saturation in temperate forest ecosystems (hypotheses revisited), *BioScience*, 48, 921–934, 1998.

Aber, J.D., Nadelhoffer, K.J., Stendler, P., and Melillo, J.M., Nitrogen saturation in northern forest ecosystems: excess fuel combustion may stress the biosphere, *BioScience*, 39, 378–386, 1989.

Alloway, B.J., *Heavy Metals in Soils*, John Wiley & Sons, New York, 1995.

Anderson, D.M., Gilbert, P.M., and Burkholder, J.M., Harmful algal blooms and eutrophication: nutrient sources, composition, and consequences, *Estuaries*, 25(4b), 704–726, 2002.

Baldigo, B.P. and Lawrence, G.B., Effects of stream acidification and habitat on fish populations of a North American river, *Aquat. Sci.*, 63, 196–222, 2001.

Barnes, B.V., Zak, D.R., Denton, S.R., and Spurr, S.H., *Forest Ecology*, 4th ed., John Wiley & Sons, New York, 1998.

Berti, W.R. and Jacobs, L.W., Chemistry and phytotoxicity of soil trace elements from repeated sewage sludge applications, *J. Environ. Qual.*, 25, 1025–1032, 1996.

Binkley, D., Patterns and processes of variation in nitrogen and phosphorus concentrations in forested streams, Technical Bulletin 836, National Council for Air and Stream Improvement, Research Triangle Park, NC, 2001.

Binkley, D. and Brown, T.C., Forest practices as nonpoint sources of pollution in North America, *Water Resourc. Bull.*, 29, 729–740, 1993.

Binkley, D., Ice, G.G., Kaye, J., and Williams, C.A., Nitrogen and phosphorus concentrations in forest streams of the United States, *J. Am. Water Resourc. Assoc.*, 40, 1277–1291, 2004.

Blanchard, P.E. and Lerch, R.N., Watershed vulnerability to losses of agricultural chemicals: interactions of chemistry, hydrology, and land-use, *Environ. Sci. Technol.*, 34, 3315–3322, 2000.

Boring, L.R., Swank, W.T., and Monk, C.D., Dynamics of early successional forest structure and processes in the Coweeta Basin, in *Forest Hydrology and Ecology at Coweeta*, Swank, W.T. and Crossley, D.A. (eds.), Springer-Verlag, New York, 1988, pp. 161–179.

Bormann, F.H. and Likens, G.E., *Pattern and Process in a Forested Ecosystem*, Springer-Verlag, New York, 1994.

Bormann, F.H., Likens, G.E., Siccama, T.E., Pierce, R.S., and Eaton, J.S., The export of nutrients and recovery of stable conditions following deforestation at Hubbard Brook, *Ecol. Monogr.*, 44, 255–277, 1974.

Bowen, J.L. and Valiela, I., The ecological effects of urbanization of coastal watersheds: historical increases in nitrogen loads and eutrophication of Waquoit Bay estuaries, *Can. J. Fish. Aquat. Sci.*, 58, 1489–1500, 2001.

Brady, N.C. and Weil, R.R., *The Nature and Properties of Soils*, 13th ed., Macmillan, New York, 2002.

Breault, R.F. and Harris, S.L., Geographical distribution and potential for adverse biological effects of selected trace elements and organic compounds in streambed sediment in the Connecticut, Housatonic, and Thames River basins, 1992–1994, Water Resources Investigations Report 97-4169, U.S. Geological Survey, Reston, VA, 1997.

Burkholder, J.M. and Glasgow, H.B., History of toxic *Pfiesteria* in North Carolina estuaries from 1991 to the present, *Bioscience*, 51, 827–841, 2001.

Burns, D.A., Murdoch, P.S., Lawrence, G.B., and Michel, R.L., Effect of groundwater springs on NO_3^- concentrations during summer in Catskill Mountain streams, *Water Resourc. Res.*, 34, 1987–1996, 1998.

Charlton, M.N., Milne, J.E., Booth, W.G., and Chiocchio, F., Lake Erie offshore in 1990: restoration and resilience in the central basin, *Great Lakes Res.*, 19, 291–309, 1993.

Chen, H. and Cutright, T.J., Preliminary evaluation of microbially mediated precipitation of cadmium, chromium, and nickel by rhizosphere consortium, *J. Environ. Eng.*, 129, 4–9, 2003.

Chesapeake Bay Foundation, The State of the Bay Report, 2003, http://www.cbf2003.org/; accessed July 2006.

Clow, D.W. and Mast, M.A., Trends in precipitation and stream-water chemistry in the northeastern United States, water years 1984–96, Fact Sheet 117-99, U.S. Geological Survey, Reston, VA, 1999.

Cooper, S.R., Chesapeake Bay watershed historical land use: impact on water quality and diatom communities, *Ecol. Applic.*, 5, 703–723, 1995.

Cooper, S.R. and Brush, G.S., A 2,500-year history of anoxia and eutrophication in Chesapeake Bay, *Estuaries*, 16(3B), 617–626, 1993.

Cooter, E.J. and Hutzell, W.T., A regional atmospheric fate and transport model for atrazine. 1. Development and implementation, *Environ. Sci. Technol.*, 36, 4091–4098, 2002.

Correll, D.L., The role of phosphorus in the eutrophication of receiving waters: a review, *J. Environ. Qual.*, 27, 261–266, 1998.

Creed, I.F. and Band, L.E., Exploring functional similarity in the export of nitrate-N from forested catchments: a mechanistic modeling approach, *Water Resourc. Res.*, 34, 3079–3093, 1998a.

Creed, I.F. and Band, L.E., Export of nitrogen from catchments within a temperate forest: evidence for a unifying mechanism regulated by variable source area dynamics, *Water Resourc. Res.*, 34, 3105–3120, 1998b.

Cronin, T.M. and Vann, C.D., The sedimentary record of climatic and anthropogenic influence on the Patuxent estuary and Chesapeake Bay ecosystems, *Estuaries*, 26(2A), 196–209, 2003.

Dolan, D.M., Point source loadings of phosphorus to Lake Erie: 1986–1990, *J. Great Lakes Res.*, 19, 212–223, 1993.

Donald, D.B., Guprasad, N.P., Quinnett-Abbott, L., and Cash, K., Diffuse geographic distribution of herbicides in northern prairie wetlands, *Environ. Toxicol. Chem.*, 20, 273–279, 2001.

Downing, J.A., Watson, S.B., and McCauley, E., Predicting cyanobacteria dominance in lakes, *Can. J. Fish. Aquat. Sci.*, 58, 1905–1908, 2001.

Driscoll, C.T., Driscoll, K.M., Roy, K.M., and Mitchell, M.J., Chemical response of lakes in the Adirondack region of New York to declines in acid deposition, *Environ. Sci. Technol.*, 37, 2036–2042, 2003.

Driscoll, C.T., Lawrence, G.B., Bulger, A.J., Butler, T.J., Cronan, C.S., Eagar, C., Lambert, K.F., Likens, G.E., Stoddard, J.L., and Weathers, K.C., *Acid Rain Revisited: Advances in Scientific Understanding Since the Passage of the 1970 and 1990 Clean Air Act Amendments*, Hubbard Brook Research Foundation, Hanover, NH, 2001.

Edmondson, W.T., Phosphorus, nitrogen, and algae in Lake Washington after diversion of sewage, *Science*, 169, 690–691, 1970.

Edwards, A.C. and Withers, P.J.A., Soil phosphorus management and water quality: a UK perspective, *Soil Use Manage.*, 14, 124–130, 1998.

EXTOXNET (Extension Toxicology Network), a pesticide information project of Cooperative Extension Offices of Cornell University, Oregon State University, University of Idaho, and the University of Davis and the Institute for Environmental Toxicology, Michigan State University, 1996, http://extoxnet.orst.edu/pips/ghindex.html; accessed July 2006.

Falconer, I.R. (ed.), *Toxins in Seafood and Drinking Water*, Academic Press, London, 1993.

Feger, K.H., Nitrogen cycling in two Norway spruce (*Picea abies*) ecosystems and effects of a $(NH_4)_2SO_4$ addition, *Water Air Soil Pollut.*, 61, 295–307, 1992.

Fenelon, J.M., Water Quality in the White River Basin, Indiana, 1992–96, Circular 1150, U.S. Geological Survey, Reston, VA, 1998.

Fenn, M.E., Poth, M.A., Aber, J.D., Baron, J.S., Bormann, B.T., Johnson, D.W., Lemly, A.D., McNulty, S.G., Ryan, D.F., and Stottlemyer, R., Nitrogen excess in North American ecosystems: predisposing factors, ecosystem responses, and management strategies, *Ecol. Applic.*, 8, 706–733, 1998.

Fisk, M.C., Zak, D.R., and Crow, T.R., Nitrogen storage in old- and second-growth northern hardwood forests, *Ecology*, 83, 73–87, 2002.

Freedman, B., *Environmental Ecology: The Impacts of Pollution and Other Stresses on Ecosystem Structure and Function*, Academic Press, San Diego, 1989.

Froelich, P.N., Kinetic control of dissolved phosphate in natural rivers and estuaries: a primer on the phosphate buffer mechanism, *Limnol. Oceanogr.*, 33(4 pt. 2), 649–668, 1988.

Gächter, R. and Müller, B., Why the phosphorus retention of lakes does not necessarily depend on the oxygen supply to their sediment source, *Limnol. Oceanogr.*, 48, 929–933, 2003.

Gaynor, J.D., MacTavish, D.C., and Findlay, W.I., Atrazine and metolachlor loss in surface and subsurface runoff from three tillage treatments in corn, *J. Environ. Qual.*, 24, 284–305, 1995.

Golterman, H.L., Vertical movement of phosphate in freshwater, *TNO nieuws: orgaan van de Organisatie voor Toegepast- Natuurwetenschappelijk Onderzoek*, 27, 96–101, 1972.

Good, J.C., Roof runoff as a diffuse source of metals and aquatic toxicity in stormwater, *Water Sci. Technol.*, 28(3–5), 317–321, 1993.

Gorham, E., Vitousek, P.M., and Reiners, W.A., The regulation of chemical budgets over the course of terrestrial ecosystem succession, *Annu. Rev. Ecol. Syst.*, 10, 53–84, 1979.

Grattan, L.M., Oldach, D., and Morris, J.G., Human health risks of exposure to *Pfiesteria piscicida*, *Bioscience*, 51, 853–858, 2001.

Groffman, P.M. and Tiedje, J.M., Denitrification in north temperate forest soils: relationships between denitrification and environmental factors at the landscape scale, *Soil Biol. Biochem.*, 21, 621–626, 1989a.

Groffman, P.M. and Tiedje, J.M., Denitrification in north temperate forest soils: spatial and temporal patterns at the landscape and seasonal scales, *Soil Biol. Biochem.*, 21, 613–620, 1989b.

Gunderson, P., Callescen, I., and deVries, W., Nitrate leaching in forest ecosystems is related to forest floor C/N ratios, *Environ. Pollut.*, 102, 403–407, 1998.

Haitzer, M., Aitken, G.R., and Ryan, J.N., Binding of mercury (II) to aquatic humic substances: influence of pH and source of humic substances, *Environ. Sci. Technol.*, 37, 2436–2441, 2003.

Harris, G.P., Comparison of the biogeochemistry of lakes and estuaries: ecosystem processes, functional groups, hysteresis effects and interactions between macro- and microbiology, *Mar. Freshwater Res.*, 50, 791–811, 1999.

Hedin, L.O., Armesto, J.J., and Johnson, A.H., Patterns of nutrient loss from unpolluted, old-growth temperate forests: evaluation of biogeochemical theory, *Ecology*, 76, 493–509, 1995.

Helvey, J.D. and Kunkle, S.H., Input-output budgets of selected nutrients on an experimental watershed near Parsons, West Virginia, Research Paper NE-584, USDA Forest Service, Northeastern Forest Experiment Station, Broomall, PA, 1986.

Hobbie, J.E. and Likens, G.E., Output of phosphorus, dissolved organic carbon and fine particulate carbon from Hubbard Brook watersheds, *Limnol. Oceanogr.*, 18, 734–742, 1973.

Holmgren, G.G.S., Meyer, M.W., Chaney, R.L., and Daniels, R.B., Cadmium, lead, zinc, copper, and nickel in agricultural soils of the United States of America, *J. Environ. Qual.*, 22, 335–348, 1993.

Hornbeck, J.W., Bailey, S.W., Buso, D.C., and Shanley, J.B., Stream water chemistry and nutrient budgets for forested watersheds in New England: variability and management implications, *For. Ecol. Manage.*, 93, 73–89, 1997.

Howarth, R.W., Marino, R., and Cole, J.J., Nitrogen fixation in freshwater estuaries. 2. Biochemical controls, *Limnol. Oceanogr.*, 33(4 pt. 2), 688–701, 1988a.

Howarth, R.W., Marino, R., Lane, J., and Cole, J.J., Nitrogen fixation in freshwater estuaries. 1. Rates and importance, *Limnol. Oceanogr.*, 33(4 pt. 2), 669–687, 1988b.

Howell, P. and Simpson, D., Abundance of marine resources in relation to dissolved oxygen in Long Island Sound, *Estuaries*, 17, 394–402, 1994.

Hurd, T.M., Raynal, D.J., and Schweitzer, C.R., Symbiotic N_2 fixation of *Alnus incana* spp. *rugosa* in shrub wetlands of the Adirondack Mountains, New York, USA, *Oecologia*, 126, 94–103, 2001.

Jaworski, N.A., Howarth, R.W., and Heitling, L.J., Atmospheric deposition of nitrogen oxides onto the landscape contributes to coastal eutrophication in the northeast United States, *Environ. Sci. Technol.*, 31, 1995–2004, 1997.

Jefts, S., Fernandez, I.J., Rustad, L.E., and Dail, D.B., Decadal responses in soil N dynamics at the Bear Brook watershed in Maine, USA, *For. Ecol. Manage.*, 189, 189–205, 2004.

Kelty, M.J., Menalled, F.D., and Carlton, M.M., Nitrogen dynamics and red pine growth following application of palletized biosolids in Massachusetts, USA, *Can. J. For. Res.*, 34, 1477–1487, 2004.

Kiekens, L., Zinc, in *Heavy Metals in Soils*, Alloway, B.J. (ed.), John Wiley & Sons, New York, 1990, pp. 261–279.

Kimmerer, T.W., Structure and function of forest trees, in *Introduction to Forest Science*, Young, R.A. and Geise, R.L. (eds.), John Wiley & Sons, New York, 1990, pp. 67–85.

Kozlowski, T.T. and Pallardy, S.G., *Physiology of Woody Plants*, 2nd ed., Academic Press, San Diego, 1997.

Landis. M.S., Vette, A.F., and Keeler, G.J., Atmospheric mercury in the Lake Michigan basin: influence of the Chicago/Gary urban area, *Environ. Sci. Technol.*, 36, 4508–4517, 2002.

Larson, S.J., Capel, P.D., and Majewski, M.S., Pesticides in surface waters: distribution, trends, and governing factors, in *Pesticides in the Hydrologic System*, vol. 3, Gilliom, R.J. (series ed.), Ann Arbor Press, Chelsea, MI, 1997.

Latty, E.F., Canham, C.D., and Marks, P.L., The effects of land-use history on soil properties and nutrient dynamics in northern hardwood forests of the Adirondack Mountains, *Ecosystems*, 7, 193–207, 2004.

Lawrence, G.B. and Huntington, T.G., Soil-calcium depletion linked to acid rain and forest growth in the eastern United States, Water Resources Investigations Report 98-4267, U.S. Geological Survey, Reston, VA, 1999.

Lawrence, G.B., David, M.B., and Shortle, W.C., A new mechanism for calcium loss in forest-floor soils, *Nature*, 378, 162–164, 1995.

Leita, L. and De Nobili, M., Heavy metals in the environment, *J. Environ. Qual.*, 20, 73–78, 1991.

Leopold, A., *A Sand County Almanac, and Sketches Here and There*, Oxford University Press, London, 1949.

Leu, C., Singer, H., Stamm, C., Müller, S., and Schwarzenbach, R.P., Simultaneous assessment of sources, processes, and factors influencing herbicide losses to surface waters in a small agricultural catchment, *Environ. Sci. Technol.*, 38, 3827–3834, 2004a.

Leu, C., Singer, H., Stamm, C., Müller, S., and Schwarzenbach, R.P., Variability of herbicide losses from 13 fields to surface water within a small catchment after a controlled herbicide application, *Environ. Sci. Technol.*, 38, 3835–3841, 2004b.

Lewis, G.P. and Likens, G.E., Low stream nitrate concentrations associated with oak forests on the Allegheny high plateau of Pennsylvania, *Water Resourc. Res.*, 36, 3091–3094, 2000.

Likens, G.E., Bormann, F.H., Johnson, N.M., Fisher, D.W., and Pierce, R.S., Effects of forest cutting and herbicide treatment on nutrient budgets in the Hubbard Brook watershed-ecosystem, *Ecol. Monogr.*, 40, 23–47, 1970.

Likens, G.E., Bormann, F.H., Johnson, N.M., and Pierce, R.S., The calcium, magnesium, potassium, and sodium budgets for a small forested ecosystem, *Ecology*, 48, 772–785, 1967.

Likens, G.E., Bormann, F.H., Pierce, R.S., Eaton, J.S., and Johnson, N.M., *Biogeochemistry of a Forested Ecosystem*, Springer-Verlag, New York, 1977 [2nd ed. published in 1995].

Likens, G.E., Driscoll, C.T., and Buso, D.C., Long-term effects of acid rain: response and recovery of a forested ecosystem, *Science*, 272, 244–246, 1996.

Lin, Y.-J., Karuppiah, M., Shaw, A., and Gupta, G., Effect of simulated sunlight on atrazine and metolachlor toxicity of surface waters, *Ecotoxicol. Environ. Saf.*, 43, 35–37, 1999.

Lindsey, B.D., Breen, K.J., Bilger, M.D., and Brightbill, R.A., Water quality in the lower Susquehanna River basin, Pennsylvania and Maryland, 1992–95, Circular 1168, U.S. Geological Survey, Reston, VA, 1998.

Long, G.R., Chang, M., and Kennen, J.G., Trace elements and organochlorine compounds in bed sediment and fish tissue at selected sites in New Jersey streams—sources and effects, Water Resources Investigations Report 99-4235, U.S. Geological Survey, Reston, VA, 2000.

Long Island Sound Study, *Sound Health* 2003, http://www.longislandsoundstudy.net/pubs/reports/soundhealth2003.htm; accessed July 2006.

Lorimer, C.G., Causes of the oak regeneration problem, in *Oak Regeneration: Serious Problems, Practical Recommendations*, Loftis, D.L. and McGee, C.E. (eds.), Gen. Tech. Rep. SE-84, USDA Forest Service, Southeast Forest Experiment Station, Asheville, NC, 1992, pp. 14–39.

Lovett, G.M. and Rueth, H., Soil nitrogen transformations in beech and maple stands along a nitrogen deposition gradient, *Ecol. Applic.*, 9, 1330–1344, 1999.

Lovett, G.M., Weathers, K.C., and Arthur, M.A., Control of nitrogen loss from forested watersheds by soil carbon: nitrogen ratio and tree species composition, *Ecosystems*, 5, 712–718, 2002.

Lynch, J.A. and Corbett, E.S., Nitrate export from managed and unmanaged forested watersheds in the Chesapeake Bay watershed, in *Towards a Sustainable Coastal Watershed: The Chesapeake Experiment, Proceedings of a Conference*, Publication 149, Chesapeake Research Consortium, Solomons, MD, 1994, pp. 656-664.

Magill, A.H., Aber, J.D., Bernston, G.M., McDowell, W.H., Nadelhoffer, K.J., Melillo, J.M., Long-term nitrogen additions and nitrogen saturation in two temperate forests, *Ecosystems*, 3, 238–253, 2000.

Magill, A.H., Aber, J.D., Hendricks, J.J., Bowden, R.D., Melillo, J.M., and Steudler, P.A., Biogeochemical response of forest ecosystems to simulated chronic nitrogen deposition, *Ecol. Applic.*, 7, 402–415, 1997.

Magill, A.H., Downs, M.R., Nadelhoffer, K.J., Hallett, R.A., and Aber, J.D., Forest ecosystem response to four years of chronic nitrate and sulfate additions at Bear Brooks watershed, Maine, USA, *For. Ecol. Manage.*, 84, 29–37, 1996.

Magnien, R.E., The dynamics of science, perception, and policy during the outbreak of *Pfiesteria* in the Chesapeake Bay, *Bioscience*, 51, 843–852, 2001.

Makarewicz, J.C., Phytoplankton biomass and species composition in Lake Erie, 1970 to 1987, *J. Great Lakes Res.*, 19, 258–274, 1993.

McClelland, J.W. and Valiela, I., Linking nitrogen in estuarine producers to land-derived sources, *Limnol. Oceanogr.*, 43, 577–585, 1998.

McDowell, R.W. and Sharpley, A.N., Phosphorus losses in subsurface flow before and after manure application to intensively farmed land, *Sci. Total Environ.*, 278, 113–125, 2001.

Meyer, J.L. and Likens, G.E., Transport and transformation of phosphorus in a forest stream ecosystem, *Ecology*, 60, 1255–1269, 1979.

Mitchell, M.J., Foster, N.W., Shepard, J.P., and Morrison, I.K., Nutrient cycling in Huntington Forest and Turkey Lakes deciduous stands: nitrogen and sulfur, *Can. J. For. Res.*, 2, 457–464, 1992.

Mullaney, J.R. and Zimmerman, M.J., Nitrogen and pesticide concentrations in an agricultural basin in north-central Connecticut, Water Resources Investigations Report 97-4076, U.S. Geological Survey, Reston, VA, 1997.

Müller, R. and Stadelmann, P., Fish habitat requirements as the basis for rehabilitation of eutrophic lakes by oxygenation, *Fish. Manage. Ecol.*, 11, 251–260, 2004.

Myers, D.N., Thomas, M.A., Frey, J.W., Rheaume, S.J., and Button, D.T., Water quality in the Lake Erie–Lake Saint Clair drainages: Michigan, Ohio, Indiana, New York, and Pennsylvania, 1996–98, Circular 1203, U.S. Geological Survey, Reston, VA, 2000.

NADP (National Atmospheric Deposition Program), Ammonium ion wet deposition, 2004, http://nadp.sws.uiuc.edu/isopleths/maps2004/; accessed July 2006.

National Research Council, *Clean Coastal Waters: Understanding and Reducing the Effects of Nutrient Pollution*, National Academies Press, Washington, D.C., 2000.

National Research Council, *Assessing the TMDL Approach to Water Quality Management*, National Academies Press, Washington, D.C., 2001.

Nixon, S.W., Coastal marine eutrophication: a definition, social causes, and future concerns, *Ophelia*, 41, 199–219, 1995.

Officer, C.B., Biggs, R.B., Taft, J.L., Cronin, L.E., Tyler, M.A., and Boynton, W.R., Chesapeake Bay anoxia: origin, development, and significance, *Science*, 223, 22–27, 1984.

Oviatt, C., Doering, P., Nowicki, B., Reed, L., Cole, J., and Frithsen, J., An ecosystem level experiment on nutrient limitation in temperate coastal marine environments, *Mar. Ecol. Prog. Ser.*, 116, 171–179, 1995.

Paerl, H.W., Dennis, R.L., and Whitall, D.R., Atmospheric deposition of nitrogen: implications for nutrient over-enrichment of coastal waters, *Estuaries*, 25(4b), 677–693, 2002.

Palmer, S.M. and Driscoll, C.T., Decline in mobilization of toxic aluminum, *Nature*, 417, 242, 2002.

Pardo, L.H., Driscoll, C.T., and Likens, G.E., Patterns of nitrate losses from a chronosequence of clear-cut watersheds, *Water Air Soil Pollut.*, 85, 1659–1664, 1995.

Persky, J.H., The relation of ground-water quality to housing density, Cape Cod, Massachusetts, Water Resources Investigations Report 86-4093, U.S. Geological Survey, Reston, VA, 1986.

Peterjohn, W.T., Foster, C.J., Christ, M.J., and Adams, M.B., Patterns of nitrogen availability within a forested watershed exhibiting symptoms of nitrogen saturation, *For. Ecol. Manage.*, 119, 247–257, 1999.

Pregitzer, K.S., Zak, D.R., Burton, A.J., Ashby, J.A., and MacDonald, N.W., Chronic nitrate additions dramatically increase the export of carbon and nitrogen from northern hardwood ecosystems, *Biogeochemistry*, 68, 179–197, 2004.

Raven, P.H., Evert, R.F., and Eichhorn, S.E., *Biology of Plants*, 6th ed., Worth Publishers, New York, 1999.

Rheaume, S.J., Button, D.T., Myers, D.N., and Hubbell, D.L., Areal distribution and concentrations of contaminants of concern in surficial streambed and lakebed sediments, Lake Erie–Lake Saint Clair drainages, 1990–1997, Water Resources Investigations Report 00-4200, U.S. Geological Survey, Reston, VA, 2001.

Richards, B.K., Steenhuis, T.S., Peverly, J.H., and McBride, M.B., Metal mobility at an old, heavily loaded sludge application site, *Environ. Pollut.*, 99, 365–377, 1998.

Ridley, M.K., Wesolowski, D.J., Palmer, D.A., Bénézeth, P., and Kettler, R.M., Effect of sulfate on the release rate of Al^{3+} from gibbsite in low-temperature acidic waters, *Environ. Sci. Technol.*, 31, 1922–1925, 1997.

Robinson, G.R. and Ayuso, R.A., Use of spatial statistics and isotopic tracers to measure the influence of arsenical pesticide use on stream sediment chemistry in New England, USA, *Appl. Geochem.*, 19, 1097–1110, 2004.

Sawhney, B.L., Bugbee, G.J., and Stilwell, D.E., Leachability of heavy metals from growth media containing source-separated municipal solid waste compost, *J. Environ. Qual.*, 23, 718–722, 1994.

Schindler, D.W., Eutrophication and recovery in experimental lakes: implications for lake management, *Science*, 184, 897–899, 1974.

Schindler, D.W., Evolution of phosphorus limitation in lakes, *Science*, 195, 260–262, 1977.

Seastedt, T.R., Mass, nitrogen, and phosphorus dynamics in foliage and root detritus of tallgrass prairie, *Ecology*, 69, 59–65, 1988.

Sharpley, A.N., Chapra, S.C., Wedepohl, R., Sims, J.T., Daniel, T.C., and Reddy, K.R., Managing agricultural phosphorus for protection of surface waters: issues and options, *J. Environ. Qual.*, 23, 437–451, 1994.

Sharpley, A.N., Menzel, R.G., Smith, S.J., Rhoades, E.D., and Olness, A.E., The sorption of soluble phosphorus by soil material during transport in runoff from cropped and grassed watersheds, *J. Environ. Qual.*, 10, 211–215, 1981.

Smith, R.L., *Elements of Ecology*, 3rd ed., Harper Collins, New York, 1992.

Smith, V.H., Eutrophication of freshwater and coastal marine ecosystems: a global problem, *Environ. Sci. Pollut. Res.*, 10, 126–139, 2003.

Squillace, P.J. and Thurman, E.M., Herbicide transport in rivers: importance of hydrology and geochemistry in nonpoint source contamination, *Environ. Sci. Technol.*, 26, 538–545, 1992.

Staver, K.W. and Brinsfield, R.B., Agriculture and water quality on the Maryland eastern shore: where do we go from here?, *Bioscience*, 51, 859–868, 2001.

Stevenson, F.J., *Cycles of Soil: Carbon, Nitrogen, Phosphorus, Sulfur, Micronutrients*, John Wiley & Sons, New York, 1986.

Stoddard, J.L., Jeffries, D.S., Lükewille, A., Clair, T.A., Dillon, P.J., Driscoll, C.T., Forsius, M., Johannessen, M., Kahl, J.S., Kellogg, J.H., Kemp, A., Mannio, J., Monteith, D.T., Murdoch, P.S., Patrick, S., Rebsdorf, A., Skjelkvale, B.L., Stainton, M.P., Traaen, T., van Dam, H., Webster, K.E., Wieting, J., and Wilander, A., Regional trends in aquatic recovery from acidification in North America and Europe, *Nature*, 401, 575–578, 1999.

U.S. Environmental Protection Agency, *National Air Pollutant Emission Trends 1900–1998*, 2000-EPA-454-R-00-002, U.S. Government Printing Office, Washington, D.C., 2000, http://www.epa.gov/ttn/chief/trends/trends98/; accessed July 2006.

U.S. Geological Survey, The quality of our nation's waters—nutrients and pesticides, Circular 1225, U.S. Geological Survey, Reston, VA, 1999.

U.S. Geological Survey, Pesticides in stream sediment and aquatic biota: current understanding of distribution and major influences, Fact Sheet 092-00, U.S. Geological Survey, Reston, VA, 2000.

Valiela, I. and Bowen, J.L., Nitrogen sources to watersheds and estuaries: role of land cover mosaics and losses within watersheds, *Environ. Pollut.*, 118, 239–248, 2002.

Valiela, I., Collins, G., Kremer, J., Lajtha, K., Geist, M., Seely, B., Brawley, J., and Sham, C.H., Nitrogen loading from coastal watersheds to receiving estuaries, new method and application, *Ecol. Applic.*, 7, 358–380, 1997.

Verry, E.S., Effects of an aspen clearcutting on water yield and quality in northern Minnesota, in *National Symposium on Watersheds in Transition*, June 19–22, 1972, Fort Collins, Colorado, Csallany, S.C., McLaughlin, T.G., and Striffler, W.D. (eds.), American Water Resources Association, Middleburg, VA, 1972, pp. 276–284.

Vitousek, P.M. and Reiners, W.A., Ecosystem succession and nutrient retention: a hypothesis, *BioScience*, 25, 376–381, 1975.

Vitousek, P.M., Aber, J., Howarth, R.W., Likens, G.E., Matson, P.A., Schindler, D.W., Schlesinger, W.H., and Tilman, G.D., Human alteration of the global nitrogen cycle: causes and consequences, *Issues Ecol.*, 1, 1–15, 1997.

Vitousek, P.M., Cassman, K., Cleveland, C., Crews, T., Field, C.B., Grimm, N.B., Howarth, R.W., Marino, R., Martinelli, L., Rastetter, E.B., and Sprent, J.I., Towards an ecological understanding of biological nitrogen fixation, *Biogeochemistry*, 57/58, 1–45, 2002.

Vogelbein, W.K., Lovko, V.J., Shields, J.D., Reece, K.S., Mason, P.L., Haas, L.W., and Walker, C.C., *Pfiesteria shumwayae* kills fish by micropredation not exotoxin secretion, *Nature*, 418, 967–970, 2002.

Wall, G.R., Riva-Murray, K., and Phillips, P.J., Water quality in the Hudson River basin, New York and adjacent states, 1992–95, Circular 1165, U.S. Geological Survey, Reston, VA, 1998.

Walter, L.E.F., Hartnett, D.C., Hetrick, B.A.D., and Schwab, A.P., Interspecific nutrient transfer in a tallgrass prairie plant community, *Am. J. Bot.*, 83, 180–184, 1996.

Watt, M.K., A hydrologic primer for New Jersey watershed management, Water Resources Investigations Report 00-4140, U.S. Geological Survey, Reston, VA, 2000.

Wetzel, R.G., *Limnology*, 2nd ed., Saunders, Philadelphia, 1983.

Whitney, G.G., *From Coastal Wilderness to Fruited Plain: A History of Environmental Change in Temperate North America, 1500 to the Present*, Cambridge University Press, Cambridge, 1994.

Yoh, M., Soil C/N ratio as affected by climate: an ecological factor of forest NO_3^- leaching, *Water Air Soil Pollut.*, 130, 661–666, 2001.

Zak, D.R., Pregitzer, K.S., Holmes, W.E., Burton, A.J., and Zogg, G.P., Anthropogenic N deposition and the fate of $^{15}NO_3^-$ in a northern hardwood ecosystem, *Biogeochemistry*, 69, 143–157, 2004.

Zimmerman, M.J., Grady, S.J., Todd Trench, E.C., Flanagan, S.M., and Nielsen, M.G., Water-quality assessment of the Connecticut, Housatonic, and Thames River basins study unit: analysis of available data on nutrients, suspended sediments, and pesticides, 1972–92, Water Resources Investigations Report 95-4203, U.S. Geological Survey, Reston, VA, 1996.

4 The Stream Ecosystem

> ...a healthy river is a living river.
>
> **James R. Karr, 1999, *Defining and Measuring River Health***

4.1 INTRODUCTION

Watershed disturbance, beginning with the conversion of forest to other land uses, affects the physical and chemical condition of streams with inevitable consequences for stream life. Water quality concerns should extend to the entire stream ecosystem, not only because fish are an important food and economic resource, but also because the condition (health, abundance, and diversity) of aquatic biota is a reflection of the health or condition of the aquatic ecosystem as a whole. Biotic communities can serve as a measure of current stream conditions and provide a means of testing the effectiveness of watershed protection and restoration measures (Karr, 1981, 1991; Kerans and Karr, 1994; Krieger and Ross, 1993).

Streams in undisturbed settings provide a dynamic and heterogeneous habitat. Aquatic species have evolved over millennia in stream environments. Their morphology, physiology, and feeding habits are suited to specific habitat niches. (A niche is defined as "the sum total of an organisms utilization of the biotic and abiotic resources of its environment" [Campbell, 1990], not just where it lives, but also what it eats and the conditions required for reproduction.) Many organisms use two or more niches (e.g., Atlantic salmon [*Salmo salar*]), requiring different conditions at different stages of their life cycle (Aadland, 1993; Nislow et al., 1999). Conversion of forested watersheds to other uses can change established patterns of streamflow, sediment delivery, water temperature, and water quality in ways that directly affect living organisms and tend to reduce habitat variability and degrade the quality of the habitat that remains. In this chapter we describe the stream ecosystem and the interactions between the biota and the physical characteristics and processes that control the stream environment.

4.2 AQUATIC BIOTA

Stream life includes colonies of single-celled organisms (bacteria, diatoms, and algae), fungi, plants, and animals (Table 4.1). Single-celled organisms may be found floating in the water column or attached to material located on the streambed. They include autotrophs (photosynthesizing algae and bacteria) and decomposers (bacteria and fungi). Attached stream microorganisms are collectively referred to as periphyton, biofilm, or aufwuchs, terms that refer to single-celled organisms covering plants and rocks, woody debris, and sediment (Hawkins et al., 1982; Lock, 1993). The term aquatic macrophyte (larger plants) is used to refer to multicellular filamentous algae, mosses, and ferns adapted to aquatic habitat, and true angiosperms or flowering plants such as water lilies (*Nuphar* and *Nymphaea*) (Wetzel, 1983). Animal species that inhabit streams can be broadly divided into invertebrate and vertebrate groups (Allan, 1995). The term "benthic macroinvertebrate" refers to organisms without backbones that are large enough to be seen with the naked eye (larger than 0.05 mm) and spend at least part of their life cycle on the streambed or benthos. Benthic macroinvertebrates include insects, aquatic worms, crustaceans, and mollusks. Frequently, insect species spend the majority of their life cycle as larvae on the streambed, emerging as adult flying insects for only a brief period of time during which they reproduce and then die. The most common stream

TABLE 4.1
Aquatic Biota

Producers/consumers	Biological kingdoms	Energy resources	Examples	
Autotrophs (photosynthesizing single-celled organisms and plants)	Monera and Protista: unicellular periphyton (microbiota)	Light and nutrients	Cyanobacteria (blue-green algae), golden algae, diatoms, green algae	
	Protists and plants: multicellular (macrophytes)	Light and nutrients	Large filamentous algae, mosses, aquatic ferns, angiosperms (flowering plants)	
Decomposers (single-celled organisms that break down organic matter)	Monera: prokaryotes	Organic matter: falling leaves and other forest litter; detritus	Bacteria	
	Fungi	Dead leaves	Hyphomycetes	
Heterotrophs (organisms that consume plants and other heterotrophic organisms)	Animals: macroinvertebrates	Dead leaves, woody debris, fungi; drifting bacteria and periphyton; surface bacteria; macrophytes and periphyton, particularly diatoms; other invertebrates	Insects	Ephemeroptera (mayflies), Plecoptera (stoneflies), Trichoptera (caddisflies), Diptera (true flies, mosquitoes, midges), Odonata (dragonflies), Coleoptera (beetles)
			Crustaceans; Mollusks Annelids (oligochaete worms)	Gammarus (freshwater shrimp), freshwater clams
	Vertebrates	Algae and plants; macroinvertebrates; fish	Fish	Trout, salmon, bass, sunfish, etc.

Allan (1995), Campbell (1990), Giller and Malmqvist (1998), Hynes (1970), Margulies and Schwartz (1999).

vertebrates are fish. Gravel deposits, cobbles, boulders, woody debris, and macrophytes, as well as streambed sediments, provide habitat for benthic macroinvertebrates and fish. The health and diversity of communities of both fish and macroinvertebrates are used as measurements of stream condition as a whole (Karr, 1981; Kerans and Karr, 1994; NRCS Watershed Science Institute, 2000; Rosenberg and Resh, 1993).

4.2.1 AQUATIC FOOD WEBS

Aquatic biota form complex food webs (Figure 4.1). External energy inputs include light, which fuels photosynthesis in algae and plants, and organic matter (mostly dead leaves) from riparian areas. Together, these two resources form the basis of the aquatic food web. Microbial communities of decomposers break down coarse particulate organic matter (CPOM) into fine (FPOM) and very fine particulate organic matter (VFPOM), releasing nutrients that can then be used by plants and macroinvertebrates, which in turn become food for other macroinvertebrates and fish. Food in the stream is generally categorized by its origin. Allochthonous material originates outside the stream—leaves from overhanging trees, for example. Autochthonous material—algae, diatoms, and macrophytes—are the result of in-stream production (Allan, 1995).

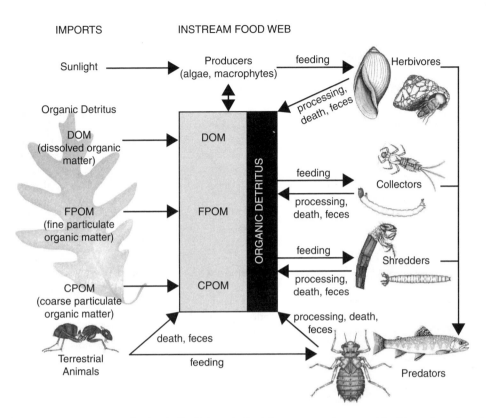

FIGURE 4.1 The in-stream food web. Sunlight and allochthonous organic matter (dead leaves) are the primary inputs and form the base of the food web. The stream supports a variety of organisms that have specialized feeding behaviors and a variety of food sources. (Original drawing by Ethan Nedeau, 2006.)

4.2.1.1 Trophic Classifications

Macroinvertebrates and fish species are frequently classified in terms of their feeding behaviors. While some are omnivores and generalists that consume a wide variety of foods in several different habitats, others are highly specialized and concentrate on particular food resources that occur only in specific habitat types. The term "trophic guild" refers to a group of species of macroinvertebrates or fish that consume similar types of food in similar habitats. The detail and complexity of classification systems varies among research studies.

Table 4.2 illustrates a classification system dividing macroinvertebrates into five trophic guilds: shredders, filterers/collectors, scrapers/grazers, generalists, and predators. Shredders consume CPOM such as leaves and associated bacteria and fungi that fall into the stream. Filterers/collectors eat drifting FPOM and VFPOM, bacteria, and drifting periphyton suspended in the water column. Scrapers/grazers feed on periphyton in the streambed. Generalists consume food from a variety of sources. Predators feed on other macroinvertebrates (Hawkins et al., 1982; Kerans and Karr, 1994; Vannote et al., 1980).

Different fish species are classified as herbivores, insectivores, piscivores (preying on other fish), or omnivores. Fish also may be grouped according to feeding behaviors. Some fish species (benthic feeders) search streambeds for insect larvae, periphyton, or detritus; some consume surface prey and insects and bacteria drifting in the water column (Allan, 1995). Again, classification systems vary among individual research studies. Table 4.3 is based on the classification system used in a study that took place in Illinois, Missouri, and southern Ohio (Horwitz, 1978). (The

TABLE 4.2
Macroinvertebrate Trophic Guilds

Trophic guild	Food	Location	Source	Examples
Shredders	Nonwoody CPOM, primarily leaves, and associated microbiota, especially fungi; woody CPOM and microbiota, fungi	Water surface	Riparian forests	Several families of Trichoptera, Plecoptera, and Crustacea, some Diptera, snails
Filterers/collectors	FPOM and VFPOM, bacteria, and sloughed periphyton in the water column	Water column	In-stream processing of CPOM; in-stream production	Net-spinning Trichoptera, some Diptera; some Ephemeroptera
Scrapers/grazers	Periphyton, especially diatoms	Benthos, streambed surfaces	In-stream production	Several families of Trichoptera, some Diptera, Coleoptera
Generalists	All of the above			
Predators	Other macroinvertebrates	Surface and water column	In-stream production	Odonata, some Plecoptera, Trichoptera, Diptera, and Coleoptera

CPOM, coarse particulate organic matter; FPOM, fine particulate organic matter; VFPOM, very fine particulate organic matter.

Allan (1995), Hawkins et al. (1982).

interested reader is referred to one of several texts in aquatic ecology and biology [Allan, 1995; Giller and Malmqvist, 1998; Hynes, 1970] for a detailed description of the intricate trophic relationships in the stream ecosystem.) Because of these specialized feeding behaviors, and because aquatic species also require suitable habitats for reproduction and juvenile growth, streams with more diverse physical habitats usually support greater aquatic biodiversity (Vannote et al., 1980).

4.2.1.2   Carbon Dynamics and Trophic Guild Distribution

4.2.1.2.1   The River Continuum Concept

The river continuum concept (RCC) (Figure 4.2) proposed by Vannote et al. (1980) describes a pattern of consistent variation in carbon dynamics and trophic guilds along a longitudinal and size gradient. According to the RCC, leaves (CPOM) from overhanging riparian trees are the primary source of carbon in narrow headwater streams. Proceeding from upstream to downstream reaches, tributaries combine to form larger streams and rivers. In general, the channel or bed slope decreases and the volume of flow increases. As a result, flow velocity declines and the width and depth of the channel increases ($Q = AV$). The riparian zone becomes progressively less important as a carbon source. Downstream communities depend more on in-stream production—algae and periphyton, for example—and the transfer of decomposed organic matter (FPOM) from upstream areas. Changes in the carbon supply lead to predictable changes in community composition. Thus, in macroinvertebrate communities, we would expect to find shredders in shaded upstream areas and an increased proportion of filterers/collectors and scrapers/grazers in downstream regions.

Since the publication of the RCC, many studies have attempted to test its predictions in a variety of stream and river environments (Hawkins et al., 1982; Hawkins and Sedell, 1981; Minshall et

TABLE 4.3
Fish Feeding Guilds

Feeding guild	Description	Species examples
Piscivores	Fish primarily eating fish and crayfish, with smaller amounts of terrestrial and aquatic insects	Herrings (*Alosa*) Pikes (*Esox*) Eel (*Anguilla*) Larger catfish (*Ictalurus catus*) Basses (*Morone*) Some sunfish (*Micropterus*, *Ambloplites*)
Snail eaters		Redear sunfish (*Lepomis microlophus*)
Generalized invertebrate feeders	Fish take a variety of invertebrate food from the bottom, surface, and water column, including terrestrial and aquatic insects, zooplankton, and benthic microcrustacea	Trout Some minnows (*Notropis*) Pirate perch (*Aphredoderus*) Sunfish (*Lepomis*)
Bottom feeders	Fish take a variety of benthic invertebrates; insect larvae are probably the main food type	Sturgeons Minnows Suckers Darters (*Percina*, *Etheostoma*, *Ammoncrypta*) Sculpins (*Cottus*)
Surface and water column feeders	Fish feed primarily on terrestrial insects, drift, and zooplankton, with some aquatic insect larvae as well	Mooneye and goldeye (*Hiodon*) Minnows (*Notropis*) Silversides (*Labidesthes*) Killifish (*Fundulus*)
Detritivores and herbivores	Bottom feeders taking in detritus, benthic diatoms, and attached algae	Minnows (*Campostoma*, *Phoxinus*, and others) Suckers (*Catostomus platyrhynchus*)
Planktivores	Midwater fishes that strain zooplankton and phytoplankton from the water	Paddlefish (*Polydon*) Shad (*Dorosoma*)
Omnivores	Consume a wide range of animal and plant food, including terrestrial and aquatic insects, fish detritus, and mollusks	Native minnows Carp and goldfish (*Cyprinus* and *Carassius*) Bullheads and channel catfish (*Ictalurus*)
Lampreys	Parasitic and nonparasitic	
Unknown		

Modified from Horwitz (1978).

al., 1983; Poole, 2002; Rosi-Marshall and Wallace, 2002; Schlosser, 1982). This has led to a wealth of new information on broad-scale habitat-community variation and to the development of several new theories designed to address river environments not adequately described by the RCC. Minshall et al. (1983) sampled streams with forested headwaters at four sites in Idaho, Michigan, Oregon, and Pennsylvania. Their data supported the "gradual change in ecosystem structure and function" proposed by the RCC in general, but they noted variations from the predicted pattern based on "(1) watershed climate and geology, (2) riparian conditions, (3) tributary patterns, and (4) location-specific lithology and geomorphology." One of the conclusions of this study and others (Hawkins et al., 1982; Schlosser, 1982) is that while the RCC describes large-scale trends and processes, local variations in habitat (riparian condition) and community processes (e.g., predatory relationships) can produce ecological patterns that, not surprisingly, are at odds with RCC predictions.

Large rivers with relatively unconfined floodplains do not fit the RCC paradigm. In these systems, processed carbon carried from upstream is of little consequence and the interaction of the river channel with its floodplain is far more important in determining river habitat and community

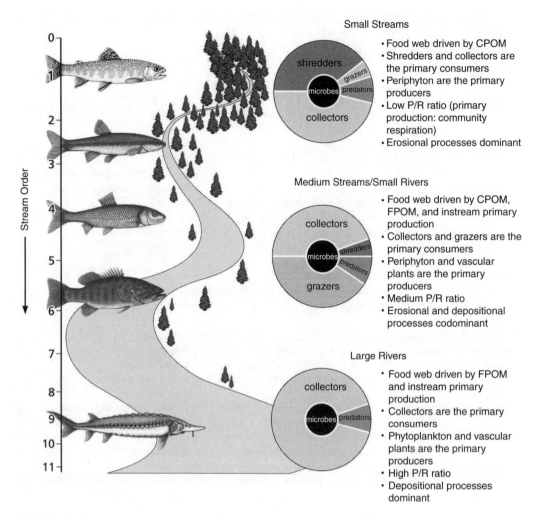

FIGURE 4.2 The river continuum concept (redrawn from Vannote et al., 1980). (Redrawn and used with the permission of the National Research Council Canada.)

response (Sedell et al., 1989). The flood pulse concept (Junk et al., 1989) describes these environments—in which habitat and biotic communities are shaped by periodic flooding—more accurately. In some settings, geologic factors may produce habitat patterns that are more pronounced than those occurring along the length of a stream. The process domains concept (PDC) (Montgomery, 1999) describes river systems in which geology and geomorphology are controlling factors, identifying discrete sections of rivers in which "community structure and dynamics respond distinctively to different disturbance regimes." Researchers in Georgia (Walters et al., 2003) sampled 31 streams and found that geomorphic variables (stream slope, bed texture, and bed mobility) were better predictors of species composition in a particular area than the longitudinal patterns identified by the RCC.

4.3 STREAM HABITAT

Within individual stream reaches there are microhabitat variations among pools, riffles, and runs (also called raceways), and between the edges and center areas of the stream channel. Riffles and pools are created by variations in the slope of the streambed and the width of the channel (Dunne and Leopold, 1978). Pools are deeper, with slower currents and more fine sediment in the bed material. In shallower riffles, the gradient and current velocity are greater, washing away fine sediments, leaving gravel streambeds with high levels of dissolved oxygen preferred by some fish species for spawning. Runs are transitional areas between pools and riffles (Berkman and Rabeni, 1987; Chapman, 1988; Dunne and Leopold, 1978). Schlosser (1982) concluded that the addition of deep pools in downstream reaches led to an increase in the number of fish species in downstream areas and that fish species diversity was significantly correlated with habitat diversity. Aadland (1993) categorized fish species in six Minnesota streams based on habitat preferences of species and during different life stages. In this study, riffles were the preferred habitat for spawning and deep pools were inhabited exclusively by adult fish.

4.3.1 Determining Factors in Stream Habitat

The physical environment of streams and river channels at any given point is controlled by streamflow and sediment (Leopold, 1997; Leopold et al., 1964). In most areas (at stream edges, in the benthos, and in the water column) it is the interaction of these two factors that determines habitat characteristics. Turbidity (light attenuation caused by suspended particulate matter) is a function of streamflow, sediment load, and sediment particle size. Rapidly flowing water keeps more sediment in suspension rather than allowing it to settle to the streambed (Hamblin, 1982). Sediment loading determines the amount of sediment in the channel, and small particles remain in suspension longer than larger ones. Habitat conditions are also affected by water temperature, dissolved oxygen concentrations, and chemical water quality. Persistent changes in streamflow and sediment conditions alter the stream environment. This may decrease habitat diversity and the availability of particular niches. This in turn affects community richness (the number of different species) and abundance (the total number of individuals per unit area) and species abundance (the number of individuals per species per unit area). Persistent changes in streamflow and sediment are often the result of changes in upstream and riparian land use.

4.3.1.1 Streamflow Variability and Aquatic Communities

Flowing water provides benefits by delivering resources to and removing wastes from stream communities. At the same time, successful stream inhabitants have evolved to live within a particular current velocity range and not be swept away (drift loss) (Allan, 1995). Microbes in streams attach to rocks, woody debris, macrophytes, and sediments and form biofilms using chemical glues (Lock, 1993). Macroinvertebrates seek refuge from strong currents under rocks, in woody debris, and in interstitial spaces of the stream substrate (Waters, 1995). Various species of Diptera (black fly larvae and some midges) and Trichoptera (caddisfly larvae) spin silk threads that attach to rocks in the streambed. Many fish species seek refuge in quieter areas at the edge of the stream and in places where coarse woody debris and rocks provide shelter from fast currents. Woody debris also provides a refuge from predators for younger, smaller fish (Everett and Ruiz, 1993). The flattened and elongated body shapes of many stream animals are thought to be adaptations that allow for controlled movements in flowing water (Allan, 1995; Hynes, 1970).

As discussed in Chapter 2, streamflow varies throughout the year and within different reaches of the river channel. The streamflow velocity is higher in steep, narrow headwater regions and decreases in the deeper, more level downstream sections of the channel. Current velocities also differ across the width of the stream and from the stream surface to the streambed. In general, the velocity is highest at the surface in the center of the stream and lowest at the streambanks and at the interface between the water and the streambed (Dunne and Leopold, 1978). Hence the species of fish and insects that inhabit the center sections of streams where currents are faster are often not the same species that inhabit areas nearer streambanks where currents are slower (Allan, 1995; Bain et al., 1988). Periods of increased streamflow and turbulence will increase the drift loss and mortality of stream organisms; however, populations can recover if the disturbance is temporary and not part of a long-term trend (Roghair et al., 2002; Waters, 1995). Streamflow variability, the difference in amplitude between the peak and minimum discharge at a particular point in a stream channel, helps to determine the type and number of species that can survive in that area. Horwitz (1978) examined the diversity of fish species as it related to the variability of streamflow in 15 river systems in southern Illinois, southern Ohio, the Missouri Ozarks, and Wyoming. In general there was far greater flow variability in stream headwaters than in downstream reaches. In all the rivers studied, the number of fish species increased from upstream to downstream reaches as the variability of streamflow diminished. It also was true that the streams with the most variable flows in headwater regions had the fewest number of fish species compared to the headwater regions of other streams. Horwitz speculated that areas subject to a more constant flow allow greater niche or habitat specialization and as a result support a greater diversity of species.

Bain et al. (1988) examined the differences in fish species diversity in the Deerfield River in western Massachusetts and the West River in southern Vermont. Both are fifth-order tributaries of the Connecticut River. The West River is unregulated; there are four hydroelectric dams on the Deerfield River, causing dramatic variability between minimum and peak discharges. Nine species representing more than 90% of all fish were restricted to habitats at the stream margins where the water was shallow and the current was slow. These species were reduced in abundance in the regulated Deerfield River and absent at the site that showed the greatest variation in streamflow. Fish in this group included small (length ≤ 100 mm) smallmouth bass (*Micropterus dolomieui*), rock bass (*Ambloplites rupestris*), white suckers (*Catostomus commersoni*), bluegills (*Lempomis macrochris*), pumpkinseeds (*Lepomis gibbosus*), blacknose dace (*Rhinichthys atratulus*), fallfish (*Semotilus corporalis*), mimic shiner (*Notropis volucellus*), and tesselated darter (*Etheostoma olmstedi*). The swift current habitat group—large rock bass (length > 100 mm), longnose dace (*Rhinichthys cataractae*), largemouth bass (*Micropterus salmoides*), and large white suckers—used deep areas of the river with swift currents, while American eels (*Anquilla rostrata*) and larger smallmouth bass were habitat generalists using various areas of the stream. The number of fish belonging to species in these two groups (swift current habitat and habitat generalists) increased in the Deerfield River.

The availability of habitats with slower currents appears to increase the survival of juvenile fish of some species. Nislow et al. (1999) found the greatest numbers of juvenile Atlantic salmon in stream habitats with a narrow range of relatively slow streamflow velocities. At the same time, juvenile fish of other species benefit from habitat diversity if predators are a part of the stream environment. Schlosser (1987) found that juvenile hornyhead chub (*Nocomis biguttatus*), white sucker, and smallmouth bass preferred pools with slow moving water when predators were absent. When predators, in this case adult smallmouth bass, were introduced to the pools, the juveniles moved to the refugia of riffles and runs.

Flowing water makes food available to fish that feed in the water column by dislodging a certain percentage of attached algae and benthic macroinvertebrates, placing them in the water column and causing them to drift downstream. Higher streamflow velocity increases the shear stress on the algae and invertebrates, dislodging more organisms, but some fish species have less success capturing food in areas with higher velocity streamflow (Borchardt, 1993; Hill and Grossman, 1993). Hill and Grossman (1993) developed and field-tested models showing that rainbow trout (*Oncorhynchus mykiss*) were found in greater numbers in areas where their prey capture success was the highest, and where the least energy was required to get the most food. Current velocity ranged from near zero to more than 40 cm/sec in the sampled streams. Prey capture success decreased as current velocity increased above a specific range. Small trout (53 to 70 mm in length) used areas with current velocities between 13 and 18 cm/sec. Larger trout were most commonly found in areas with current velocities between 18 and 21 cm/sec.

An increase in the frequency of high velocity streamflows can result in reductions in periphyton and benthic macroinvertebrate populations and consequently limits the fish food supply. Clausen and Biggs (1997), investigating streams in New Zealand, found that periphyton diversity was lowest in streams with the most frequent peak flows. This study also found that the species diversity of benthic macroinvertebrates was highest at intermediate levels of peak flow frequency and that invertebrate density decreased in frequently flooded streams. Townsend et al. (1997), also working in New Zealand, correlated bed disturbance, measured by the movement of stones on the streambed, with variations in invertebrate populations. In this study, invertebrate species abundance and richness also were at their highest levels at intermediate levels of disturbance. Macroinvertebrate orders identified in this study included Ephemeroptera (mayflies), Plecoptera (stoneflies), and Trichoptera (caddisflies) (collectively referred to as EPT), Coleoptera (beetles), Diptera (flies, mosquitoes, and midges), Hemiptera (true bugs), Odonata (dragonflies), and mollusks, crustaceans, and worms. The lowest levels of invertebrate richness and abundance were found in both the least disturbed and in the most disturbed stream environments.

The presence of woody debris and other habitat features that can provide refugia for invertebrates and fish during periods of high flows can be important in protecting populations and community structure (Lenat, 1988). Borchardt (1993) found that population losses decreased for the crustacean *Gammarus pulex*, a type of freshwater shrimp, and *Ephemerella ignita*, a mayfly, as woody debris was added to an experimental channel. Increasing woody debris from 800 g/m^2 to 1600 g/m^2 reduced the loss of *G. pulex* by half and the loss of *E. ignita* by one-third. Larger amounts of woody debris (>1600 g/m^2) had little effect.

4.3.1.2 Sediment Variability and Aquatic Communities

Accelerated soil and channel erosion leading to excess sediment has been recognized as the most important pollutant of stream environments by both the U.S. Fish and Wildlife Service and the Environmental Protection Agency (Waters, 1995). The addition of sediment to a streambed can reduce the survival and productivity of stream species. Increased sediment loading can cause (1) abrasion and clogging of respiratory structures, (2) reduced feeding efficiency, (3) exposure to sediment-associated toxins, and (4) higher drift rates of benthic organisms (Ward, 1992). Excess sediment also degrades habitat, causing long-term declines in stream populations and particularly affecting species that feed or spawn on rocky, gravelly substrates (lithophilic species) (Chapman, 1988; Ellis, 1936).

Fine sediments suspended and carried in streamflow increase turbidity. If turbidity is high, less light passes through the water column and the water appears cloudy or opaque. Lower light levels can limit photosynthesis in algae and macrophytes, reducing the food supply for herbivorous macroinvertebrates that serve as food for other macroinvertebrates and fish (Allan, 1995; Ryan, 1991; Waters, 1995). Higher turbidity also reduces visibility, making it more difficult for fish to find food (Waters, 1995). Studies of larval striped bass (Breitburg, 1988) and bluegills in North Carolina (Gardner, 1981) both reported lower prey capture success when turbidity levels were high. In contrast, one study in California (Boehlert and Morgan, 1985) found that larval Pacific herring actually captured more food at intermediate levels of turbidity than at low levels. Turbidity may have increased the visual contrast at this small scale, highlighting prey for the tiny larvae and enhancing capture success.

Fine sediments deposited on the streambed can fill interstitial spaces between larger rock and sand particles, reducing oxygen levels for macroinvertebrates that burrow in the streambed and eliminating refugia for these organisms during storm events. Lenat et al. (1981) examined benthic communities in two streams in North Carolina, portions of which had been contaminated with sediment from highway construction. In general, macroinvertebrate density was lowest at sampling sites directly downstream from construction areas and highest at stream sites with rocky substrates. During low flow periods, periphyton and a distinct community of sediment-tolerant benthic grazers rapidly colonized sandy areas. However, these sandy areas were not suitable for any invertebrate species during stormflow. The density of benthic organisms was greatly reduced at sandy sites at these times.

Culp et al. (1986) found that adding sediment to areas where flow velocity was low (5 to 6 cm/sec) had little immediate effect on the benthic community (long-term effects of habitat change were not the subject of this study). A significant decrease in density was observed for one taxa only, *Paraleptophlebia*, a type of mayfly. In areas where the flow velocity was high enough (35 to 38 cm/sec) to move the sediment, benthic macroinvertebrate density was reduced by more than 50% in 24 hours. Saltation (bouncing, sliding, and drifting of sedimentary particles on the streambed) of the added sediment, created by turbulent flow, dislodged benthic macroinvertebrates and greatly increased drift loss of the organisms. Mayflies and stoneflies (*Baetis tricaudatus*, *Cinygmula*, and *Zapada*), and midges (family Chironomidae) residing in the upper 3 cm of the substrate were immediately affected. Drift loss rates of mayfly and stonefly species residing in the lower substrate (*Paraleptophlebia* and *Alloperla*), 3 to 6 cm below the surface, increased 6 to 9 hours later. Both species showing a delayed response to sediment additions are known to come to the surface at night to feed. It is also possible that sediments had filtered down into the deeper substrate, filling interstices, thus lowering oxygen availability in this microhabitat and forcing organisms up to the surface of the substrate. Here they became vulnerable to gravel and cobbles that roll and bounce (saltation) when flow velocity and shear stress increases.

Berkman and Rabeni (1987) studied the effect of sedimentation on fish communities in streams in northeast Missouri. Fish species were grouped by both feeding and reproductive behaviors. The feeding classifications were (1) omnivores—consuming plant, animal, and detrital material; (2) general insectivores—consuming surface, midwater, and benthic insects; (3) benthic insectivores; and (4) herbivores. (Piscivores constituted less than 1% of the fish species captured and were not included in the study.) Species were further classified as "simple spawners" or "complex spawners" based on "preparation of spawning site, territorial defense, and pre-spawning social organization." Complex spawners built nests and guarded nesting sites. Complex spawners were also divided into groups that exhibited parental care and those that did not. Simple spawners were divided into groups that required clean gravel versus those that could spawn on sand, silt, or submerged vegetation. Riffles with the least amount of sediment in the substrate had unique communities dominated by "riffle species," benthic insectivorous and herbivorous fish species and species requiring clean gravel for spawning. As the amount of sand in the riffle substrate increased, the relative abundance of riffle species decreased, while the relative abundance of the general insectivores and species that spawned on a variety of substrates increased. There was little change in the relative abundance of complex spawners. The increase in sediment tended to homogenize the stream environment. As riffle habitat became more similar to that of runs and pools, the unique biological riffle community was lost as well. The addition of sediment buried benthic food resources and eliminated suitable spawning sites for these species.

It should be noted that aquatic communities (periphyton, macrophyton, benthic macroinvertebrates, and fish) are all accustomed to disturbance—peak flows during rain and snowmelt events and low flows in periods between storms, as well as variations in turbidity and sediment inputs corresponding to variations in streamflow. Typically populations experience temporary declines in response to these events. Recovery and recolonization are fairly rapid, however, even after disturbances such as a severe drought (Caruso, 2002; Ryan, 1991). It is only when the overall pattern of disturbance—the dynamic equilibrium of a stream—changes for extended periods of time that there are major and persistent changes in the density and composition of aquatic communities (Waters, 1995).

4.3.1.3 Temperature (Dissolved Oxygen and Biological Metabolism)

Under natural conditions, stream water temperature varies seasonally and geographically and generally increases from shaded, low-order headwater streams to wider, higher-order, downstream areas, where riparian shading has relatively little influence. The temperature range varies as well; shallow upstream waters show the greatest diurnal and season variation, while downstream areas with greater discharge (thermal inertia) are generally more stable (Galli and Dubose, 1990).

Aquatic organisms are ectotherms; their body temperature equilibrates with their surroundings (Campbell, 1990; Galli and Dubose, 1990). They are classified as stenothermal (cold water: 0°C to 20°C; upper lethal temperature <26°C), mesothermal (intermediate: 20°C to 28°C; upper lethal temperature 28°C to 34°C), or euthermal (warm water: >28°C; upper lethal temperature >34°C) species based on the optimal temperature conditions required for population survival, growth, and reproduction (Galli and Dubose, 1990). Examples of coldwater fish species include many trout and salmon species; warmwater fish include the Centrarchidae (sunfish and bass). Orders of insects requiring coldwater environments include the Plecoptera (stoneflies) and Trichoptera (caddisflies) and many species in the order Ephemeroptera (mayflies). An example of an insect order preferring warmwater conditions is the Odonata (dragonflies and damselflies) (Galli and Dubose, 1990).

Water temperature directly influences the metabolic rate of stream organisms. Respiration rates in aquatic animals can increase by 10% or more for every 1°C increase in water temperature (Giller and Malmqvist, 1998). Animals respiring at higher rates require more oxygen, but dissolved oxygen decreases as water temperature rises. Pure water at standard pressure contains 13.1 mg/L of dissolved oxygen at 4°C and only 7.55 mg/L at 30°C (Wetzel, 1983). Simply put, "while aquatic organisms require more oxygen at higher temperatures, less is available" (Ward, 1992).

In general, turbulent, fast flowing streams are at 100% saturation because oxygen is mixed into the water through diffusion from the air. Laboratory studies have shown that lotic species (species living in moving water or streams) have higher metabolic rates and use more oxygen than species of similar families in still (lentic) waters. This difference has been observed for several groups of insect species. In one study, anesthetized stream species of mayfly nymphs consumed four times more oxygen than anesthetized pond species at 10°C. The insects were anesthetized to ensure that activity levels did not influence metabolic rates (Ward, 1992). In addition, insects in still water are able to alter respiratory behaviors to increase oxygen uptake. Because stream species have evolved in an environment where dissolved oxygen is generally abundant, there has been little selective pressure to develop physiologic adaptations that conserve or increase oxygen uptake. This increases the vulnerability of stream species to even slight changes in dissolved oxygen concentrations (Ward, 1992). Hynes (1970) proposed that the inability of coldwater species such as brook trout to survive in warmer temperature regimes is primarily a response to reductions in the concentration of dissolved oxygen.

Water temperature also provides cues that control the life cycles of stream organisms. At each stage of the life cycle there is a species-specific optimal temperature range for growth and development. Studies of insect communities in White Clay Creek, Pennsylvania, showed that insects exposed to temperatures warmer than their optimal range used more energy in maintenance metabolism than in growth (Vannote and Sweeney, 1980). Adult maturation began earlier in development. These changes resulted in decreased adult body size and egg production. Movement to cooler conditions also resulted in reduced growth and reproduction from reduced metabolic rates and reduced assimilation of nutrient resources. Other studies have shown that while growth rates of fish increase initially with increasing temperatures, growth falls off drastically as temperatures approach the upper limits of the species ideal temperature range (Allan, 1995).

Aquatic communities can tolerate and recover from episodic temperature variation as long as larger scale temperature regimes (maximum and minimum temperatures and range of temperature variation over extended time periods) within individual habitat areas remain relatively stable. Human alteration of the landscape (e.g., the removal of riparian trees, urbanization, or the loss of forests and wetlands) can change long-term water temperature regimes, causing population reductions and the elimination of species that are unable to adapt. (The ramifications of global climate change, while potentially substantial [Morgan et al., 2001; Schindler, 2001], are beyond the scope of this review.)

4.3.1.4 Chemical Water Quality

As noted in the discussion of eutrophication in Chapter 3, excess loading of nitrogen and phosphorus brings about major changes in stream community composition and can ultimately cause the creation of anoxic "dead zones" where few aquatic organisms survive. Herbicides and pesticides have deleterious effects on stream life as well. The effects of chemical compounds associated with agricultural and urban land use are discussed in Part II, Chapters 7 and 8.

4.4   BIOLOGICAL MONITORING

The health, diversity, and abundance of the biotic community are often used as indicators of the general condition of a stream and its surrounding watershed (Karr and Chu, 1999). Living organisms experience the full range of stream conditions: flow, sediment, temperature, and pollutants (Figure 4.3). Indications of deterioration in stream conditions are often reflected in the biotic community

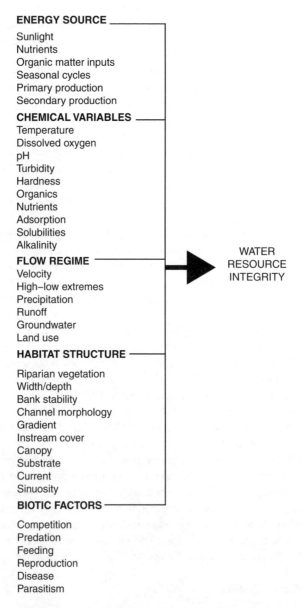

ENERGY SOURCE
Sunlight
Nutrients
Organic matter inputs
Seasonal cycles
Primary production
Secondary production

CHEMICAL VARIABLES
Temperature
Dissolved oxygen
pH
Turbidity
Hardness
Organics
Nutrients
Adsorption
Solubilities
Alkalinity

FLOW REGIME
Velocity
High–low extremes
Precipitation
Runoff
Groundwater
Land use

HABITAT STRUCTURE

Riparian vegetation
Width/depth
Bank stability
Channel morphology
Gradient
Instream cover
Canopy
Substrate
Current
Sinuosity

BIOTIC FACTORS

Competition
Predation
Feeding
Reproduction
Disease
Parasitism

WATER
RESOURCE
INTEGRITY

FIGURE 4.3 Factors that influence the integrity of streams (NRCS 1998; modified from Karr, 1986; http://www.nrcs.usda.gov/technical/ECS/aquatic/svapfnl.pdf).

long before a clear signal of a change in water chemistry can be detected (Horner et al., 1996; Karr, 1981). Monitoring biotic populations can provide an early warning signal and alert watershed managers and others to the need for enhanced river protection.

Both fish and benthic macroinvertebrates have been used in biotic monitoring. The use of fish was initially preferred because, until recently, taxonomy was better known, information about species life history was more readily available, and it was thought to be less difficult to generate public concern about fish than about insects, worms, and snails (Karr, 1981). Benthic macroinvertebrates are now commonly used as well. Macroinvertebrates are ubiquitous and knowledge of the taxonomy of these organisms has grown. (There was a 60% increase in the number, 231 to 371, of valid genera classified in the order Ephemeroptera from 1976 to 1990 [Giller and Malmqvist, 1998].) In contrast with fish, macroinvertebrates are basically sedentary. Thus the benthic macroinvertebrate community can serve as a continuous monitor of conditions at a specific stream site, allowing for long-term analysis of change (Krieger and Ross, 1993; Rosenberg and Resh, 1993).

Biological monitoring designs have been based on the following assumptions: (1) undisturbed stream habitats have characteristic biotic communities; (2) biotic communities in stream habitats adversely affected by human activities differ from those in undisturbed streams; and (3) some aquatic species are more susceptible to human induced disturbance than others (Figure 4.4). Thus

Pollution Intolerant

- Sensitive to pollution such as low oxygen, warm temperatures, eutrophication, and contaminants
- Usually have external gills, partly or wholly aquatic life stages, and relatively long life cycles
- Usually confined to coldwater streams and rivers with good to excellent water quality and unimpaired habitat
- Common examples: stoneflies (Plecoptera), mayflies (Ephemeroptera), and caddisflies (Trichoptera)

Somewhat Pollution Tolerant

- Capable of withstanding moderate amounts of environmental stress and pollution
- Found in a wide range of habitats in both lotic and lentic waterbodies
- Not particularly good indicator species because of their wide tolerance
- Common examples: dragonflies and damselflies (Odonata), aquatic beetles (Coleoptera), crayfish (Decapoda), scuds (Amphipoda), and certain caddisflies (Trichoptera)

Pollution Tolerant

- Can withstand severely polluted conditions and naturally stressful environments such as anoxic mud, lagoons, or waterbodies that dry periodically (e.g., vernal pools)
- Though they can be present anywhere, they are often numerically dominant in polluted environments where other species cannot persist
- Common examples: aquatic worms (Oligochaeta, esp. Tubificidae), air-breathing snail (Gastropoda), fly larvae (Diptera, esp. Chironomidae), and leeches (Hirudinea)

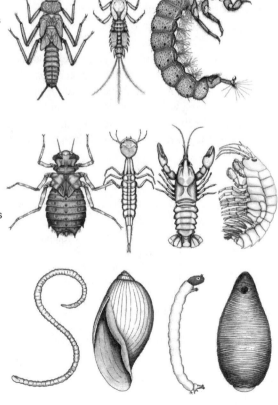

FIGURE 4.4 Stream invertebrates: examples of pollution tolerance characteristics and groupings. (Original drawing by Ethan Nedeau, 2006.)

measurements of biotic communities in similar channel environments of disturbed and undisturbed streams (within areas of similar terrain, soils, geology, and climate, sometimes referred to as ecoregions [Omernik, 1987]) can reflect the degree of disturbance to the stream and aquatic habitat.

In the United States, most biological monitoring has been done using the multimetric approach (Norris and Hawkins, 2000). This requires the development of a biological index. The biological index identifies stream community measurements, or metrics, that correlate significantly with varying degrees of stream disturbance and water quality in a particular region (Fore et al., 1996; Karr, 1981; Kerans and Karr, 1994). Stream community metrics may include the total number of taxa (taxa richness, a reflection of biological diversity) and the number and abundance of pollution tolerant and intolerant species and measurements of community processes such as the proportions of different trophic guilds (e.g., shredders, grazers, etc.) in the community—a reflection of habitat diversity (Karr, 1981; Kerans and Karr, 1994). The index is developed on a set of sample streams and can then be used to characterize the condition of other streams throughout the region.

The development of the index of biotic integrity (IBI) was described by Karr in 1981. Karr used measurements of fish communities in streams in Indiana and northeastern Illinois to develop a system that "accurately reflected the status of fish communities and the environment supporting them." The IBI has subsequently been refined and modified to fit local conditions in other areas (Steedman, 1988).

Karr (1981) used 12 parameters (metrics) in assessing stream health (Table 4.4). Metrics were scored based on how closely the sample measurements approximated comparable data from similar habitats (pools or riffles) in undisturbed streams. Scores of 1, 3, and 5 were assigned based on the relation of a particular sample measurement to degraded aquatic conditions. For example, a high percentage of green sunfish (*Lepomis cyanellus*) resulted in a score of 1 for that metric because green sunfish are considered very tolerant of degraded conditions. The presence of species hybrids was thought to be the result of habitat degradation and increased interaction among species. If hybrid fish were present at a site, the site received a 1 for that metric. A high proportion of omnivores also received a score of 1. An abundance of omnivores indicates degradation of specialized habitats that would support a wide variety of species at lower trophic levels and allow for specialized trophic behaviors. Relatively healthy, trophically diverse communities are able to support top level predators, thus a metric score of 5 was given if these species were present. The site score for each metric

---

TABLE 4.4
Parameters Used in the Assessment of Fish Communities

Species composition and richness
1. Number of species
2. Presence of intolerant species
3. Species richness and composition of darters
4. Species richness and composition of suckers
5. Species richness and composition of sunfish (except green sunfish)
6. Proportion of green sunfish
7. Proportion of hybrid individuals

Ecological factors
8. Number of individuals in the sample
9. Proportion of omnivores (individuals)
10. Proportion of insectivorous cyprinids
11. Proportion of top carnivores
12. Proportion with disease, tumors, fin damage, and other anomalies

---

Karr (1981); reprinted with the permission of the American Fisheries Society.

TABLE 4.5
Biotic Integrity Classes Used in the Assessment of Fish Communities Along with General Descriptions of Their Attributes

Class	Attributes
Excellent	Comparable to the best situations without the influence of man; all regionally expected species for the habitat and stream size, including the most intolerant forms, are present with a full array of age and sex classes; balanced trophic structure
Good	Species richness somewhat below expectation, especially due to loss of the most intolerant forms; some species with less than optimal abundances or size distribution; trophic structure shows some signs of stress
Fair	Signs of additional deterioration include fewer intolerant forms, more skewed trophic structure (e.g., increasing frequency of omnivores); older age classes of top predators may be rare
Poor	Dominated by omnivores, pollution-tolerant forms, and habitat generalists; few top carnivores; growth rates and condition factors commonly depressed; hybrids and diseased fish often present
Very Poor	Few fish present, mostly introduced or very tolerant forms; hybrids common; disease, parasites, fin damage, and other anomalies
No Fish	Repetitive sampling fails to turn up any fish

Karr (1981); reprinted with permission of the American Fisheries Society.

was added to produce an IBI score ranging from 0 to 60. Sites were then ranked from very poor (IBI < 23) to excellent (IBI 57 to 60) (Table 4.5). Excellent conditions would be comparable to streams in an undisturbed landscape. Karr stressed the necessity of having aquatic biologists who were familiar with the technique adapt the IBI to local conditions in other areas.

The development of an index of biotic integrity using benthic macroinvertebrates (B-IBI) is described in a study performed in the Tennessee Valley (Kerans and Karr, 1994). Eighteen attributes or metrics, 10 measuring community composition and 8 measuring ecological factors or community processes, were evaluated. The goal was to identify metrics that consistently produced significantly different values among sites with known variations in stream conditions. Stream condition and water quality indicators included mean, minimum, and maximum measurements of streamflow, temperature, dissolved oxygen, pH, suspended and dissolved solids, nutrient (nitrogen and phosphorus) concentrations, and fecal coliform concentrations. Thirteen of the original 18 metrics proved to be valuable in discriminating sites. Discarded metrics failed to discriminate among sites by water quality, were strongly correlated with other measurements, or showed excessive variability over the 3 year sampling period. Among the community composition metrics, total taxa richness and the presence or absence of pollution intolerant species, including certain species of snails and mussels, mayflies, caddisflies, and stoneflies, proved to be good measures of stream condition (Table 4.6).

Lenat (1988) describes the development of a B-IBI for coastal regions in North Carolina. The area being sampled in this case contained three distinct ecoregions: mountains, Piedmont, and Coastal Plain. Rankings were developed based on total taxa richness and the taxa richness for EPT for stressed and unstressed sites within each ecoregion. Water quality parameters used as indicators of "stress" included concentrations of fluorides, phenols, sulfates, dissolved oxygen, metals, fecal coliform, and nutrients, and temperature, pH, and turbidity. Sites were scored based on the number of times state water quality standards were exceeded and by how much. Stressed sites received higher scores, indicating impaired water quality. Unstressed sites were defined as "areas with no known chemical/physical alterations having a high diversity of invertebrates and/or fish." It was found that while total taxa richness did not differ among unstressed sites in the three ecoregions, EPT taxa richness was significantly higher in the mountains than in the Coastal Plain. Thus different criteria were developed for evaluating water quality based on invertebrate sampling for each of the three regions. An EPT score greater than 27 was classified as excellent in the Coastal Plain region. In comparison, an EPT score greater than 41 was required for an excellent rating in the mountains.

TABLE 4.6
Biological Attributes Used in the Benthic Index of Biotic Integrity (B-IBI)

Metric	Hypothesized response to degraded stream condition
Measurements of community structure	
Total taxa richness	Decline
Number of intolerant snails and mussels	Decline
Ephemeropteran taxa richness	Decline
Trichopteran taxa richness	Decline
Plecopteran taxa richness	Decline
Proportion of individuals as *Corbicula**	Increase
Proportion of individuals as oligochaetes	Increase
Proportion of individuals as chironomids	Increase
Proportion of individuals in the two most abundant taxa	Increase
Measurements of community processes	
Proportion of individuals as omnivores and scavengers	Increase
Proportion of individuals as detritivores	Increase
Proportion of individuals as shredders	Decline
Proportion of individuals as collector-gatherers	Increase
Proportion of individuals as collector-filterers	Increase
Proportion of individuals as grazers-scrapers	Decline
Proportion of individuals as strict predators (excluding chironomids and flatworms)	Decline
Total abundance	Decline

* Corbicula: clams.

Slightly modified from Kerans and Karr (1994). (Reprinted with permission of the Ecological Society of America.)

4.4.1 APPLICATIONS OF BIOTIC INDICATORS IN AQUATIC HABITATS

4.4.1.1 Olentangy River, Central Ohio

Gatz and Harig (1993) used a variation of Karr's (1981) IBI to quantify the condition of the Delaware Run, a tributary of the Olentangy River in central Ohio. This is an unusual study since earlier records of fish species richness and abundance were available. In 1940 an undergraduate student at Ohio Wesleyan University working with an employee of the Ohio State Department of Conservation made eight collections of fishes spaced along the entire stream. Records of species collected, preserved specimens, and verbal descriptions of species abundance had been retained from the 1940s project. The human population of the watershed grew from less than 9,000 to more than 20,000 from the time of the first to that of the second study. As the proportion of human land use increased, riparian vegetation was damaged and destroyed, streams were channelized, and the pollutant load increased. In 1992 Gatz and Harig collected fish at three of the same sites used in 1940. They used IBI rankings to compare stream conditions in 1992 to those in 1940. It showed a major shift from pollution intolerant to pollution tolerant fish species. The species collected in 1940 included several species of shiners (*Luxilus chrysocephalus*, *Lythrurus umbratilis*, *Notropis photogenis*, *Notropis rubellus*, and *Notropis stramineus*), pollution intolerant species that were absent in 1992. Species such as carp (*Cyprinus carpio*) and green sunfish (*Lepomis cyanellus*), absent in 1940, were present in large numbers in 1993. Creek chubs (*Semotilus atromaculatus*), white suckers (*Catostomus commersoni*), and green sunfish, all habitat generalists, made up 70% of the fish population in 1992, an increase from 9% in 1940. In general there had been a decline in both the

TABLE 4.7
IBI Metric Scores for Delaware Run, Ohio, 1940 and 1992

	Metric score in 1940	Metric score in 1992
Number of		
1. Native fish	5	5
2. Darter and sculpin species	3 (1)	5 (1)
3. Headwater species	1	1
4. Minnow species	3 (5)	1 (1)
5. Sensitive species	5 (5)	3 (1)
6. Individuals excluding tolerant species	3	1
7. Simple lithophilic spawning species	5 (5)	3 (1)
Percentage of individuals that are		
8. Tolerant species	3 (5)	1 (1)
9. Omnivores	1	3
10. Insectivores	5 (3)	3 (1)
11. Pioneering species	3	1
12. Diseased or have anomalies	5	3
SUM = IBI	42 (42)	30 (20)

Scores using Karr (1981) IBI are listed first. Modified scores (see text) and the total score using modified scores are shown in parentheses (Gatz and Harig, 1993).

number and abundance of species (such as the shiners mentioned above) that require clean rocky gravel substrates for spawning. There also was an increase in the number and abundance of pioneering species (fish that move into new habitats first). In addition, there was an increase in the number of fish with "deformities, eroded fins, lesions, and tumors." Scores were modified from Karr's original IBI to reflect the characteristics of local species within the major species groups. This application of the IBI confirmed that land use changes in the watershed during the 52-year period from 1940 to 1992 had added sediment and pollutants to stream waters and made stream habitat generally more unstable (Table 4.7).

4.4.1.2 Lake Erie

Macroinvertebrate sampling has been used to monitor the effectiveness of pollution control measures intended to improve the condition of Cleveland Harbor and the adjacent nearshore areas of Lake Erie in Cleveland, Ohio (Krieger, 1984; Krieger and Ross, 1993). The Cuyahoga River, which enters Lake Erie at Cleveland Harbor, became the focus of national attention in 1968 when debris and oil, floating on the river, caught fire. (In fact, the river first ignited in 1936 and had been the site of several fires before gaining widespread public attention [EPA, 2000]). In addition, inadequate wastewater treatment facilities frequently released untreated domestic sewage into the river and harbor areas, causing organic enrichment and oxygen depletion (Krieger, 1984). Outrage over the state of the river helped to motivate passage of the Clean Water Act in 1972.

Sampling in 1978 and 1979 revealed, as expected, that aquatic habitat in the harbor was severely degraded. For this reason, using macroinvertebrates as biotic indicators involved measuring and comparing the diversity and abundance of a small group of pollution tolerant species. The relative abundance of the most pollution tolerant taxa, oligochaete worms, was used as a measure of the condition of aquatic habitat. Less than 60% oligochaete worms was defined as a good condition, 60% to 80% implied "doubtful conditions," and more than 80% indicated seriously degraded conditions caused by organic enrichment or industrial pollution. The median abundance of oligochaete worms in the harbor in 1978 was measured at 29,500 individuals/m^2 (98% of total macroinvertebrate abundance) compared with 1,400 individuals/m^2 (66%) of total abundance in areas of open water outside the harbor.

During the 1980s, more than $750 million was spent improving wastewater treatment facilities that dealt both with sewage and industrial waste. The concentration of metals passing into the river and harbor was reduced by more than 80% (Krieger and Ross, 1993). Benthic sampling in 1989 revealed that habitat conditions, while still reflecting high levels of organic enrichment, had improved compared to 1978. In Cleveland Harbor, the relative abundance of oligochaete worms had declined to 85.6%. Outside the harbor, the percentage was reduced to 30%. In Cleveland Harbor, the median number of taxa per sample had increased from 11 in 1978 to 19 in 1989. In open water, the increase was even greater, 9 in 1978 to almost 24 in 1989. Nonoligochaete taxa included species of Sphaeriidae (fingernail clams) belonging to the family Corbiculacea and Chironomidae (midges). In comparison to EPT species, all of these groups are considered pollution tolerant (Kerans and Karr; 1994; Yandora, 1998). Nevertheless, in the highly degraded environment in and around Cleveland Harbor, an increase in biological diversity even among these species was seen as an indication of improved water and aquatic habitat quality. The presence of caddisfly larvae (*Oecetis* sp.) in a number of open water areas outside the harbor was a most encouraging sign of recovery.

4.5 SUMMARY

4.5.1 AQUATIC BIOTA

Stream and river ecosystems are inhabited by single-celled organisms (bacteria and algae), fungi, plants, and animals. Animal communities are divided into invertebrate (insects, crustaceans, mollusks, and worms) and vertebrate taxa (primarily fish).

4.5.2 AQUATIC FOOD WEBS

Living organisms in streams are part of complex food webs. External inputs include light, nutrients, and CPOM (allochthonous material) from riparian areas. Organic matter is produced in the stream (autochthonous material) from the growth and reproduction of photosynthesizing algae, bacteria, and plants. Decomposing bacteria and fungi process CPOM into FPOM. CPOM, FPOM, and in-stream bacteria and algae are consumed by macroinvertebrates. Various species of fish consume plants, macroinvertebrates, and other fish.

Macroinvertebrate and fish species have adapted to specific niches. Macroinvertebrate and fish taxa have been grouped into trophic guilds based on preferred food sources and feeding behaviors. The river continuum concept describes the large-scale distribution of trophic guilds based on differences in carbon inputs from upstream to downstream. Other theories place more emphasis on geologic factors as key determinants of species distribution in streams.

4.5.3 STREAM HABITAT

Different small-scale habitats are used by species at different stages of the life cycle. Smaller, younger fish often prefer river edges and shallow pools with lower streamflow velocities. Riffles and runs and coarse woody debris provide refugia for smaller, younger organisms seeking to avoid larger predators in deep pools. Shallow riffles with fast flowing water and gravelly substrates tend to be preferred sites for spawning.

4.5.3.1 Determining Factors in Stream Habitat

4.5.3.1.1 Streamflow and Sediment

Long-term changes in streamflow and sediment conditions will change the stream environment, affecting habitat diversity and the availability of particular niches. This in turn affects species distribution and community composition. Changes in streamflow and sediment supply are often the result of changes in watershed land use.

4.5.3.1.2 Temperature

Stream water temperature is a critical feature of the aquatic environment. Optimal temperature ranges vary for different fish and macroinvertebrate species. Individual species require specific temperature ranges during various life cycle stages for optimal development and growth. Temperature controls metabolic rates in stream organisms and the concentration of dissolved oxygen. As water temperature rises, metabolic rates go up, increasing the demand for dissolved oxygen. At the same time, the dissolved oxygen availability is reduced in warmer waters.

4.5.4 BIOLOGICAL MONITORING

This technique uses macroinvertebrates and fish as indicators of stream condition. The IBI and the B-IBI identify measurements (metrics) of aquatic communities that accurately reflect a stream's condition (water quality, habitat diversity, etc.) within a certain area. Using these indices, scientists and managers can determine the relative condition of other streams in the same area by sampling stream organisms.

REFERENCES

Aadland, L.P., Stream habitat types: their fish assemblages and relationship to flow, *N. Am. J. Fish. Manage.*, 13, 790–806, 1993.

Allan, D.J., *Stream Ecology: Structure and Function of Running Waters*, Chapman & Hall, London, 1995.

Bain, M.B., Finn, J.T., and Brooke, H.E., Streamflow regulation and fish community structure, *Ecology*, 69, 382–392, 1988.

Berkman, H.E. and Rabeni, C.F., Effect of siltation on fish communities, *Environ. Biol. Fish.*, 18, 285–294, 1987.

Boehlert, G.W. and Morgan, J.B., Turbidity enhances feeding abilities of larval Pacific herring, *Clupea harengus pallasi*, *Hydrobiologia*, 123, 161–170, 1985.

Borchardt, D., Effects of flow and refugia on drift loss of benthic macroinvertebrates—implications for habitat restoration in lowland streams, *Freshwater Biol.*, 29, 221–227, 1993.

Breitburg, D.L., Effects of turbidity on prey consumption by striped bass larvae, *Trans. Am. Fish. Soc.*, 117, 72–77, 1988.

Campbell, N.A., *Biology*, Benjamin/Cummings, Redwood City, CA, 1990.

Caruso, B.S., Temporal and spatial patterns of extreme low flows and effects on stream ecosystems in Otago, New Zealand, *J. Hydrol.*, 257, 115–133, 2002.

Chapman, D.W., Critical review of variables used to define effects of fines in redds of large salmonids, *Trans. Am. Fish. Soc.*, 117, 1–21, 1988.

Clausen, B. and Biggs, B.J.F., Relationships between benthic biota and hydrologic indices in New Zealand streams, *Freshwater Biol.*, 38, 327–342, 1997.

Culp, J.M., Wrona, F.J., and Davies, R.W., Response of stream benthos and drift to fine sediment deposition versus transport, *Can. J. Zool.*, 64, 1345–1351, 1986.

Dunne, T. and Leopold, L.B., *Water in Environmental Planning*, W.H. Freeman, New York, 1978.

Ellis, M.M., Erosion silt as a factor in aquatic environments, *Ecology*, 17, 29–42, 1936.

EPA (U.S. Environmental Protection Agency), Cuyahoga River area of concern, U.S. Environmental Protection Agency, Washington, D.C., http://www.epa.gov/glnpo/aoc//cuyahoga.html; accessed July 2006.

Everett, R.A. and Ruiz, G.M., Coarse woody debris as a refuge from predation in aquatic communities: an experimental test, *Oecologia*, 93, 475–486, 1993.

Fore, L.S., Karr, J.R., and Wisseman, R.W., Assessing invertebrate responses to human activities: evaluating alternative approaches, *J. N. Am. Benthol. Soc.*, 15, 212–231, 1996.

Galli, J. and Dubose, R., Water temperature and freshwater stream biota: an overview, Maryland Department of the Environment, Baltimore, MD, 1990.

Gardner, M.B., Effects of turbidity on feeding rates and selectivity of bluegills, *Trans. Am. Fish. Soc.*, 110, 446–450, 1981.

Gatz, A.J. and Harig, A.L., Decline in the index of biotic integrity of Delaware Run, Ohio, over 50 years, *Ohio J. Sci.*, 93, 95–100, 1993.

Giller, P.S. and Malmqvist, B., *The Biology of Streams and Rivers*, Oxford University Press, New York, 1998.

Hamblin, W.K., *The Earth's Dynamic Systems*, Burgess Publishing, Minneapolis, MN, 1982.

Hawkins, C.P. and Sedell, J.R., Longitudinal and seasonal changes in functional organization of macroinvertebrate communities in four Oregon streams, *Ecology*, 62, 387–397, 1981.

Hawkins, C.P., Murphy, M.L., and Anderson, N.H., Effects of canopy, substrate composition, and gradient on the structure of macroinvertebrate communities in Cascade Range streams of Oregon, *Ecology*, 63, 1840–1856, 1982.

Hill, J. and Grossman, G.D., An energetic model of microhabitat use for rainbow trout and rosyside dace, *Ecology*, 74, 685–698, 1993.

Horner, R.R., Booth, D.B., Azous, A., and May, C.W., Watershed determinants of ecosystem functioning, in *Effects of Watershed Development and Management on Aquatic Ecosystems*, Roesner, L.A. (ed.), American Society of Civil Engineers, New York, 1996, pp. 251–274.

Horwitz, R.J., Temporal variability patterns and the distributional patterns of stream fishes, *Ecol. Monogr.*, 48, 307–324, 1978.

Hynes, H.B.N., *The Ecology of Running Waters*, University of Toronto Press, Toronto, 1970.

Junk, W.J., Bayley, P.B., and Sparks, R.E., The flood pulse concept in river-floodplain systems, in *Proceedings of the International Large River Symposium*, Dodge, D.P. (ed.), Canadian Special Publication of Fisheries and Aquatic Sciences, Department of Fisheries and Oceans, Ottawa, Ontario, Canada, 1989, pp. 110–127.

Karr, J.R., Assessment of biological integrity using fish communities, *Fisheries*, 6(6), 21–27, 1981.

Karr, J.R., Biological integrity: a long-neglected aspect of water resource management, *Ecol. Applic.*, 1, 66–84, 1991.

Karr, J.R., Defining and measuring river health, *Freshwater Biol.*, 41, 221–234, 1999.

Karr, J.R. and Chu, E.W., *Restoring Life in Running Waters*, Island Press, Washington, D.C., 1999.

Kerans, B.L. and Karr, J.R., A benthic index of biotic integrity (B-IBI) for rivers of the Tennessee Valley, *Ecol. Applic.*, 4, 768–785, 1994.

Krieger, K.A., Benthic macroinvertebrates as indicators of environmental degradation in the southern nearshore zone of the central basin of Lake Erie, *J. Great Lakes Res.*, 10, 197–209, 1984.

Krieger, K.A. and Ross, L.S., Changes in the benthic macroinvertebrate community of the Cleveland Harbor area of Lake Erie from 1978–1989, *J. Great Lakes Res.*, 19, 237–249, 1993.

Lenat, D.R., Water quality assessment of streams using a qualitative collection method for benthic macroinvertebrates, *J. N. Am. Benthol. Soc.*, 7, 222–233, 1988.

Lenat, D.R., Penrose, D.L., and Eagleson, K.W., Variable effects of sediment addition on stream benthos, *Hydrobiologia*, 79, 187–194, 1981.

Leopold, L.B., *Water, Rivers, and Creeks*, University Science Books, Sausalito, CA, 1997.

Leopold, L.B., Wolman, M.G., and Miller, J.P., *Fluvial Processes in Geomorphology*, W.H. Freeman, San Francisco, 1964.

Lock, M.A., Attached microbial communities in rivers, in *Aquatic Microbiology: An Ecological Approach*, Ford, T.E. (ed.), Blackwell Scientific, Boston, 1993, pp. 113–138.

Margulies, L. and Schwartz, K.V., *Five Kingdoms: An Illustrated Guide to the Phyla of Life on Earth*, 3rd ed., W.H. Freeman, New York, 1999.

Minshall, G.W., Petersen, R.C., Cummins, K.W., Bott, T.L., Sedell, J.R., Cushing, C.E., and Vannote, R.L., Interbiome comparison of stream ecosystem dynamics, *Ecol. Monogr.*, 53, 1–25, 1983.

Montgomery, D.R., Process domains and the river continuum, *J. Am. Water Resourc. Assoc.*, 35, 397–410, 1999.

Morgan, I.J., McDonald, D.G., and Wood, C.M., The cost of living for freshwater fish in a warmer, more polluted world, *Global Change Biol.*, 7, 345–355, 2001.

Nislow, K.H., Folt, C.L., and Parrish, D.L., Favorable foraging locations for young Atlantic salmon: application to habitat and population restoration, *Ecol. Applic.*, 9, 1085–1099, 1999.

Norris, R.H. and Hawkins, C.P., Monitoring river health, *Hydrobiologia*, 435, 5–17, 2000.

NRCS (Natural Resources Conservation Service), Stream visual assessment protocol, NWCC-TN-99-1, National Water and Climate Center, Portland, OR, 1998, http://www.nrcs.usda.gov/technical/ECS/aquatic/svapfnl.pdf; accessed July, 2006.

NRCS (Natural Resources Conservation Service) Watershed Science Institute, Watershed Condition Series, Technical Notes 1, 2, and 3, Draft Copy, USDA NRCS Watershed Science Institute, c/o North Carolina State University, Room 1210B, Williams Hall, Raleigh, NC 27695-7619, 2000.

Omernik, J.M., Map supplement: ecoregions of the conterminous United States, *Ann. Assoc. Am. Geog.*, 77, 118–125, 1987.

Poole, G.C., Fluvial landscape ecology: addressing uniqueness within the river discontinuum, *Freshwater Biol.*, 47, 641–660, 2002.

Roghair, C.N., Dolloff, C.A., and Underwood, M.K., Response of a brook trout population and instream habitat to a catastrophic flood and debris flow, *Trans. Am. Fish. Soc.*, 131, 718–730, 2002.

Rosenberg, D.M. and Resh, V.H., Introduction to freshwater biomonitoring and benthic macroinvertebrates, in *Freshwater Biomonitoring and Benthic Macroinvertebrates*, Rosenberg, D.M. and Resh, V.H. (eds.), Chapman & Hall, New York, 1993, pp. 1–9.

Rosi-Marshall, E.J. and Wallace, J.B., Invertebrate food webs along a stream resource gradient, *Freshwater Biol.*, 47, 129–141, 2002.

Ryan, P.A., Environmental effects of sediment on New Zealand streams: a review, *N. Z. J. Mar. Freshwater Res.*, 25, 207–221, 1991.

Schindler, D.W., The cumulative effects of climate warming and other human stresses on Canadian freshwaters in the new millennium, *Can. J. Fish. Aquat. Sci.*, 58, 18–29, 2001.

Schlosser, I.J., Fish community structure and function along two habitat gradients in a headwater stream, *Ecol. Monogr.*, 52, 395–414, 1982.

Schlosser, I.J., The role of predation in age- and size-related habitat use by stream fishes, *Ecology*, 68, 651–659, 1987.

Sedell, J.R., Richey, J.E., and Swanson, F.J., The river continuum concept: a basis for the expected ecosystem behavior of very large rivers?, in *Proceedings of the International Large River Symposium*, Dodge, D.P. (ed.), Canadian Special Publication of Fisheries and Aquatic Sciences, Department of Fisheries and Oceans, Ottawa, Ontario, Canada, 1989, pp. 49–64.

Steedman, R.J., Modification and assessment of an index of biotic integrity to quantify stream quality in southern Ontario, *Can. J. Fish. Aquat. Sci.*, 45, 492–501, 1988.

Townsend, C.R., Scarsbrook, M.R., and Dolédec, S., The intermediate disturbance hypothesis, refugia, and biodiversity in streams, *Limnol. Oceanogr.*, 42, 938–949, 1997.

Vannote, R.L. and Sweeney, B.W., Geographical analysis of thermal equilibria: a conceptual model for evaluating the effect of natural and modified thermal regimes on aquatic insect communities, *Am. Nat.*, 115, 667–695, 1980.

Vannote, R.L., Minshall, G.W., Cummins, K.W., Sedell, J.R., and Cushing, C.E., The river continuum concept, *Can. J. Fish. Aquat. Sci.*, 37, 130–137, 1980.

Ward, J.V., *Aquatic Insect Ecology: 1. Biology and Habitat*, John Wiley & Sons, New York, 1992.

Walters, D.M., Leigh, D.S., Freeman, M.C., Freeman, B.J., and Pringle, C.M., Geomorphology and fish assemblages in a Piedmont River basin, U.S.A., *Freshwater Biol.*, 48, 1950–1970, 2003.

Waters, T.F., *Sediment in Streams: Sources, Biological Effects, and Control*, Monograph 7, American Fisheries Society, Bethesda, MD, 1995.

Wetzel, R.G., *Limnology*, 2nd ed., Saunders, Philadelphia, 1983.

Yandora, K., Rapid bioassessment of benthic macroinvertebrates illustrates water quality in small order urban streams in a North Carolina Piedmont city, in *Proceedings of the National Water Quality Monitoring Council (NWQMC)*, July 7–9, 1998, Reno, NV, http://www.nwqmc.org/98proceedings/Papers/40-YAND .html; accessed July, 2006.

5 Riparian Areas and Riparian Buffers

When they go, we will lose not only storage basins for water and final refuges for birds and wildlife, but museums of the past, examples to shifting humans of the stability of an ecological community that has stood the test of time. We will lose much of beauty if we trade them for farmland, and we will lose our chance to study a primeval environment, its interrelationships and dependencies.

**Sigurd Olson, 1956, *The Singing Wilderness* (from "Smoky Gold"
...an essay on tamarack swamps, the riparian wetlands that border
many rivers in the northern Lake States)**

5.1 INTRODUCTION

Riparian areas are the physical and ecological link between upland and aquatic ecosystems (Figure 5.1). Riparian buffers are defined by laws, regulations, or voluntary guidelines to maintain or restore vegetation and limit some forms of land use in order to protect water quality and aquatic ecosystems. Understanding the structure and function of riparian areas is essential to developing strategies to protect stream channels through the conservation and management of riparian buffers. Two recently published books, by the National Research Council (2002) and Verry et al. (2000), provide a comprehensive overview of the current state of the science regarding riparian areas and riparian area management. Key reviews of riparian buffer structure and function include Lowrance et al. (1995a, 1997a) and Klapproth and Johnson (2000a,b). Verry et al. (2004) propose a "linked sequence

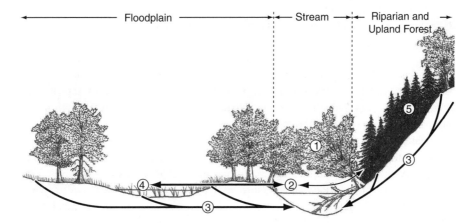

FIGURE 5.1 (1) Streamside vegetation shades the stream channel, moderating water temperature. Roots stabilize streambanks. Fallen leaves supply organic matter. Coarse woody debris stabilizes the channel and provides habitat. (2) Nutrients and sediment enter the stream channel from riparian areas and are returned to riparian areas during flood events. (3) Subsurface flow from the uplands passes through riparian areas and enters the stream as baseflow. (4) Riparian areas include wetlands that process nutrients (denitrification) and provide storage and infiltration sites for floodwaters. (5) Riparian and upland forests minimize erosion and sediment transport and store nutrients. (Original drawing by Ethan Nedeau.)

of definition, delineation, and riparian sampling" that can be used to "accurately assess riparian resources on the ground."

5.2 RIPARIAN AREAS

5.2.1 Definitions

Scientists have developed many different definitions of riparian areas (Verry et al., 2004). These definitions have been based on vegetation, soils, and geomorphology. Until recently, most riparian studies have identified the location of the stream channel and then sampled at various distances from the stream channel to study riparian characteristics. A new definition developed by Verry et al. (2004) provides a method of delineation and sampling that uses the stream valley rather than the channel to locate the riparian area.

The following definitions of riparian areas are a few of many that have been used by scientists and federal agencies to assist in the study and delineation of riparian areas:

1. USDA Natural Resources Conservation Service (NRCS) (1991)
   Riparian areas are ecosystems that occur along watercourses and water bodies. They are distinctly different from the surrounding lands because of unique soil and vegetation characteristics that are strongly influenced by free or unbound water in the soil. Riparian ecosystems occupy the transitional area between the terrestrial and aquatic ecosystems. Typical examples include floodplains, streambanks, and lakeshores.
2. U.S. Fish and Wildlife Service (1997)
   Riparian areas are plant communities contiguous to and affected by surface and subsurface hydrologic features of perennial or intermittent lotic and lentic water bodies (rivers, streams, lakes, or drainage ways). Riparian areas have one or both of the following characteristics: (1) distinctly different vegetative species than adjacent areas, and (2) species similar to adjacent areas but exhibiting more vigorous or robust growth forms. Riparian areas are usually transitional between wetland and upland.
3. USDA Forest Service (2004)
   Riparian areas: Geographically delineable areas with distinctive resource values and characteristics that are comprised of the aquatic and riparian ecosystems.
   Riparian ecosystems: A transition area between the aquatic ecosystem and the adjacent terrestrial ecosystem; identified by soil characteristics or distinctive vegetation communities that require free or unbound water.
4. National Research Council (2002)
   Riparian areas are transitional between terrestrial and aquatic ecosystems and are distinguished by gradients in biophysical conditions, ecological processes, and biota. They are areas through which surface and subsurface hydrology connect water bodies with their adjacent uplands. They include those portions of terrestrial ecosystems that significantly influence exchanges of energy and matter with aquatic ecosystems (i.e., zones of influence). Riparian areas are adjacent to perennial, intermittent, and ephemeral streams, lakes, and estuarine-marine shorelines.
5. Verry et al. (2004)
   Riparian ecotones are a three-dimensional space of interaction that include terrestrial and aquatic ecosystems, that extend down into the groundwater, up above the canopy, outward across the floodplain, up the near-slopes that drain to the water, laterally into the terrestrial ecosystem, and along the water course at a variable width. Riparian ecotones include the

flood-prone area and 30 m landward along the valley. Obvious landslide or slump areas are added with a 15 m band around their edge.

Because the riparian area has been described as a "collection of ecosystems," Verry et al. (2004) contend that the correct term for the riparian area should be "riparian ecotone," where ecotone is defined as a gradient across ecosystems. The flood-prone area is the distance to the terrace slopes, measured at an elevation twice the bankfull depth of the stream (the approximate water surface elevation of the 50 year recurrence interval flood) (Verry et al., 2004). The shape of the stream valley and adjacent upland slopes define the riparian area.

The physical relationship of the stream to its valley can be quantified using the "entrenchment ratio." The entrenchment ratio is "the ratio of the width of the flood-prone area to the bankfull surface width of the channel" (Rosgen, 1994). Rosgen (1994) used the entrenchment ratio as a means of categorizing natural streams. The riparian areas of highly entrenched streams flowing through V-shaped valleys are narrower than the riparian areas of only slightly entrenched streams that meander through well-developed floodplains and broad valleys (Figure 5.2 and Figure 5.3).

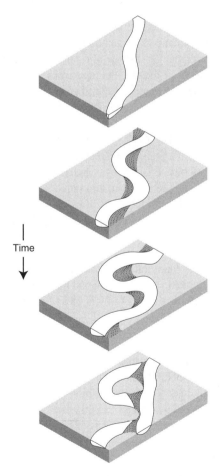

Time

FIGURE 5.2 Over time the path of the stream channel moves across the stream valley forming meander bends and cutting new channels, leaving the oxbows of former meander bends behind. The widths of the meander belt and the riparian area are determined by the physical characteristics of the stream valley. (Original drawing by Ethan Nedeau.)

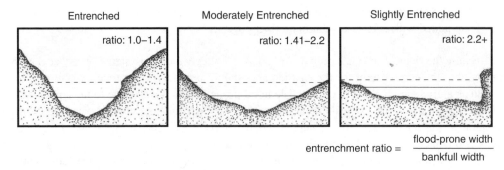

FIGURE 5.3 Examples of stream channels with different entrenchment ratios. (Modified from Rosgen, 1994.)

5.2.2 RIPARIAN AREA FUNCTIONS

5.2.2.1 Streams and Rivers

Hydrologic processes and biological activity in the riparian area have a significant influence on the quantity and quality of streamflow and stream channel habitat. These effects operate at different scales, from the immediate channel and streamside vegetation to the riparian area as a whole (Figure 5.1). Within the stream channel, the presence or absence of riparian vegetation can influence channel shape and width (Sweeney et al., 2004). Near the channel, tree roots help to stabilize streambanks (Beeson and Doyle, 1995). The overhanging tree canopy shades the stream channel and can substantially reduce temperature fluctuations (Beschta and Taylor, 1988; Patric, 1980). As noted in Chapter 4, riparian trees and shrubs provide allochthonous input—organic matter (leaves, branches, etc.)—that forms the base of aquatic food webs (Allan, 1995). The riparian area contributes coarse woody debris—fallen tree trunks and branches. These help to maintain channel stability and also provide habitat for insects and fish (Beeson and Doyle, 1995; Booth, 1991; Martin and Hornbeck, 1994). Following storm events, infiltration and subsequent water storage in the riparian area slow the movement of water to stream channels and help to sustain streamflow. In the event of overbank flows, riparian areas provide natural storage for floodwaters. Nutrient cycling in riparian areas, including riparian wetlands, can reduce nutrient inputs to streams (Cooper et al., 1986). Nitrogen and phosphorus are stored in riparian soils and riparian vegetation. Riparian wetlands provide sites for denitrification. Riparian areas also provide specialized and diverse habitats for many plant and animal species (Klapproth and Johnson, 2000a; Naiman et al., 1993). During periodic floods, sediment and nutrients are returned to the floodplain from the stream channel. Over time, natural stream channels meander through the floodplain (Figure 5.2).

5.2.2.2 Littoral Zones of Lakes

There have been fewer attempts to define the area between a lake and the uplands in its watershed. Still, it is clear that these transitional regions fulfill many of the same functions as stream riparian areas, acting as "filters, barriers, sources, and sinks for numerous biological, physical, and chemical processes at diverse space and time scales" (Bolgrien and Kratz, 2000). The traditional term for the areas nearest the lake shoreline is the littoral zone. Detailed descriptions of biological and chemical processes and plant and animal communities in the littoral zone may be found in Wetzel (1983). It is widely recognized that the littoral zone and associated wetlands are exceedingly important in regulating the processing of nutrients and organic carbon, thereby influencing the trophic status of lakes (Wetzel, 1983).

5.3 RIPARIAN BUFFERS

Riparian areas and riparian buffers are not the same thing. As we have seen, the riparian area is the stream valley and the part of the terrace slopes and upland that interact with the stream. Riparian buffers—established by laws, regulations, or voluntary programs—are areas bordering rivers and streams (and lakes) that are protected and managed in order to mitigate the effects of human activity. Riparian buffers have been shown to have many beneficial effects to the extent that they partially preserve the ecological functions of the larger riparian area. However, riparian buffers are usually expected to "do more with less"—to process increased amounts of sediment, nutrients, and new pollutants in a smaller space and less time than that provided by the undisturbed (predevelopment) riparian area.

Prior to European settlement, uplands and large areas of river valleys were typically covered with forests (Verry and Dolloff, 2000). Water flowed through undisturbed riparian areas, that often included extensive wetlands, prior to entering stream and river channels. At the present time, upland and riparian ecosystems are often altered by human activity—timber harvesting, agriculture, and urban development—and only a small percentage of presettlement natural wetland area remains (McCorvie and Lant, 1993). Nutrients, in the form of chemical fertilizers and animal manure, have been added, often in excess of what can be retained by soils or is required by crops and perennial vegetation (Sharpley et al., 2001). Riparian areas often have fertile soils and level topography. As a result, they have been favored for agricultural and more recently for residential and urban development (National Research Council, 2002). In addition, many new pollutants—pesticides and complex mixes of industrial and urban chemicals—have been introduced (Garabedian et al., 1998; Myers et al., 2000; Wall et al., 1998).

The width of these buffers is often established somewhat arbitrarily, based on political or administrative decisions rather than scientific judgments (Castelle et al., 1994). For example, the New Jersey Department of Environmental Protection (2000) defines riparian buffers as "an area of trees and shrubs, 35–300 feet wide located upgradient, adjacent, and parallel to the edge of a water feature." Clearly the boundaries of the true riparian area may lie far beyond or well within that of the protected buffer.

5.3.1 RIPARIAN BUFFER FUNCTIONS

The streamside functions of riparian areas may be the ones most easily provided by a riparian buffer—shading and inputs of large woody debris. These can be provided by buffers that are 30 m wide (Castelle et al., 1994). Sweeney et al. (2004) found that removing streamside trees and the shade provided by the forest canopy encouraged the growth of grasses at the edge of and in stream channels. This caused stream channels to become narrower and resulted in the loss of habitat and reduced biodiversity and abundance in populations of insects and fish. Beyond the stream edge, the ability of riparian buffers to provide the ecosystem functions necessary to maintain streamflow patterns and protect water quality vary with buffer area and local environmental conditions. Riparian buffers can trap sediment generated from upland disturbance (Lee et al., 2000; Uusi-Kämppä et al., 1997) and reduce the concentration of nutrients (nitrogen and phosphorus) in water that reaches the stream channel (Lowrance et al., 1995a,b, 1997a; Peterjohn and Correll, 1984). Buffers can also reduce pesticide concentrations in overland and shallow subsurface flow through vegetative uptake and by trapping pesticide-contaminated sediments (Lowrance et al., 1997b).

5.3.1.1 Sources of Variability in Buffer Functions

The effectiveness of riparian buffers depends upon the characteristics of the watershed (e.g., topography, hydrologic setting, soil, and vegetation) and regional climate (e.g., the magnitude and frequency of rainfall and snowmelt events). Watershed characteristics and climate, in turn, influence the residence time of water, nutrients, sediment, and pollutants such as pesticides in the buffer. The thickness of the unsaturated zone (depth to the water table) and its seasonal variability are also important attributes. In sum, the balance between pollutant loading from the adjacent uplands and the assimilative capacity of the riparian buffer, or lack thereof, largely determine its long-term effectiveness (Addy et al., 1999; Barling and Moore, 1994; Lowrance et al., 1995a).

5.3.1.1.1 Residence Time

Many of the processes governing the storage and conversion of nutrients and pesticides in riparian buffers are biological processes carried out by plants and soil microbes. Residence time is obviously important because the longer substances remain in the buffer, the greater the likelihood that they will be immobilized or transformed by plants or bacteria (Groffman et al., 1996). Since pollutants (including sediment) are transported by water, residence time is a function of hydrologic processes. An increase in the rate of overland flow and subsurface flow from uplands to streams will decrease residence time.

The first factor controlling residence time is climate—temperature and precipitation—and corresponding changes in the variable source area (VSA) (Chapter 2). As the VSA increases due to rain or snowmelt, flow velocity increases and reduces the residence time of water, sediment, and pollutants in the riparian area. Expansion and contraction of the VSA introduces variability in residence time at two time scales. First, residence time varies during storm events. Increased flow velocity produces a flush of pollutants during and after a storm event. (The effectiveness of riparian buffers also varies seasonally in relation to rates of biological activity.) Second, residence time may vary annually with variations in annual precipitation. During exceptionally dry years, increases in available soil moisture storage may reduce pollutant loading that might otherwise occur during storms. In contrast, in wet years, saturated soils may sustain the transport of pollutants from uplands to streams with little time available for processing in the riparian area.

Topography is another factor controlling residence time. Steep slopes cause the acceleration of overland and subsurface flow through the buffer, thereby decreasing residence time (Snyder et al., 1998). Soil permeability is a key attribute. Pollutants move more rapidly through the highly permeable sandy soils found in glacial outwash deposits and coastal regions than through poorly drained clay soils.

5.3.1.1.2 Depth to the Water Table

The processes that reduce pollutants in water passing through riparian buffers take place at or near the soil surface. Sediment and sediment-associated nutrients and pesticides are trapped in surface vegetation. Dissolved nutrients and pesticides are retained, taken up by plants, and broken down in the vegetative root zone. The root zone in temperate forests is generally limited to between 1 and 2 m below the soil surface (Raven et al., 1999). Geological conditions that channel groundwater and associated upland pollutants below the root zone will limit the effectiveness of riparian buffer processes (Gilliam et al., 1997; Jordan et al., 1993; Lowrance et al., 1995a, 1997a; Snyder et al., 1998) (Figure 5.4). Nutrients may also bypass the biologically active root zone if there is a persistent lowering of the water table (saturated zone) because of land use change. This can occur in urban and suburban areas, where increased impervious surface area accelerates overland flow and decreases infiltration (Groffman et al., 2002). Obviously, riparian buffers will be far less effective when overland flow and nonpoint source pollutants are channeled directly to streams. This can occur in agriculture areas where tile drains have been installed (Cooper et al., 1986; Osborne and Kovacic, 1993) and in urban areas with storm drains.

5.3.1.1.3 Upland Land Use

Questions have been raised about the relative importance of riparian buffers vis-à-vis overall patterns of watershed land use. Numerous short-term, small-scale edge-of-field experimental studies (Clausen et al., 2000; Cooper et al., 1986; Jordan et al., 1993; Lee et al., 2000; Lowrance et al., 1997b; Peterjohn and Correll, 1984) document the ability of vegetative buffers to retain sediment and prevent pollutants from reaching streams. These are studies that measure nutrient and pesticide concentrations in soil water, first at the field edge of the buffer and then at streamside after water has passed through the buffer. Meador and Goldstein (2003) found that impaired riparian areas

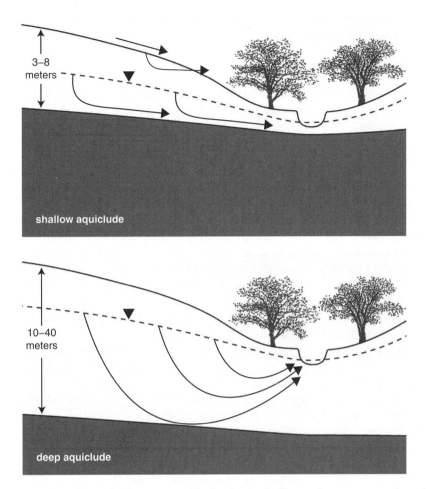

FIGURE 5.4 In order to be retained or processed in the riparian buffer, nutrients must pass through the root zone. In areas where the water table is far below the root zone, nutrients in subsurface flow may bypass the riparian buffer and enter the stream channel directly. (Redrawn from Lowrance et al., 1995a.)

were associated with impaired stream conditions, higher levels of total nitrogen, and suspended sediment across large geographic areas. On the other hand, Omernik et al. (1981) examined the nutrient outputs of 80 large watersheds (average area 40 km²) affected by only nonpoint source pollution. They found that "the proximity of forest and agriculture to main streams [did] not bear a significant relationship to stream nutrient concentrations beyond that attributable to the proportion of the total watershed that these land uses comprise." In other words, land use patterns across the entire watershed may be as important or more important than the presence or absence of riparian buffers in determining water quality.

More recently, Harding et al. (1998) found that the whole watershed land use 40 years earlier was the best predictor of current biological diversity in selected streams in western North Carolina. This study included (1) forested watersheds, (2) agricultural watersheds, and (3) agricultural watersheds in 1950 that were reforested by 1990. Streams in reforested agricultural watersheds (92% forest in the 1990s with forested riparian areas) had invertebrate populations that were typical of streams in watersheds that had been farmed continuously. Reforestation of both the watershed and the riparian buffer had not yet been sufficient to overcome the influence of earlier agricultural land use.

Studies in urban areas (discussed in Chapter 8) find that stream condition, as measured by biotic indices, is consistently poor once urban land use or total impervious area exceeds approximately

40% of the total watershed area. This is true even in areas where riparian buffers have been conserved or restored (Booth et al., 2004).

5.3.2 TESTS OF RIPARIAN BUFFER FUNCTIONS

The following sections provide summaries of key studies that have investigated the effectiveness of riparian buffers in the storage and removal of various pollutants and the protection of fish communities.

5.3.2.1 Sediment

Riparian buffers can be particularly effective in trapping and storing sediment that is being transported in overland flow. Sediment retention rates greater than 80% are frequently reported (Table 5.1). Large particles of sand are deposited first, while smaller silt and clay particles can be carried further into the buffer area. A study in Iowa (Lee et al., 2000) found that a switchgrass (*Panicum virgatum* L.) buffer 7.1 m wide retained 82% to 89% of the larger sand particles and 72% to 76% of the silt particles, but only 15% to 49% of clay particles in overland flow from bare cropland. A 16.3 m wide buffer of grass and trees (7.1 m of switchgrass with an additional 9.2 m wide strip of woody vegetation) removed 52% to 89% of the clay particles. While deep-rooted forest trees play a more important role in nutrient uptake, the increase in sediment retention may primarily be the result of the increased buffer width (and infiltration capacity). Grass strips between agricultural fields and riparian forests increase the likelihood that sediment will be kept out of stream channels in two ways. First, the dense cover of grass stems may be more effective in trapping larger sand particles than forest floor vegetation, especially if overstory shade limits plant growth on the forest floor (Cooper et al., 1986; Klapproth and Johnson, 2000b). Second, grass strips help to spread overland flow over a wide area. This dissipates concentrated flow and prevents sediment-laden water from forming gullies and ephemeral stream channels. Sediment retention is greatly reduced if the water carrying the sediment is flowing in channels before it reaches the riparian forest buffer (Daniels and Gilliam, 1996). Concentrated flow and channeling is more apt to occur during severe storms and in steep areas where flow rates are greater. Current U.S. Department of Agriculture (USDA) specifications for riparian buffers in agricultural areas call for the establishment of grass filter strips between croplands and riparian forests (Lowrance et al., 2000).

Riparian vegetation also prevents bank erosion. Beeson and Doyle (1995) assessed 748 river bends in four stream reaches in British Columbia following major floods in 1990. Thirty-five river bends showed evidence of major erosion from the floods. Analysis of pre-flood vegetation from aerial photographs showed that 34 of the 35 river bends were not vegetated. The remaining river bend was protected by only a single row of trees.

5.3.2.2 Nitrogen

Riparian buffers have repeatedly been shown to act as sinks for nitrogen. Nitrogen associated with sediment and organic matter (ammonium [NH_4^+] and organic nitrogen) in overland flow is retained when that flow is slowed and the sediment is deposited (Daniels and Gilliam, 1996; Gilliam et al. 1997). Figure 5.5 illustrates nitrogen pathways and riparian buffer mechanisms involved in nitrogen retention. However, riparian buffers may afford little protection during high intensity rainstorms, when fast-flowing water carries more organic matter and associated nitrogen directly into streams (Lowrance et al., 1997a). As discussed earlier, loss of the forest, combined with excessive nitrogen inputs from human activity, increases the amount of dissolved nitrogen, nitrate (NO_3^-) in shallow groundwater. Retention of forested riparian buffers can mitigate these impacts because a large portion of the dissolved nitrate can be absorbed and removed by processes that occur in streamside forests—vegetative uptake and denitrification (Jordan et al., 1993).

TABLE 5.1
Sediment Reduction Across Riparian Buffers and Vegetative Filter Strips

Location	Adjacent land use	Buffer vegetation	Width (m)	Sediment reduction (%)	Reference
Riparian buffers					
Maryland	Agriculture (row crops)	Deciduous forest	50	94	Peterjohn and Correll (1984)
Connecticut	Agriculture (corn)	Grass (fine-leafed fescue [*Festuca* spp.])	35	92	Clausen et al. (2000)
North Carolina (Coastal Plain)	Agriculture	Bottomland forest	Variable	88	Cooper et al. (1986)
Vegetated filter strips					
Iowa	Bare cropland	Switchgrass (*Panicum virgatum*)	7.1	Sand > 82 Silt > 71 Clay > 15	Lee et al. (2000)
		Switchgrass, woody plants	7.1 (switchgrass) + 9.2 (woody plants) = 16.3	Sand > 98 Silt > 93 Clay > 52	
Nebraska	Grain sorghum and soybeans	Grass	7.5 15	76 87	Schmitt et al. (1999)
		Grass, trees, and shrubs	7.5 15	79 88	
North Carolina (Piedmont)	Agriculture	Grass	6	50–60	Daniels and Gilliam (1996)
		Grass + riparian forest	13–18	80	
Virginia	Bare cropland	Orchard grass (*Dactylis glomerata*)	4.6 9.1	70[a] 84[a]	Dillaha et al. (1989)

[a] Average of 6 trials; effectiveness of vegetative filter strips in retaining sediment decreased over time.

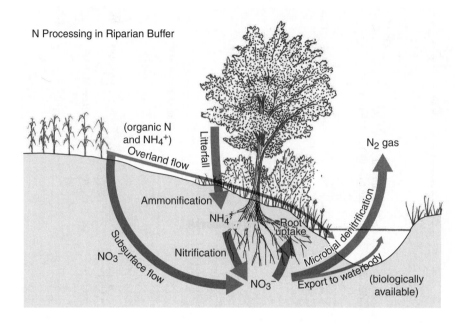

FIGURE 5.5 Nitrogen is primarily transported as dissolved nitrate in subsurface flow. Much of the nitrogen retention and loss occurs through denitrification in saturated soils in the riparian buffer. (Original drawing by Ethan Nedeau.)

Vegetative uptake is an important part of nutrient retention in riparian buffers (Clausen et al., 2000). There are limits, however, to the quantities of nitrogen that vegetation can absorb. Furthermore, much of the nitrogen taken up during the growing season is returned to the forest floor when leaves and twigs fall and decompose. Peterjohn and Correll (1984) estimate that forest trees in an approximately 50 m wide riparian buffer took up 77 kg/ha/yr of nitrogen, and that 62 kg/ha/yr or 80% of vegetative uptake was returned to the soil in litter fall. Many studies conclude that the most significant pathway for nitrogen loss in riparian buffers is denitrification (Addy et al., 1999; Cooper et al., 1986; Groffman et al., 1996; Gilliam et al., 1997; Lowrance et al., 1997a). Riparian forests and riparian wetlands provide ideal environments for denitrification (Barnes et al., 1998). The saturated zone (water table) in these areas is near the soil surface, but rises and falls seasonally (higher in spring and fall, lower during the growing season when evapotranspiration is highest). During drier periods, oxygen required by microbes to decompose forest litter is readily available in the soil. Decomposition of litter releases organic carbon compounds. When the water table rises, the environment near the soil surface becomes anaerobic and there is an ample supply of organic carbon and denitrification can occur (Simmons et al., 1992). In general, denitrification occurs in anaerobic environments in the root zone or in sediments just below the root zone (Groffman et al., 1992; Lowrance et al., 1995b). Denitrification is also possible at lower depths if organic carbon supplies are adequate (Hill, 1996).

Within riparian buffers, poorly drained soils provide the most favorable environment for denitrification. Little or no denitrification takes place at sites where soils are moderately well drained, well drained, or excessively well drained. Because of this, nitrate removal is not uniformly distributed throughout the riparian area. Most denitrification occurs in patches of poorly drained soils, referred to as denitrification hot spots (Gold et al., 1998; Hill, 1996). Table 5.2 summarizes key information from several nitrogen retention studies.

TABLE 5.2
Nitrate (NO$_3^-$-N) Reduction Across Riparian Buffers and Vegetated Filter Strips

Location	Adjacent land use	Buffer vegetation	Width (m)	Nitrate reduction (%)	Reference
Riparian buffers					
Maryland	Agriculture (row crops)	Deciduous forest	50	Overland flow: 79 Subsurface flow: 90	Peterjohn and Correll (1984)
Connecticut	Agriculture (corn)	Grass (fine-leafed fescue [*Festuca* spp.])	35	Overland flow: 83 Subsurface flow: 70	Clausen et al. (2000)
Maryland	Agriculture (cropland)	Deciduous forest	60	Subsurface flow: 95	Jordan et al. (1993)
Illinois	Agriculture (corn and soybeans)	Grass Forest	39 16	Subsurface flow: 85 Subsurface flow: 95	Osborne and Kovacic (1993)
Rhode Island	Urban (unsewered residential development)	Forest	25–60	Subsurface flow: >80	Simmons et al. (1992)
Virginia (Coastal Plain)	Agriculture	Riparian wetlands	10–40	48[a]	Snyder et al. (1998)
North Carolina (Coastal Plain)	Agriculture	Bottomland forest	Variable	86	Cooper et al. (1986)
Vegetated filter strips					
Iowa	Bare cropland	Switchgrass (*Panicum virgatum*) Switchgrass, woody plants	7.1 7.1 (switchgrass) + 9.2 (woody plants) = 16.3	61[b] 92[b]	Lee et al. (2000)
Nebraska	Grain sorghum and soybeans	Grass Grass, trees, and shrubs	7.5 15 7.5 15	25 40 25 40	Schmitt et al. (1999)
North Carolina (Piedmont)	Agriculture	Grass Grass + riparian forest	6 13–18	50 50	Daniels and Gilliam (1996)
Virginia	Bare cropland	Orchard grass (*Dactylis glomerata*)	4.6 9.1	27 57	Dillaha et al. (1989)

[a] Site characterized by steep riparian slopes and saturated zone 5 to 10 m below the surface to within 20 to 30 m of the stream channel.
[b] Measurements taken during a 2 hour rainfall simulation.

5.3.2.3 Phosphorus

While riparian forest buffer sites are well suited for the storage and removal of nitrogen, they are less effective in reducing the transport of phosphorus from watershed soils to streams (Lowrance et al., 1995a; Snyder et al. 1998). In addition, the capacity of riparian forest buffers to retain phosphorus is highly variable (Table 5.3). This variability is apparent among watersheds in different locations with different site conditions and within the same watershed both seasonally and over longer time periods (Cooper et al., 1986; Osborne and Kovacic, 1993).

There are several reasons for the limited capacity of riparian forest buffers to prevent phosphorus loading to streams. A large proportion of the phosphorus entering a riparian buffer is bound to sediment particles. Phosphorus binds preferentially to silts, clay, and organic matter, which are the first to erode during rainstorms and travel farther than larger sand particles. Hence it is more likely that these fine particles will be transported through the buffer directly to streams (Calhoun et al., 2002; Whigham et al., 1986). Whether these particles are transported or retained depends on ambient soil water content, infiltration capacity and the rate of the overland flow, the presence of rills or gullies, and the width of the buffer, as noted in Section 5.3.2.1 on sediment retention.

Phosphorus in the dissolved form is immediately available for plant uptake and growth. However, there often is more phosphorus present in soil solution (due to excess inputs from agricultural sources) than riparian vegetation can use. Phosphorus entering riparian buffers in dissolved form or as sediment-associated particles may only be stored temporarily before it is transported to the stream or other receiving water (Figure 5.6).

As with nitrogen, steep slopes and high volumes or rates of precipitation increase the likelihood that dissolved phosphorus and sediment-associated phosphorus will flow through riparian buffers. There is, however, no biochemical equivalent to denitrification that removes phosphorus from the soil. As discussed in Chapter 3, the anoxic conditions that occur when riparian forests are flooded can promote the release of dissolved, biologically available phosphorus from ferric-phosphate compounds (Golterman, 1972). If phosphorus continues to be deposited and retained in the buffer, riparian forest soils can gradually develop higher phosphorus concentrations than upland soils. Varying amounts of this particle-associated phosphorus may be converted to dissolved phosphate, depending on the water content, temperature, pH, and soil type in the forest. Under these conditions, it is possible that riparian forests can eventually become a source of (rather than a sink for) phosphorus loading to streams and lakes (Cooper et al., 1986; Ng et al., 1993; Uusi-Kämppä et al., 1997; Vanek, 1991).

Peterjohn and Correll (1984) studied the Rhode River basin in the Chesapeake Bay watershed. They found that in a single year, the approximately 50 m wide riparian forest retained almost 85% of the sediment-associated phosphorus (4.8 mg/L entering the forest versus 0.74 mg/L leaving the forest), but that dissolved phosphorus in overland flow was essentially unchanged (0.208 mg/L entering versus 0.217 mg/L leaving). Phosphorus concentrations in shallow groundwater increased from 0.015 to 0.062 mg/L for one transect and from 0.13 to 0.247 mg/L for the second transect. Since the greatest amount of phosphorus entering the riparian forest is associated with sediment, the retention rate for total phosphorus was still high (80%).

Cooper et al. (1986) studied sediment and nutrient processing and transport in an 1860 ha watershed in the coastal area of North Carolina. Approximately half of the watershed area was in cropland while the rest remained as irregularly shaped riparian forests and swamps. More than 80% of the sediments exported from the agricultural fields during the previous 20 years had been retained in the riparian areas. As a result, phosphorus concentrations were higher in the riparian buffers than in the agricultural fields and the researchers noted that it was likely that the riparian area had become a source, leaking dissolved phosphorus to the stream.

A study of restored grass riparian buffers adjacent to cornfields in Connecticut (Clausen et al., 2000) used a paired-watershed approach to study nutrient retention in riparian buffers. The riparian buffer decreased the concentration of total phosphorus in overland flow by 73% compared with a control area with no buffer. However, the concentration of dissolved phosphorus in groundwater

TABLE 5.3
Phosphorus Reduction Through Storage Across Riparian Buffers and Vegetated Filter Strips

Location	Adjacent land use	Buffer vegetation	Width (m)	Phosphorus reduction (%)	Reference
Riparian buffers					
Maryland	Agriculture (row crops)	Deciduous forest	50	Particulate: 85 / Dissolved: (114)	Peterjohn and Correll (1984)
Connecticut	Agriculture (corn)	Grass (fine-leafed fescue [Festuca spp.])	35	Overland flow: 73 / Subsurface flow: (122[a])	Clausen et al. (2000)
Maryland	Agriculture (cropland)	Deciduous forest	60	Subsurface flow / Increase in phosphorus at streamside[a]	Jordan et al. (1993)
Illinois	Agriculture (corn and soybeans)	Grass / Forest	39 / 16	Both grass and forest retained phosphorus in the growing season and released phosphorus in the dormant season	Osborne and Kovacic (1993)
Virginia (Coastal Plain)	Agriculture	Riparian wetlands	10–40	No trend	Snyder et al. (1998)
Vegetated filter strips					
Iowa	Bare cropland	Switchgrass (Panicum virgatum) / Switchgrass, woody plants	7.1 / 7.1 + 9.2 = 16.3	Total phosphorus: 46 / Total phosphorus: 81	Lee et al. (2000)
Nebraska	Grain sorghum and soybeans	Grass; grass/trees and shrubs (different vegetation types performed similarly)	7.5 / 15 / 7.5 / 15	Particulate: 60 / Dissolved: 20 / Particulate: 75 / Dissolved: 30	Schmitt et al. (1999)
North Carolina (Piedmont)	Agriculture	Grass / Grass and riparian forest	6 / 13–18	Total phosphorus: 50–60	Daniels and Gilliam (1996)
Virginia	Bare cropland	Orchard grass (Dactylis glomerata)	4.6 / 9.1	Total phosphorus: 61 / Total phosphorus: 79	Dillaha et al. (1989)

[a] Dissolved phosphate increased from barely detectable to approximately 0.07 mg/L from the cornfield to the streamside, possibly due to the release of sediment-associated phosphorus.

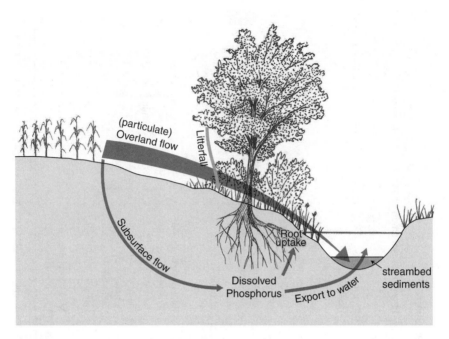

FIGURE 5.6 Phosphorus tends to adsorb to sediments and is primarily exported in particulate form with overland flow. Riparian buffers trap and store sediments and associated phosphorus. Phosphorus is also stored in riparian vegetation. (Original drawing by Ethan Nedeau.)

was not reduced. It is probable that manure applications in the upland fields following the calibration period of this study caused substantial increases in the concentration of dissolved phosphorus in shallow groundwater (measured in a water table well, depth 2.4 m) that passed through both the buffered and unbuffered areas and beneath the root zone.

Osborne and Kovacic (1993), working in Illinois, found that phosphorus concentrations in shallow groundwater were greater in riparian forest buffers than in grass buffers or cropped areas during the dormant season. They concluded that phosphorus in the cropped area was washed straight into the stream through tile drains and that leaching of phosphorus to groundwater from surface organic matter might result in higher concentrations of groundwater phosphorus in the forested areas. Both grass and forested buffers released phosphorus to groundwater during the dormant season. Osborne and Kovacic suggested that regular cropping and removal of organic matter from grass buffers and careful selective harvesting of trees in forest buffers would encourage the growth of new vegetation and increase the vegetative uptake of dissolved phosphorus.

A study in Sweden (Vanek, 1991) found that a riparian forest buffer was actually the source of high levels of phosphorus entering a small (12 ha) lake. This lake received water from direct precipitation and groundwater. It had never received point discharges of wastewater or other pollutants and was separated from nearby cropland and developed areas by a 30 to 50 m strip of riparian forest. Vanek concluded that phosphorus accumulated over time in the riparian forest and was currently being released into the lake with groundwater flow, causing algal blooms and eutrophication.

5.3.2.4 Vegetative Filter Strips

The term vegetative buffer or filter strip may also refer to narrow strips (usually less than 15 m) of vegetation located in the middle of agriculture fields in relatively level areas (Daniels and Gilliam, 1996) or as midslope contour buffers intended to slow overland flow, trap sediment, and prevent gully formation (Dosskey, 2001). Riparian buffers and midfield vegetated filter strips perform many similar functions and research in one area is often used as a source of information for the other

(Schmitt et al., 1999). Recommended widths for riparian buffers are usually greater, up to 90 m (Lowrance et al., 2000), and vegetated filter strips located away from the stream side are less likely to provide sites for denitrification. Narrow grass strips may fill with coarse sediment after several precipitation and erosion events and evince a decrease in effectiveness. An experimental study of a 4.6-m grass filter strip (Dillaha et al., 1989) showed that sediment retention was reduced from 90% to 5% after six simulated rainfall events. Smaller clay and silt particles may be carried through these grass strips during high intensity storms.

5.3.2.5 Pesticides

Pesticides differ in their persistence (half-life), solubility, and tendency to adhere to soil particles (see Table 3.1A,B). Therefore the ability of riparian buffers to prevent contamination of streams and other water bodies by pesticides depends on the nature of the particular substances involved as well as the soil and vegetation in the buffer (Burken and Schnoor, 1996; Harris and Forster, 1997). As we have discussed earlier, riparian buffers function particularly well as sediment traps, so the pesticides that are adsorbed to the coarse-textured sediment particles can also be trapped and held in riparian buffers. Trees, litter, and herbaceous vegetation in riparian buffers combine to decrease the volume and rate of overland flow. In addition to allowing time for sediment to settle, this increases the likelihood that pesticides will be retained in the riparian buffer either as sediment-associated particles or dissolved in soil solution or shallow groundwater. Pesticides that enter the soil are more likely to break down if they remain in the unsaturated zone where fluctuating moisture and oxygen levels increase rates of microbial decomposition (Misra et al., 1996; Paterson and Schnoor, 1992). Once again, residence time and those factors that increase or decrease residence time have a dominant influence on the transport and fate of pesticides.

Lowrance et al. (1997b) investigated the role of riparian buffers in preventing stream contamination by pesticides in Georgia. The buffers averaged 50 m in width and consisted of a grass strip adjacent to a cornfield, a managed pine forest downslope of the grass buffer, and a hardwood forest adjacent to the stream. They found that atrazine concentrations decreased from an average of 34.1 μg/L at the field edge to ≤1 μg/L near the edge of the stream during a period of high herbicide transport in the spring. Alachlor concentrations were reduced from 9.1 μg/L to ≤1 μg/L by the riparian buffers.

Naturally the pattern of precipitation during pesticide application has a substantial influence on the transport of these chemicals through riparian buffers. Arora et al. (1996) investigated the movement of three herbicides (atrazine, metolachlor, and cyanazine) through six 1.52 m wide vegetative buffer strips in Iowa. The primary process for herbicide retention was infiltration. Even these very narrow vegetated strips retained 40% to 100% of the total sediment moving through the buffer. However, only 5% of the herbicide stored in the buffer was adsorbed to sediment particles. Retention of the pesticides via infiltration was highly variable: 16% to 100% for metolachlor, 8% to 100% for cyanazine, and 11% to 100% for atrazine. High intensity rainstorms shortly after herbicide application increased the probability that more herbicides would be transported to streams.

Research into phytoremediation has shown that the presence of certain species of trees and grass in riparian buffers and filter strips increases the decomposition rate for some pesticides (Burken and Schnoor, 1996, 1997; Karthikeyan et al., 2004). Using atrazine labeled with C^{14} radioactive tracers, Burken and Schnoor (1996, 1997) demonstrated that hybrid poplar trees (*Populus deltoids nigra* DN34) can "uptake, hydrolyze and dealkylate atrazine to less toxic metabolites." The breakdown of atrazine took place in all parts of the trees—roots, stems, and leaves—and became more complete as the chemicals remained in the plant tissue over time. It should be noted that the poplar trees displayed no adverse effects from the uptake of atrazine. The presence of nontarget (for the herbicide) vegetation also increases pesticide breakdown rates in soils by adding organic matter and oxygen which enhance the growth of decomposing microbial populations (Burken and Schnoor, 1996).

5.3.2.6 Coliform Bacteria

Entry et al. (2000a,b) investigated the movement and survival of swine wastewater in water and soil by measuring total and fecal coliform concentrations as the effluent moved through riparian buffers of various compositions and sizes. Three buffer types were tested: (1) 20 m grass and 10 m forest, (2) 10 m grass and 20 m forest, and (3) 10 m grass and 20 m maidencane (*Panicum hemitomon* Schult.), a wetland herbaceous species. Each buffer was 4 m wide and 30 m long with the longer axis of the rectangular buffer oriented orthogonally to the hill slope. In each case the grass portions were located upslope of the forest or maidencane portions. The results of this study indicate that the movement of coliform bacteria through the riparian buffer treatments was limited (less than 15 m) and that time was the primary controlling factor in the decrease in concentration. Total and fecal coliform numbers decreased approximately 10-fold every 7 days for the first 14 days, regardless of the season or vegetation type. Total fecal coliform mortality in shallow (1.5 and 2 m) wells correlated ($r^2 = 0.89$) with decreasing soil water content and increasing soil temperature. At 90 to 120 days after the wastewater had been applied, there was no difference between treated and untreated soils. The authors concluded that vegetated filter strips did not reduce concentrations of total or fecal coliform bacteria and recommended holding systems (tanks or ponds) for waste storage. The waste can then be applied to fields during periods of dry, warm weather, when transport to streams is minimized and coliform bacteria mortality is most rapid.

One of the important functions of protected riparian buffers may simply be to limit the access of farm animals to streams. Meals (2001) investigated phosphorus and bacteria loadings in tributary streams of Lake Champlain in Vermont. "Dairy cows in Vermont traditionally spend half of the year on pasture and livestock commonly have free access to streams and streambanks" (Meals, 2001; Meals and Hopkins, 2002). Riparian protection in experimental watersheds in this project area included livestock exclusion, streambank protection (erosion control), and riparian restoration. Their initial results demonstrated significant reductions in phosphorus concentrations and bacterial concentrations in streams in the treatment watersheds. In the reference watersheds, pollutant concentrations in streams correlated positively with temperature and negatively with streamflow. This is consistent with the observation that direct deposition of animal waste in streams is a primary pollutant source. Higher streamflow may dilute concentrations of pollutants in streams. In addition, dairy cows are most likely to be in the streams during the summer when the temperatures are warmer. In agricultural watersheds without livestock, nutrient concentrations tend to increase in the winter (dormant season), when vegetative uptake is limited or negligible.

5.3.3 Riparian Area Condition and Aquatic Biota

Changes in fish and benthic macroinvertebrate communities provide a measure of the in-stream disturbance caused by changes in the riparian environment and other watershed conditions (Chapter 4). Jones et al. (1999) completed a study of the effects of the loss of riparian trees on fish communities in southern Appalachian streams. All 12 of the study watersheds were at least 95% forested. Sampling sites in each watershed were located 1 to 2 km downstream of riparian areas that varied in size and condition from entirely forested to deforested gaps up to 5.6 km in length. The width of the deforested gaps varied from 0.2 to 1.3 km. Trees were removed and the rest of the vegetation remained in place. This study differs from those described above, in that, aside from the deforested patches, the watersheds and riparian areas were undisturbed. It also is unique in its focus on the streamside length and area of the gaps in riparian forests rather than the width of preserved or restored riparian buffers.

Habitat sampling indicated that longer deforested riparian gaps were associated with an increase in stream sediment downstream. This caused a decline in habitat diversity because riffles filled with sediment. Longer deforested gaps were also positively correlated with an overall decrease in fish abundance. In some cases, benthic-dependent species were replaced by more sediment-tolerant and invasive species. While there was no correlation between upstream gap length and fish species diversity, there were statistically significant changes in community structure when fish species were grouped according to reproductive behavior. In general, fish species were categorized in two groups: nonguarders and guarders. Nonguarding fish species either broadcast their eggs and abandon them or hide their eggs in benthic sediments with only slight or no modification of the sediment. Guarding fish species spawned eggs underneath rock surfaces, constructed pebble piles for spawning and then guarded the eggs, or constructed pit nests for eggs in soft bottom areas. Longer deforested gaps upstream were significantly ($p < 0.05$) associated with an increase in the relative abundance of pebble pile builders and pit spawners—species that cleaned silt from their nests and guarded their eggs. At the same time, species that did not clean silt from nests decreased in density.

The authors conclude that "loss of riparian forests leads to a decrease in species that are dependent on rapidly flowing shallow water in relatively sediment-free stream reaches, or that hide but do not guard their eggs." Loss of riparian forest led to an increase in exotic nonsalmonid species that favored slow deep-water habitats. Jones et al. (1999) also found that there was a negative correlation between fish density and gap length ($r^2 = 0.458$, $p < 0.02$), but no significant relation with gap width. This indicates that much of the increased sediment load may have originated from the erosion of exposed streambanks.

Destruction of riparian vegetation has also been associated with declines in macroinvertebrate populations. A study describing the effect of timber harvesting in streamside areas on macroinvertebrates is described in Chapter 6 (Culp and Davis, 1983).

5.3.4 RIPARIAN BUFFER WIDTH, DESIGN, AND MANAGEMENT

Much time and effort have been invested by regulatory agencies in trying to determine the optimal width of riparian buffers (Wenger, 1999). Optimal buffer widths allow reasonable use of the land while protecting the stream channels, water quality, aquatic ecosystems, and riparian habitat. Clearly the effectiveness of riparian buffers depends upon several interrelated factors—topography, surface and groundwater hydrology, and upland land use (Castelle et al., 1994). As noted earlier, steep slopes accelerate the movement of water, sediment, and pollutants to streams. Soil erosion typically begins with fine textured soils and progresses to coarse textured soils (Calhoun et al., 2002). In order to maintain the same effectiveness, the riparian buffer width should be increased in proportion to site characteristics. The Massachusetts Department of Conservation and Recreation (2003), for example, specifies that riparian buffers should be increased by 40 feet for every 10% increment of slope greater than 10% on poorly and moderately drained soils. On well-drained outwash and till soils, the buffer is increased 40 feet for each 10% increase in slope greater than 20%. The width of riparian buffers also may be designated in response to habitat requirements of specific species or to conserve ecosystems at a particular scale (Crow et al., 2000).

Greater pollutant loading in adjacent upland areas requires wider buffers to assimilate and efficiently process those sediments and nutrients before they reach the receiving water. Low soil infiltration rates in some areas may reduce riparian buffer effectiveness and, again, wider buffers may be needed to achieve that same level of nutrient retention. Another concern centers on what kind of management, if any, should be allowed within riparian buffers. Lowrance et al. (2000) (see also Sheridan et al., 1999) tested the USDA guidelines for managed riparian buffers as applied to a site in the Coastal Plain near Tifton, Georgia. These guidelines call for a three-zone buffer system. The first zone, located next to the stream, is composed of native hardwood trees and shrubs. It is intended to provide shade and coarse woody debris, stabilization of stream habitat conditions, and a storage site for nutrients. No forest management activities are allowed in zone 1. The second zone, located upslope of zone 1, is carefully managed to maintain the assimilative capacity of the forest. The third zone, located between the agricultural field or other nonpoint pollutant source and zone 2 is a grassed filter strip designed to slow or stop overland flow, convert channel flow to sheet flow, and trap sediment.

In this particular experimental design, the entire buffer had an average width of 75 m from the streambank to the cultivated field. Zone 1 was a 15 m wide strip of trees—yellow poplar (*Lyriodendron tulipifera*) and swamp black gum (*Nyssa sylvatica* var. *biflora*). Zone 2 (a 50-year-old pine forest) ranged from 45 to 60 m wide and was divided into three 40 m sections: (1) a control, (2) clearcut, and (3) partial cut (thinning). Zone 3 was an 8 m wide strip vegetated with Bermuda grass (*Cynodon dactylon*), Bahia grass (*Paspalum notatum flugge*), and perennial rye grass (*Lolium perrene*). The average slope of the cultivated field was 2.5% and the soils were sandy loams. Sampling of nitrate concentrations in the adjacent field showed some unexpected variability. More nitrate was found in soil water above the control (12.0 mg/L) and clearcut (11.0 mg/L) than above the thinned (5.3 mg/L) sections. There were significant reductions in nitrate concentrations from field to streamside in the control and clearcut plots (96% reduction for the control plots, 88% reduction preharvest versus 90% reduction postharvest for the clearcut). There was an unexpected increase in nitrate concentrations in an area of the riparian forest near the stream in the partial-cut treatment. This may have been the result of groundwater flow patterns routing water from the field around the riparian buffer system. Tests of the effects of the silvicultural treatments were confounded by (1) groundwater flow paths, (2) the aforementioned spatial differences in nitrate concentration of soil water entering the three buffer areas, (3) seasonal and annual variability of nitrate concentrations, and (4) the amount of nitrate stored in zone 3, the grass buffer. The authors concluded that overall there was little negative short-term effect that could be attributed to timber harvesting in zone 2.

Box 5.1   Riparian Buffers in Agricultural: Prairie Grasslands

The Bear Creek Riparian Buffer National Research and Demonstration Area (Josiah and Kemperman, 1998; Schultz et al., 2004)

The Bear Creek watershed (Iowa) is a long-term research site for the study of riparian buffers in grassland agricultural ecosystems. It is located in a predominantly agricultural landscape. Row crops and intensive grazing are the most common agricultural activities. Sediment loading and channelization have caused channel instability and bank erosion in streams (Figure 5.7). In addition, fertilizers and pesticides entered streams with subsurface and overland flow, degrading water quality and aquatic habitat. Livestock grazing near and in stream channels posed an additional threat to bank stability, channel integrity, and water quality.

Researchers based at Iowa State University (Ames, Iowa) designed and constructed multi-species riparian buffers consisting of various mixtures of trees, shrubs, grasses, and forbs as an alternative to the three-zone forested buffer system commonly recommended in regions of natural forest vegetation to the east (Lowrance et al., 2000) (Figure 5.8 and Figure 5.9). These riparian buffers were installed on first- to fourth-order streams. In most cases, row cropping or intensive grazing had reached up to the stream edge for many years. The design of each buffer reflected both site characteristics and the interests of individual landowners. Some of the buffer recommendations included planting trees on steep slopes in order to stabilize soil with perennial roots. On moderately sloping sites, grass buffers were frequently sufficient. When landowners were concerned about trees falling into restored streams a combination of woody and herbaceous plants were used. Constructed wetlands within the buffers acted as receiving basins for discharges from tile drains. The buffers were extended beyond the meander belt for tightly meandering streams. Buffers adjacent to channelized streams were made wide enough to accommodate meandering if the restoration proceeded naturally or with additional engineering work. Trees provided shade and organic carbon to streams. Shrubs also added woody roots, while grasses, especially the native switchgrass, trapped sediment. The mixture of vegetation types produced a complex vertical structure that enhanced wildlife habitat.

FIGURE 5.7 A degraded agricultural stream in the Bear Creek watershed, Iowa. (Photograph by Dick Schultz, Department of Natural Resource Ecology and Management, Iowa State University, Ames, Iowa, http://www. buffer.forestry.iastate.edu/Photogallery/photogallery.htm; accessed July 2006.)

FIGURE 5.8 The same stream following riparian restoration. (Photograph by Dick Schultz, Department of Natural Resource Ecology and Management, Iowa State University, Ames, Iowa, http://www.buffer.forestry. iastate.edu/Photogallery/photogallery.htm; accessed July 2006.)

The Bear Creek Study showed that:

1. A 7 m grass buffer reduced sediment loading by 95% and nitrogen and phosphorus by more than 60%. A 9 m buffer increased sediment storage slightly, to 97%. Nitrogen and phosphorus loading was reduced by 80% and there also was a 20% reduction in the transport of dissolved nutrients to streams.
2. Water infiltration rates were up to five times greater in buffers than in row-cropped or heavily grazed pastures.

FIGURE 5.9 Aerial view of the restored riparian buffer. (Photograph by Tom Schultz, Department of Natural Resource Ecology and Management, Iowa State University, Ames, Iowa, http://www.buffer.forestry.iastate.edu/ Photogallery/photogallery.htm; accessed July 2006.)

3. Soils in riparian buffers accumulated organic carbon and immobilized nitrogen. Hybrid poplar trees sequestered 3000 kg C/ha/yr and 37 kg N/ha/yr; switchgrass sequestered 800 kg C/ha/yr and 16 kg N/ha/yr.
4. Sediment loss from stabilized streambanks was reduced by 80%.
5. Biodiversity (fish) increased in streams where riparian buffers had been restored. The number of bird species in vegetated buffers was five times greater than in row-cropped or heavily grazed riparian areas.
6. Water quality changes (1 through 4, above) were clearly apparent at the field edge, but there was no measurable water quality improvement in the stream. One-third of the watershed was located upstream of the buffered stream segments. The untreated upstream reaches contained twice the channel length of the buffered portion. This clearly demonstrated that there is a "need for buffering watersheds in a systematic manner beginning in the headwaters and moving downstream."

Buttle (2002) questions the "ribbons" and "donuts" designs of riparian buffers—strips of forest along the edges of streams and rivers, and lakes, respectively, with fixed widths that may ignore variations in topography and the hydrologic characteristics of surrounding areas. From hydrologic and forest management studies in Ontario, Canada, where disturbance from timber harvesting is the primary concern, Buttle concluded that, in some cases, protecting upland groundwater recharge areas may be as or more important than excluding timber harvesting from riparian buffers. This is especially true for the preservation of native populations of brook trout (*Salvelinus fontinalis*) that require groundwater seeps (springs) for spawning areas. Accurately identifying groundwater recharge areas has been cost prohibitive, but the availability of digital elevation models (DEMs) along with digital surficial and bedrock geology data make this type of hydrologic analysis increasingly feasible for mapping, planning, and watershed management.

5.4 SUMMARY

5.4.1 RIPARIAN AREAS

Riparian areas are defined by the stream valley and terrace slope area (Verry et al., 2004). They include the stream channel, floodplain, terrace slopes, and some part of the adjacent uplands. Riparian areas are sites of interaction between aquatic and terrestrial ecosystems. Forests in riparian areas shade stream channels. Forest litter provides organic matter to aquatic food webs. Tree roots prevent the erosion of streambanks. Coarse woody debris from riparian trees helps to maintain stream channel stability and provides habitat for aquatic biota. Riparian areas and associated wetlands act as storage sites for flood waters and as storage and processing sites for organic material and nutrients that are transported from uplands to streams and from streams to the floodplain. Littoral shoreline zones influence the processing of nutrients and organic carbon in lakes.

5.4.2 RIPARIAN BUFFERS

Riparian buffers are designated areas bordering rivers and streams (and lakes) that are protected and managed in order to mitigate the adverse impacts of human activity. The ability of riparian buffers to protect stream channels, water quality, and aquatic habitat varies in relation to site characteristics and how well the design and management of the buffer is adapted to these conditions.

Sources of variability in riparian function include (1) the residence time of pollutants in the buffer, (2) the thickness of the unsaturated zone (depth and temporal variation of the water table), and (3) upland land use. Pollutants that move quickly through buffers may escape processing, storage, or removal. When the saturated zone is below the riparian root zone, pollutants in groundwater can bypass the riparian buffer and be channeled directly to streams. In addition, pollutant loading from adjacent agricultural and urban lands may overwhelm the capacity of riparian buffers to process, store, and assimilate pollutants.

5.4.2.1 Tests of Riparian Buffer Function

5.4.2.1.1 Sediment

Riparian buffers are most effective at trapping sand particles. Silt and clay particles may pass through the buffer if overland flow reaches the receiving waterbody. The capacity of the buffer to retain sediment may diminish over time if the upland source is not controlled.

5.4.2.1.2 Nitrogen

Riparian buffers reduce the nitrogen concentration of water entering streams from agricultural fields. While vegetative uptake and transformation of nitrogen is important, especially during the growing season, the primary mechanism for nitrogen removal appears to be denitrification in poorly drained soils of riparian wetlands. This underscores the need for the landform-based delineation of buffers versus fixed width strips.

5.4.2.1.3 Phosphorus

Riparian buffers are less effective in phosphorus removal. While a high percentage of sediment-associated phosphorus may be retained, buffers are far less effective in the retention of dissolved phosphorus. Continued phosphorus loading in riparian sediments may lead to a build up of phosphorus in riparian buffer soils that will, over time, increase phosphorus concentrations in streams and lakes.

5.4.2.1.4 Pesticides

Riparian buffers can be effective sites for the storage and chemical breakdown of pesticides. Pesticide degradation is highly dependent upon residence time and soil conditions—available organic matter, adequate water content and oxygen—that support microbial populations. Some species of vegetation also may absorb and process pesticides, creating nontoxic metabolites as end products. High intensity rainstorms immediately following herbicide applications increase the probability that pesticides will be washed directly to streams.

5.4.2.1.5 Coliform Bacteria

Coliform bacteria found in animal manure declines 10-fold every 7 days during the first 14 days following deposition, regardless of the time of year or vegetation type. Storing animal waste in areas where it cannot be transported to streams and restricting animal access to riparian buffers and streams is the most effective way to reduce coliform contamination.

5.4.2.2 Riparian Buffer Condition and Aquatic Biota

Cutting streamside trees can cause channel instability and erosion that degrades aquatic habitat even when the rest of the watershed is forested. The composition and structure of fish and aquatic macroinvertebrate communities reflect this change.

5.4.2.3 Riparian Buffer Width

The required width of riparian buffers depends upon "the specific functions that a buffer needs to provide under site-specific conditions" as well as slope, soil, and plant community characteristics (Castelle et al., 1994). Narrow buffers (30 m) may be sufficient to moderate stream temperature and provide coarse woody debris. Buffers up to 60 m wide may be needed for effective sediment removal. Wider buffers (more than 60 m) are required for nutrient retention and habitat protection. Their delineation should be based on landform and forest characteristics (Verry et al., 2004). While buffers may effectively reduce sediment, nutrient, and pesticide loading at a particular site, the lack of buffers upstream and upland land use can make it difficult to achieve improvements in water quality, especially in agricultural, mixed use, and urban watersheds. Buffer continuity, providing continuous protection from headwaters to downstream areas, may be *the* key factor in achieving improved water quality (see Box 10.2, The Bronx River Restoration).

REFERENCES

Addy, K.L., Gold, A.J., Groffman, P.M., and Jacinthe, P.A., Ground water nitrate removal in subsoil of forested and mowed riparian buffer zones, *J. Environ. Qual.*, 28, 962–970, 1999.

Allan, J.D., *Stream Ecology: Structure and Function of Running Waters*, Chapman & Hall, London, 1995.

Arora, K., Michelson, S.K., Baker, J.L., Tierney, D.P., and Peters, C.J., Herbicide retention by vegetative buffer strips from runoff under natural rainfall, *Trans. Am. Soc. Agric. Eng.*, 39, 2155–2162, 1996.

Barling, R.D. and Moore, I.D., Role of buffer strips in management of waterway pollution: a review, *Environ. Manage.*, 18, 543–558, 1994.

Barnes, B.V., Zak, D.R., Denton, S.R., and Spurr, S.H., *Forest Ecology*, 4th ed., John Wiley & Sons, New York, 1998.

Beeson, C.E. and Doyle, P.F., Comparison of bank erosion at vegetated and non-vegetated channel bends, *Water Resourc. Bull.*, 31, 983–990, 1995.

Beschta, R.L. and Taylor, R.L., Stream temperature increases and land use in a forested Oregon watershed, *Water Resourc. Bull.*, 24, 19–25, 1988.

Bolgrien, D.W. and Kratz, T.K., Lake riparian areas, in *Riparian Management in Forests*, Verry, E.S., Hornbeck, J.W., and Dolloff, C.A. (eds.), Lewis Publishers, Boca Raton, FL, 2000, pp. 207–217.

Booth, D.B., Urbanization and the natural drainage system—impacts, solutions, and prognoses, *Northwest Environ. J.*, 7:93–118, 1991.

Booth, D.B., Karr, J.R., Schauman, S., Konrad, C.P., Morley, S.A., Larson, M.G., and Burges, S.J., Reviving urban streams, land use, hydrology, biology, and human behavior, *J. Am. Water Resourc. Assoc.*, 40, 1351–1364, 2004.

Burken, J.G. and Schnoor, J.L., Phytoremediation: plant uptake of atrazine and role of root exudates, *J. Environ. Eng.*, 122, 958–963, 1996.

Burken, J.G. and Schnoor, J.L., Uptake and metabolism of atrazine by poplar trees, *Environ. Sci. Technol.*, 31, 1399–1405, 1997.

Buttle, J.M., Rethinking the donut: the case for hydrologically relevant buffer zones, *Hydrol. Process.*, 16, 3093–3096, 2002.

Calhoun, F.G., Baker, D.B., and Slater, B.K., Soils, water quality, and watershed size: interactions in the Maumee and Sandusky River basins of northwestern Ohio, *J. Environ. Qual.*, 31, 47–53, 2002.

Castelle, A.J., Johnson, A.W., and Conolly, C., Wetland and stream buffer requirements—a review, *J. Environ. Qual.*, 23, 878–882, 1994.

Clausen, J.C., Guillard, K., Sigmund, C.M., and Dors, K.M., Water quality changes from riparian buffer restoration in Connecticut, *J. Environ. Qual.*, 29, 1751–1761, 2000.

Cooper, J.R., Gilliam, J.W., and Jacobs, T.C., Riparian areas as a control of nonpoint pollutants, in *Watershed Research Perspectives*, Correll, D.L. (ed.), Smithsonian Institution Press, Washington, D.C., 1986, pp. 166–190.

Crow, T.R., Baker, M.E., and Barnes, B.V., Diversity in riparian landscapes, in *Riparian Management in Forests of the Continental Eastern United States*, Verry, E.S., Hornbeck, J.W., and Dolloff, C.A. (eds.), Lewis Publishers, Boca Raton, FL, 2000, pp. 43–65.

Culp, J.M. and Davis, R.W., An assessment of the effects of streambank clear-cutting on macroinvertebrate communities in a managed watershed, *Can. Tech. Rep. Fish. Aquat. Sci.*, 1208, 1983.

Daniels, R.B. and Gilliam, J.W., Sediment and chemical load reduction by grass and riparian filters, *Soil Sci. Soc. Am. J.*, 60, 246–251, 1996.

Dillaha, T.A., Reneau, R.B., Mostaghimi, S., and Lee, D., Vegetative filter strips for agricultural nonpoint source pollution control, *Trans. Am. Soc. Agric. Eng.*, 32, 513–519, 1989.

Dosskey, M.G., Toward quantifying water pollution abatement in response to installing buffers on crop land, *Environ. Manage.*, 28, 577–598, 2001.

Entry, J.A., Hubbard, R.K., Thies, J.E., and Fuhrmann, J.J., The influence of vegetation in riparian filterstrips on coliform bacteria: I. Movement and survival in water, *J. Environ. Qual.*, 29, 1206–1214, 2000a.

Entry, J.A., Hubbard, R.K., Thies, J.E., and Fuhrmann, J.J., The influence of vegetation in riparian filterstrips on coliform bacteria: II. Survival in soils, *J. Environ. Qual.*, 29, 1215–1224, 2000b.

Garabedian, S.P., Coles, J.F., Grady, S.J., Trench, E.C.T., and Zimmerman, M.J., Water quality in the Connecticut, Housatonic, and Thames River basins, Connecticut, Massachusetts, New Hampshire, New York, and Vermont, 1992–1995, Circular 1155, U.S. Geological Survey, Reston, VA, 1998, http://ma.water.usgs.gov/projects/MA-100/MA-100_Project_Publications.htm; accessed July 2006.

Gilliam, J.W., Parsons, J.E., and Mikkelsen, R.L., Nitrogen dynamics and buffer zones, in *Proceedings – Buffer Zones Their Processes and Potential in Water Protection*, Haycock, N., Burt, T., Goulding, K., and Pinay, G. (eds.), Quest Environmental, Hertfordshire, UK, 1997, pp. 54–61.

Gold, A.J., Jacinthe, P.A., Groffman, P.M., Wright, W.R., and Puffer, R.H., Patchiness in groundwater nitrate removal in riparian forest, *J. Environ. Qual.*, 27, 146–155, 1998.

Golterman, H.L., Vertical movement of phosphate in freshwater, *TNO nieuws: orgaan van de Organisatie voor Toegepast- Natuurwetenschappelijk Onderzoek*, 27, 96–101, 1972.

Groffman, P.M., Boulware, N.J., Zipperer, W.C., Pouyat, R.V., Band, L.E., and Colosimo, M.F., Soil nitrogen processes in urban riparian zones, *Environ. Sci. Technol.*, 36, 4547–4552, 2002.

Groffman, P.M., Gold, A.J., and Simmons, R.C., Nitrate dynamics in riparian forests: microbial studies, *J. Environ. Qual.*, 21, 666–671, 1992.

Groffman, P.M., Howard, G., Gold, A.J., and Nelson, W.M., Microbial nitrate processing in shallow groundwater in a riparian forest, *J. Environ. Qual.*, 25, 1309–1316, 1996.

Harding, J.S., Benfield, E.F., Bolstad, P.V., Helfman, G.S., and Jones, E.B.D., III, Stream biodiversity: the ghost of land use past, *Proc. Natl. Acad. Sci. USA*, 95, 14843–14847, 1998.

Harris, G.L. and Forster, A., Pesticide contamination of surface waters—the potential role of buffer zones, in *Proceedings – Buffer Zones Their Processes and Potential in Water Protection*, Haycock, N., Burt, T., Goulding, K., and Pinay, G. (eds.), Quest Environmental, Hertfordshire, UK, 1997, pp. 62–69.

Hill, A.R., Nitrate removal in stream riparian zones, *J. Environ. Qual.*, 25, 743–755, 1996.

Jones, E.B.D., III, Helfman, G.S., Harper, J.O., and Bolstad, P.V., Effects of riparian forest removal on fish assemblages in southern Appalachian streams, *Conserv. Biol.*, 13, 1454–1465, 1999.

Jordan, T.E., Correll, D.L., and Weller, D.E., Nutrient interception by a riparian forest receiving inputs from adjacent cropland, *J. Environ. Qual.*, 22, 467–473, 1993.

Josiah, S.J. and Kemperman, J., Emerging agroforestry opportunities, *J. For.*, 96(11), 4–9, 1998.

Karthikeyan, R., Davis, L.C., Erickson, L.E., Al-Khatib, K., Kulakow, P.A., Barnes, P.L., Hutchinson, S.L., and Nurzhanova, A.A., Potential for plant-based remediation of pesticide-contaminated soil and water using nontarget plants such as trees, shrubs, and grasses, *Crit. Rev. Plant Sci.*, 23, 91–101, 2004.

Klapproth, J. and Johnson, J.E., Understanding the science behind riparian forest buffers: effects on plant and animal communities, Publication 420-152, Virginia Cooperative Extension Service, Virginia Polytechnic Institute and State University, Blacksburg, VA, 2000a, http://www.ext.vt.edu/pubs/forestry/420-152/420-152.html; accessed July 2006.

Klapproth, J. and Johnson, J.E., Understanding the science behind riparian forest buffers: effects on water quality, Publication 420-151, Virginia Cooperative Extension Service, Virginia Polytechnic Institute and State University, Blacksburg, VA, 2000b, http://www.ext.vt.edu/pubs/forestry/420-151/420-151.html; accessed July 2006.

Lee, K.-H., Isenhart, T.M., Schultz, R.C., and Mickelson, S.K., Multispecies riparian buffers trap sediment and nutrients during rainfall simulations, *J. Environ. Qual.*, 29, 1200–1205, 2000.

Lowrance, R., Altier, L.S., Newbold, J.D., Schnabel, R.R., Groffman, P.M., Denver, J.M., Correll, D.L., Gilliam, J.W., Robinson, J.L., Brinsfield, R.B., Staver, K.W., Lucas, W., and Todd, A.H., Water quality functions of riparian forest buffer systems in the Chesapeake Bay watershed, Publication 903-R-95-004 CBP/TRS 134/95, U.S. Environmental Protection Agency, Washington, D.C., 1995a, http://www.chesapeakebay.net/search/pubs.htm; accessed July 2006.

Lowrance, R., Aliter, L.S., Newbold, J.D., Schnabel, R.R., Groffman, P.M., Denver, J.D., Correll, D.L., Gilliam, J.W., Robinson, J.L., Brinsfield, R.B., Staver, K.W., Lucas, W., and Todd, A.H., Water quality functions of riparian forest buffers in Chesapeake Bay watersheds, *Environ. Manage.*, 21, 687–712, 1997a.

Lowrance, R., Hubbard, R.K., and Williams, R.G., Effects of a managed three zone riparian buffer system on shallow groundwater quality in the southeastern Coastal Plain, *J. Soil Water Conserv.*, 55, 212–220, 2000.

Lowrance, R., Vellidis, G., and Hubbard, R.K., Denitrification in a restored riparian forest wetland, *J. Environ. Qual.*, 24, 808–815, 1995b.

Lowrance, R., Vellidis, G., Wauchope, R.D., Gay, P., and Bosch, D.D., Herbicide transport in a managed riparian forest buffer system, *Trans. Am. Soc. Agric. Eng.*, 40, 1047–1057, 1997b.

Martin, C.W. and Hornbeck, J.W., Logging in New England need not cause sedimentation of streams, *Northern J. Appl. For.*, 11, 17–23, 1994.

Massachusetts Department of Conservation and Recreation, *Ware River Watershed Land Management Plan 2003–2012*, Massachusetts Department of Conservation and Recreation, Boston, MA, 2003.

McCorvie, M.R. and Lant, C.L., Drainage district formation and the loss of Midwestern wetlands, 1850–1930, *Agric. Hist.*, 67(4), 13–39, 1993.

Meador, M.R. and Goldstein, R.M., Assessing water quality at large geographic scales: relations among land use, water physicochemistry, riparian condition, and fish community structure, *Environ. Manage.*, 31, 504–517, 2003.

Meals, D.W., Water quality response to riparian restoration in an agricultural watershed in Vermont, USA, *Water Sci. Technol.*, 43(5), 175–182, 2001.

Meals, D.W. and Hopkins, R.B., Phosphorus reductions following riparian restoration in two agricultural watersheds in Vermont, USA, *Water Sci. Technol.*, 45(9), 51–60, 2002.

Misra, A.K., Baker, J.L., Mickelson, S.K., and Shang, H., Contributing area and concentration effects on herbicide removal by vegetative buffer strips, *Trans. Am. Soc. Agric. Eng.*, 39(6), 2105–2111, 1996.

Myers, D.N., Thomas, M.A., Frey, J.W., Rheaume, S.J., and Button, D.T., *Water Quality in the Lake Erie–Lake Saint Clair Drainages: Michigan, Ohio, Indiana, New York, and Pennsylvania, 1996–1998*, Circular 1203, U.S. Geological Survey, Reston, VA, 2000.

Naiman, R.J., Decamps, H., and Pollack, M., The role of riparian corridors in maintaining regional biodiversity, *Ecol. Applic.*, 3, 209–212, 1993.

National Research Council, *Riparian Areas: Functions and Strategies for Management*, National Academies Press, Washington, D.C., 2002.

New Jersey Department of Environmental Protection, Standard for riparian forest buffer, in *Watershed Management*, New Jersey Department of Environmental Protection, Trenton, NJ, 2000, chap. 5, http://www.state.nj.us/dep/watershedmgt/bmpmanual.htm; accessed July 2006.

Ng, H.Y.F., Mayer, T., and Marsalek, J., Phosphorus transport in runoff from a small agricultural watershed, *Water Sci. Technol.*, 28(3–5), 451–460, 1993.

Olson, S., *The Singing Wilderness*, Alfred A. Knopf, New York, 1956 [1997 reprint edition].

Omernik, J.M., Abernathy, A.R., and Male, L.M., Stream nutrient levels and proximity of agricultural and forest lands to streams: some relationships, *J. Soil Water Conserv.*, 36, 227–231, 1981.

Osborne, L.L. and Kovacic, D.A., Riparian vegetated buffer strips in water quality restoration and stream management, *Freshwater Biol.*, 29, 243–258, 1993.

Paterson, K.G. and Schnoor, J.L., Fate of alachlor and atrazine in a riparian zone field site, *Water Environ. Res.*, 64, 274–283, 1992.

Patric, J.H., Effects of wood products harvest on forest soil and water relations, *J. Environ. Qual.*, 9, 73–80, 1980.

Peterjohn, W.T. and Correll, D.L., Nutrient dynamics in an agricultural watershed: observations on the role of a riparian forest, *Ecology*, 65, 1466–1475, 1984.

Raven, P.H., Evert, R.F., and Eichhorn, S.E., *Biology of Plants*, 6th ed., Worth Publishers, New York, 1999.

Rosgen, D.L., A classification of natural rivers, *Catena*, 22, 169–199, 1994.

Schmitt, T.J., Dosskey, M.G., and Hoagland, K.D., Filter strip performance and processes for different vegetation, widths, and contaminants, *J. Environ. Qual.*, 28, 1479–1489, 1999.

Schultz, R.C., Isenhart, T.M., Simpkins, W.W., and Colletti, J.P., Riparian forest buffers in agroecosystems—lessons learned from the Bear Creek watershed, central Iowa, USA, *Agrofor. Syst.*, 61, 35–50, 2004.

Sharpley, A.N., Kleinman, P., and McDowell, R., Innovative management of agricultural phosphorus to protect soil and water resources, *Commun. Soil Sci. Plant Anal.*, 32, 1071–1100, 2001.

Sheridan, J.M., Lowrance, R., and Bosch, D.D., Management effects on runoff and sediment transport in riparian forest buffers, *Trans. Am. Soc. Agric. Eng.*, 42, 55–64, 1999.

Simmons, R.C., Gold, A.J., and Groffman, P.M., Nitrate dynamics in riparian forests—groundwater studies, *J. Environ. Qual.*, 21, 659–665, 1992.

Snyder, N.J., Mostaghimi, S., Berry, D.F., Reneau, R.B., Hong, S., McClellan, P.W., and Smith, E.P., Impact of riparian forest buffers on agricultural nonpoint source pollution, *J. Am. Water Resourc. Assoc.*, 34, 385–396, 1998.

Sweeney, B.W., Bott, T.L., Jackson, J.K., Kaplan, L.A., Newbold, J.D., Standley, L.J., Hession, W.C., and Horwitz, R.J., Riparian deforestation, stream narrowing, and loss of ecosystem services, *Proc. Natl. Acad. Sci. USA* 101, 14132–14137, 2004.

USDA Forest Service, Forest Service Manual, Title 2500, *Watershed and Air Management*, Section 2560.05, Washington, D.C., 2004.

USDA Natural Resources Conservation Service, General Manual, 190-GM, part 411, USDA Natural Resources Conservation Service, Washington, D.C., 1991.

U.S. Fish and Wildlife Service, Endangered and threatened wildlife and plants; final rule to list the northern population of the bog turtle as threatened and the southern population as threatened due to similarity of appearance, *Fed. Reg.*, 62, 59605–59623, 1997.

Uusi-Kämppä, J., Turtola, J., Hartikainen, H., and Ylaranta, T., The interactions of buffer zones and phosphorus runoff, in *Proceedings — Buffer Zones Their Processes and Potential in Water Protection*, Haycock, N., Burt, T., Goulding, K., and Pinay, G. (eds.), Quest Environmental, Hertfordshire, UK, 1997, pp. 43–53.

Vanek, V., Riparian zone as a source of phosphorus for a groundwater dominated lake, *Water Res.*, 25, 409–418, 1991.

Verry, E.S. and Dolloff, C.A., The challenge of managing for healthy riparian areas, in *Riparian Management in Forests of the Continental Eastern United States*, Verry, E.S., Hornbeck, J.W., and Dolloff, C.A. (eds.), Lewis Publishers, Boca Raton, FL, 2000, pp. 1–22.

Verry, E.S., Dolloff, C.A., and Manning, M.E., Riparian ecotone: a functional definition and delineation for resource assessment, *Water Air Soil Pollut. Focus*, 4, 67–94, 2004.

Verry, E.S., Hornbeck, J.W., and Dolloff, C.A., *Riparian Management in Forests of the Continental Eastern United States*, Lewis Publishers, Boca Raton, FL, 2000.

Wall, G.R., Riva-Murray, K., and Phillips, P.J., *Water Quality in the Hudson River Basin, New York and Adjacent States, 1992–95*, Circular 1165, U.S. Geological Survey, Reston, VA, 1998.

Wenger, S., A review of the scientific literature on riparian buffer width, extent and vegetation, Institute of Ecology, University of Georgia, Athens, GA, 1999.

Wetzel, R.G., *Limnology*, Saunders, Philadelphia, 1983.

Whigham, D.F., Chitterling, C., Palmer, B., and O'Neill, J., Modification of runoff from upland watersheds—the influence of a diverse riparian ecosystem, in *Watershed Research Perspectives*, Correll, D.L. (ed.), Smithsonian Institution Press, Washington, D.C., 1986, pp. 305–331.

6 Forest Management and Natural Disturbances

A national policy which, though considering the direct value of forests as a source of timber, fails to take full account also of their influence upon erosion, the flow of streams, and climate, may easily endanger the well-being of the whole people.

Raphael Zon, 1927, *Forests and Water in the Light of Scientific Investigation*

6.1 INTRODUCTION

Headwater streams in forests are the source of 70% to 80% of the water flowing in the rivers of the United States (Binkley et al., 2004). Watersheds within the National Forest System alone provide the public water supplies of 3400 towns and cities. The National Forests comprise only 30% of the forested land in the United States, so many more municipalities depend on water from forests owned by states, corporations, and private individuals (Dissmeyer, 2000). As discussed in the preceeding chapters, the condition of forest ecosystems was a major influence on the characteristics of headwater streams. These characteristics include the quantity and timing of baseflow and storm-flow, concentrations of sediment and dissolved nutrients, water temperature, and the stability of the stream channels.

In this chapter we discuss the effects of forest management on streams, beginning with a brief history of logging in the Northeast region in the 19th and early 20th centuries. We examine the development and growth of forest research programs and summarize the results of experiments designed to measure the effects of forest management and other disturbances on streamflow, sediment, and water chemistry. We conclude with a discussion of best management practices (BMPs) to protect water quality and current issues in northeastern forests.

6.2 LOGGING AND FOREST CONVERSION IN THE 19TH AND EARLY 20TH CENTURIES

In New England and the Mid-Atlantic regions, European settlers cleared old-growth forests from most areas that were in any way amenable to agriculture by the 1850s. Large areas of Ohio and Indiana were cleared as well. The harvested wood was used locally for fuel, construction, trans-portation, tool production, and huge volumes were simply burned. Old-growth forests survived in regions that were either too cold or too mountainous to be practical for farming: northern Maine, the White Mountains of New Hampshire, the Green Mountains of Vermont, the Adirondacks in New York, and the high plateau region of Pennsylvania. In the Lake States, old-growth forest still covered most of Michigan, the newly admitted state of Wisconsin (1848), and the territory that was to become the state of Minnesota (Greeley, 1925; Irland, 1999).

Logging expanded after 1850 in response to an increased demand for wood from growing urban centers on the East Coast and from new settlements on the treeless midwestern prairies. Until steam winches, logging railroads (late 1800s), and crawler tractors and trucks (1950s to present) came into widespread use, large volumes of logs and pulpwood could only be moved by water (Figure 6.1). Initially loggers sought large white pine and spruce trees. These species were widely used in

FIGURE 6.1 Log drive on the Connecticut River near Hanover, New Hampshire (late 19th century). (Gove, 2003; reprinted with the permission of Bondcliff Books.)

construction and, unlike hardwoods, they floated. In the late 1800s, as river drives were supplemented by railroads and pulp and paper mills were built, smaller trees, including hardwoods, were cut as well. By 1920, only 4% of the original old-growth forest remained (Greeley, 1925; Irland, 1999; Whitney, 1994) (Figure 6.2).

Logging typically took place during the winter months. Sleds drawn by horses and oxen, piled high with loads of wood, moved slowly over carefully graded and maintained ice roads to "log landings" or "log decks" on the frozen surface of lakes, streams, or rivers. While logging with horses and oxen on ice and snow caused minimal damage to the forest floor, the impacts on the rivers used for transporting the logs was considerable (Irland, 1999). During and after spring breakup, log drives on streams swollen with melting snow and early season rains conveyed enormous volumes of wood to downstream mills (Russell, 1928). Dams were built on many headwater lakes to store water and logs, raise water levels, and regulate outflow. A sluice gate was used to release water and logs as needed. On smaller streams, "splash" dams were built to store water (and energy) for the log drive. These dams were deliberately breached—by releasing blocks, levering out a key log, or with a well-placed charge of black powder—sending a torrent of water and logs downstream. When a dam is breached, it sends a flood downstream that ranges from 15% to 45% of the original dam height (Henderson, 1966). This produced the hydraulic and ecological equivalent of a low-frequency flood event (i.e., 50 to 100 year recurrence interval) every year for as long as logging operations were active in the watershed. Because waterways were used to transport logs and lumbermen were paid on the basis of the volume they cut, large trees in riparian areas were the first to go. An 1881 map of timber harvesting in Wisconsin showed that white pine had been harvested in 10-mile wide bands along all the major rivers (Whitney, 1994). This practice removed trees destined to become large woody debris as well as the root support of banks and riparian areas. The loss of both functions inevitably reduced stream channel stability and increased bed and bank erosion.

In the course of the log drive, river drivers with pike poles and peaveys were posted at falls, bends, or anywhere stranded logs could cause a jam in order to keep the wood moving downstream. In preparation for log drives, men and draft horses "snagged" streambeds (pulled out large woody debris), moved boulders, and blasted away troublesome rock outcrops that might cause a jam (Williams, 1976). On larger streams and rivers, heavy-planked jam boats shot the rapids along with tumbling masses

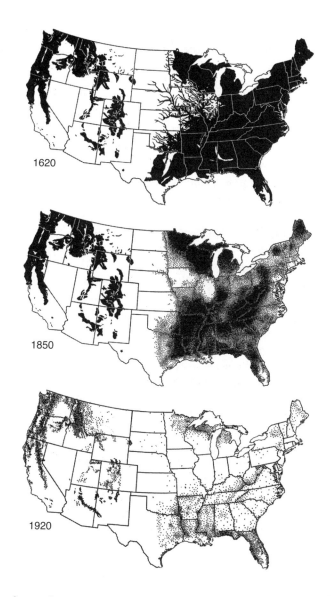

1620

1850

1920

FIGURE 6.2 Extent of uncut forest ("virgin forest") in 1620, 1850, and 1920. Areas of forests are based on estimates by states. Each dot represents 25,000 acres. Dots are not all correctly located. The Black Swamp region of northwestern Ohio, for instance, was almost a solid forest in 1850. (Greeley, 1925; reprinted with the permission of Economic Geography, Clark University, Worcester, MA.)

of logs. In calmer waters, logs were chained together in large rafts (Russell, 1928), some of which were maneuvered by steamboats. Clearly the effects of log and pulpwood drives, splash dams, and snagging on stream channel stability and aquatic habitat quality must have been devastating in some stream and river reaches. Irland (1999) describes the effects of logging in Maine as follows:

> The region's rivers were once choked with wood for several months of the year. The demands of log driving led the lumbermen to modify the natural drainage of the entire region to improve water flows.

The log drive ended at "booms"—corral-like structures floating on the river to capture and store logs for sawmills or pulp mills. This kept the logs clean and relatively free of discoloration and decay before they were processed into lumber or pulp. In so doing, booms also fouled the water and river bed with tannins, loose bark, and "sinkers." In addition, mill waste and sawdust were usually discarded directly into rivers. Before conversion to steam power, all the mill equipment was powered by water. Eventually many large mills with high dams also generated hydroelectric power. Although greatly diminished in number after 1890, log drives continued in some areas well into the 20th century. The last river drive occurred in Maine in the fall of 1976 (Irland, 1999).

This relentlessly efficient system of production was originally developed in Maine and then transferred to New Hampshire, Vermont, New York, Pennsylvania, and the Lake States. In Maine, the Penobscot, Kennebec, and Androscoggin Rivers reached hundreds of miles, via thousands of tributary streams, into the flat or gently sloping interior. This river network connected millions of hectares of forests to milling and transportation centers along the seacoast. Lumber schooners, and later railroads like the Boston & Maine, carried spruce and pine to regional, national, and international markets (Irland, 1999).

The scale of the industry in Maine presaged the fate of forests and water across the northern United States. Augusta, Bangor, Bath, Ellsworth, Orono, Old Town, Skowhegan, and Waterville all had large mill complexes in the 1800s. Bangor alone had 410 saws, and as Holbrook (1938:16) put it, "set the classic pattern that would follow the timber line West to the Pacific shore, distant by three thousand miles, one hundred years, and two trillion board feet of lumber." Gifford Pinchot (1947) described the exploitation common across the United States from about 1870 to 1920: "The American Colossus was fiercely at work turning natural resources into money. [It was] a perfect orgy of forest destruction."

Walter O'Meara's (1974) memoir of Cloquet, Minnesota, offers a poignant view of this time period. Transformed from a sleepy trading post in the 1870s, by about 1890 Cloquet had five sawmills that were supplied by more than 40 logging camps—"the supply of timber, everyone said, was inexhaustible." In less than 20 years it had displaced Bangor, Maine, as the self-proclaimed "White Pine Capital of the World"—complete with a newspaper called the *Pine Knot.* A little more than a decade later, mill owners and loggers were already casting an eye toward the western white pine forests Coeur d'Alene, Idaho. They were "the vanguard of the last trek in the long migration of rivermen and shanty boys from the Penobscot to the giant fir forests of the Pacific Coast." The end of the white pine era came abruptly, and tragically:

> In the month of October, 1918, as if in retribution for the carnage wrought by the logger's ax, the vast cutover encircling Cloquet burst into flame, and a great conflagration wiped out nearly every vestige of human habitation in our town (O'Meara, 1974).

6.3 FOREST WATERSHED RESEARCH

6.3.1 Development of Research Programs throughout the 20th Century

Federal programs for forest research and protection were initially established in response to the massive fires and floods that accompanied the destructive logging practices of the 19th century. In the 1890s and early 1900s, many people in the Northeast believed that widespread clearcutting was largely to blame for increased flooding in the spring and decreases in summer baseflow (Figure 6.3). Conservation and civic groups, including the Society for the Protection of New Hampshire Forests and the Appalachian Mountain Club, joined with industrialists, who required a reliable source of water power to run their mills, in support of federal programs for forest protection (Carlson and Ober, 2001; West, 1992). At the same time, opponents of federal funding questioned the scientific connection between forests and floods and the U.S. Army Corps of Engineers believed that flood control was fundamentally an issue involving the construction of dams and levees, not land management (Carlson and Ober, 2001; Leopold and Maddock, 1954; West, 1992).

FIGURE 6.3 Logged hillside in New Hampshire circa 1900. (Reprinted with the permission of the Society for the Protection of New Hampshire Forests Archives.)

Raphael Zon, then Chief of the U.S. Forest Service Research Office, completed an extensive review of the scientific literature on forests and water, which was submitted to Congress during this controversy (Verry, 1986; Zon, 1927). Zon concluded that, while the output of large rivers depends primarily on climatic factors (precipitation and evaporation), forest cover does affect the storage capacity of a watershed. He noted that forests cannot prevent large, low-frequency floods, but they can reduce their destructiveness. Zon (1927) felt strongly that the government should support research in this area:

> There is perhaps no other problem facing the American people to-day which demands such care in the scientific accuracy of its data and conclusions as does the relation between forests and water.

Zon encouraged Gifford Pinchot, then Chief of the U.S. Forest Service, to establish research programs that would rigorously investigate stream response to forest disturbance (West, 1992). The Weeks Act (Graves, 1911), which authorized the establishment of national forests in the eastern United States for "the conservation and improvement of the navigability of a river" was passed in 1911 (Carlson and Ober, 2001; McCullough and Robinson, 1993). The preamble of the Weeks Act reads:

> The general purpose of this law is to secure the maintenance of a perpetual growth of forest on the watersheds of navigable streams where such growth will materially aid in preventing erosion of steep slopes and the silting up of river channels, and thereby improve the flow of water for navigation.

At about the same time, the first U.S. Forest Service Research Stations were established. The USDA Forest Service Research Program has grown and expanded and been joined by a variety of other publicly and privately funded educational and research institutions. Writing in the *Journal of Hydrology* in 1993, two British researchers made the following observations:

> Although there have been many catchment [watershed] studies throughout the world, by far the majority of these have been carried out in the USA. Internationally, some of the most noteworthy contributions to catchment area research have been made in the USA... (McCullough and Robinson, 1993).

Following the initial concern with flood mitigation, researchers investigated the possibility of manipulating forest cover to increase streamflow during dry periods (Douglass, 1983; Hornbeck et al., 1993; Kittredge, 1948; Kochenderfer et al., 1990; Sopper and Lull, 1967). Removing forest cover reduces evapotranspiration, which typically increases water yield (Hornbeck, 1973). Beginning in the 1960s and 1970s, the primary focus of forest watershed research shifted from water yield to water quality. Scientists have become increasingly concerned with the effects of forest management on the entire forest and stream ecosystem, including aquatic biota, biogeochemical cycles, and soil nutrient processes (Likens et al., 1977; Swank and Crossley, 1988). Recently, work by scientists affiliated with research forests has expanded to include studies of large-scale environmental phenomena such as atmospheric deposition ("acid rain") and climate change on forests and forest streams (Aber et al., 1995; Adams et al., 1993; Hornbeck et al., 1997a). A recent book edited by Ice and Stednick (2004) is an excellent summary of this watershed research.

6.3.2 Experimental Watersheds and Research Forests

Experimental watersheds and research forests of the U.S. Forest Service have been used to provide sites for paired watershed experiments (Box 6.1) that produce long-term data on the hydrologic regime and the biogeochemistry of forest ecosystems following various experimental treatments (Hornbeck et al., 1993; Likens et al., 1977; Swank and Crossley, 1988). The first paired watershed research study in the United States was established at Wagon Wheel Gap in Colorado between 1909 and 1911 (McCullough and Robinson, 1993). The Coweeta Hydrologic Laboratory is the oldest continuously operating watershed research program in the United States. It was founded in 1934 and is located in western North Carolina (Douglass and Hoover, 1988). Other U.S. Forest Service research sites include Fool Creek in Colorado (Troendle and King, 1985), the H. J. Andrews Experimental Forest in Oregon (Anderson et al., 1976; Beschta, 1998), and in the southwest at Beaver Creek, Arizona (Baker, 1986; Lopes and Ffolliott, 1993). The Fernow Experimental Forest in Parsons, West Virginia (Reinhart et al., 1963), Hubbard Brook Experimental Forest in the White Mountains of New Hampshire (Likens et al., 1977), Leading Ridge Experimental Watershed Research Unit in Pennsylvania (Lynch and Corbett, 1990), and Marcell Experimental Forest in Minnesota (Barten, 1988; Verry, 1972) are some of the best known programs in the Northeast region (Figure 6.4). Both Coweeta and Hubbard Brook are part of the Long-Term Ecological Research (LTER) program sponsored by the National Science Foundation.

Characteristics and experimental treatments at these northeastern research forests are listed in Table 6.1. The Coweeta Hydrologic Laboratory is included in this list as well. Although it is not located in the Northeast, research contributions from Coweeta are numerous and important. These five sites evince a range of climatic and geologic conditions that in turn influence hydrologic regime and response to disturbance. Coweeta has a maritime climate and receives the largest amount of precipitation, more than twice that of the Marcell Experimental Forest located in the upper Midwest. Both Marcell and Hubbard Brook have persistent winter snowpacks, while less than 5% of the precipitation at Coweeta falls as snow. Marcell is the only area in which the soil freezes during the winter months. In all five forests, the predominant source of streamflow is subsurface flow. This is the typical flow path in forested watersheds where soil infiltration rates are high compared to other land uses and in relation to rainfall intensities and snowmelt rates.

6.3.3 Types of Forest Watershed Experiments

Forest watershed studies fall into three general categories: (1) ecological experiments, (2) natural experiments, and (3) evaluations of forest management, or in some deliberate cases, mismanagement. Ecological experiments typically employ the paired watershed approach (Box 6.1) to examine the effects of a change in a particular environmental variable on streams within the forest ecosystem. Natural experiments take advantage of a natural disturbance (wind damage caused by hurricanes, insect defoliations, wildfires, and the like) to opportunistically study the effect of that disturbance on streams.

FIGURE 6.4 Watershed 5, Hubbard Brook Experimental Forest, White Mountains of New Hampshire. (Courtesy of the USDA Forest Service, Northeastern Research Station.

The third type of study examines the effects of logging and roads on research forests or by comparing a commercial timber sale to a reference site. Examples of each type of experiment are presented below. The results of these studies and others of the same type are summarized later in the chapter.

6.3.3.1 An Ecological Experiment: Hubbard Brook Watershed 2, White Mountain National Forest, New Hampshire

The Hubbard Brook Experimental Forest consists of several small (12 to 43 ha), south-facing watersheds. A U.S. Forest Service scientist, Robert Pierce, identified the site and designed the network of weirs and meteorological stations in 1955. Streamflow and weather data collection began in 1957. In 1965, Watershed 2 (15.6 ha) was experimentally treated while Watersheds 3 and 6 were maintained as control or reference watersheds (Bormann et al., 1974; Hornbeck, 1973). The experiment at Hubbard Brook Watershed 2 was designed to

> test the effect of complete deforestation (devegetation) on (1) the hydrologic cycle, (2) intrasystem nutrient cycling, (3) nutrient export and eutrophication of stream water, and (4) erosion and export of particulate matter. A basic objective was to test the homeostatic capacity of a previously undisturbed forest ecosystem to hold on to nutrients when the main uptake pathway was blocked by destruction of vegetation and prevention of regrowth while the process of decomposition continued (Bormann et al., 1974). (Reprinted with permission of the Ecological Society of America.)

TABLE 6.1
Eastern Experimental Forests: Disturbance and Treatment History

Forest name	Disturbance history	Experimental watersheds	Treatments	Experimental watersheds	Treatments	Watershed area	References
Coweeta (1933)	Light semiannual burning and grazing prior to 1842, logging of the entire basin 1900–1923	1	Cut, burned 1956	19	Understory cut 1948–1949	3–760 ha	Douglass and Hoover (1988), Hibbert (1967)
		2	Reference	21	Reference		
		3	Cut, burned 1940	22	Strip-cut		
		6	Clearcut 1958	27	Reference		
		7	Grazed 1941–1952 Clearcut 1958	28	Multiple use demonstration		
		8	Control and treated	31	NA		
		9	Control and treated	32	Reference		
		10	Exploitive selective logging 1942–1956	34	Reference		
		13	Clearcut 1962	36	Reference		
		14	Reference	37	Clearcut 1963		
		16	Control and treated	40	Selection cut 1955		
		17	Clearcut 1940	41	Selection cut 1955		
		18	Reference	49	NA		
Fernow (1951)	Heavily cut 1905–1910	1	Commercial clearcut (all > 15 cm dbh) 1958; 1971			15–39 ha	Hibbert (1967), Hornbeck et al. (1993), Reinhart et al. (1963)
		2	Diameter limit (all > 43 cm dbh) 1958 and 1970				
		3	Intensive selection (all > 12.7 cm dbh) 1958 and 1963; clearcut 1970				
		4	Reference				
		5	Extensive selection (all > 28 cm dbh) 1957				
		6	Clearcut lower half 1964; upper half 1967; herbicides				
		7	Clearcut upper half 1963; lower half 1967; herbicides				

Site	History	Watershed	Treatment	Size	Reference
Hubbard Brook (1955)	Logged 1910–1919	W1	Reference	12–43 ha	Bormann and Likens (1994), Hornbeck et al. 1997 Martin and Hornbeck 1994
		W2	Clearcut 1965 + herbicide; 1970 regrowth		
		W3	Reference		
		W4	Progressive strip-cut beginning in 1970		
		W5	Whole-tree harvest 1983–1984		
		W6	Reference		
		W101	Commercial clearcut 1970		
Marcell (1960)	Logged 1920	S-1	Reference	9.7, 34.2, 53	Barten (1988), Verry (1972, 1987)
		S-2	Reference		
		S-3			
		S-4	Aspen-birch upland clearcut 1970		
		S-5	Strip-cut, then clearcut of black spruce in peat bog		
		S-6			
Leading Ridge	Unknown	LR1	Reference	43–123 ha	Lynch and Corbett (1990, 1994), Lynch et al. (1972)
		LR2	Lower slope clearcut 1966		
			Midslope clearcut 1972		
			Herbicide treatment 1974 on cutover areas		
			Final forest harvest 1976		
			Entire watershed herbicide treatment 1977		
		LR3	Commercial clearcut 1976–1977 44 ha of 104 ha watershed With BMPs		

Box 6.1   Paired Watershed Experiments (Brooks et al., 2003)

A paired watershed experiment is designed to measure the effect of some experimental treatment (e.g., clearcutting, strip-cuts, herbicide treatments) on streamflow and water quality. This is done by comparing streamflow in a treated or experimental watershed to that in an untreated or control watershed (Wilm, 1944). The paired watershed approach allows researchers to separate the effect of the treatment from natural variations in streamflow caused by annual variation in climate, precipitation, and temperature. Paired watershed experiments are also used to evaluate changes in water quality. Paired watershed experiments involve three steps:

1. Watershed selection and instrumentation
    Two or more watersheds are selected. The watersheds are chosen to be similar in size and within reasonable proximity of each other. Topography, bedrock and surficial geology, soils, and vegetation should also be as similar as possible. Gages to measure streamflow are then installed at the outlet of each watershed (Brooks et al., 2003). Additional measurements typically include sediment, turbidity, and water chemistry.
2. Calibration period (pretreatment)
    During the calibration, streamflow data are collected from each of the watersheds. These data are used to develop regression models that relate the streamflow of the [future] treatment watershed(s) to the streamflow of the control or reference watershed. The calibration period may require 5 to 10 years to ensure that reliable predictions can be made after the treatment is imposed (Wilm, 1949).
3. Treatment and data collection
    Following the calibration period, treatments are applied to the experimental watersheds while the control watershed remains undisturbed. Streamflow data are collected for several years following treatment. Streamflow data from the control watershed and the regression model are used to calculate the streamflow expected from the experimental watersheds absent treatment effects. The difference between this estimate and the measurement of the actual streamflow is the net effect of the treatment (Figure 6.5).

Watershed 2 was clearcut in winter with deep snow to protect the forest floor. All woody material was left on the site. This carefully controlled cutting procedure eliminated any soil disturbance so that the role of the living vegetation could be examined in isolation. Herbicides were applied for 3 years following cutting to prevent vegetative regrowth (Bormann et al., 1974; see also Bormann and Likens, 1994). It is important to note that this experimental treatment differs from conventional commercial timber harvesting in several ways. A logging operation would have removed the boles (or possibly more) of the harvested trees for sawlogs, pulp, or fuelwood. While herbicides might be selectively applied to reduce competition and encourage the growth of desirable tree species, they are not used to stop all regrowth.

6.3.3.2 A Natural Experiment: Sacandaga River Watershed, Adirondack Mountains, New York

The Sacandaga River watershed is a large (1272 km²) watershed with diverse topography. The area presented a unique opportunity to study long-term streamflow patterns in relation to land use change—streamflow, air temperature, and precipitation have been recorded for the watershed since the early 1900s. Patterns of forest stand density and composition between 1912 and 1963 were determined from increment cores, and vegetation and stump sampling. Additional information regarding land use change and increases in the beaver population was determined from the records

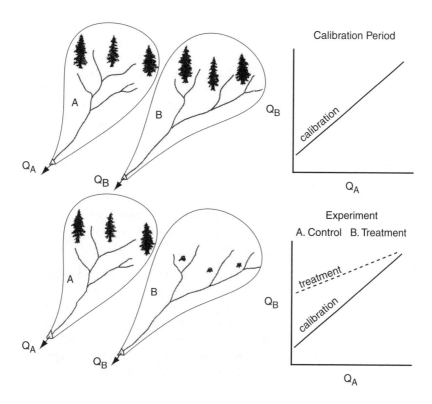

FIGURE 6.5 Paired watershed experiments. (Brooks et al., 2002; reprinted with the permission of Blackwell Publishing Professional.)

of state and local agencies. Regression analysis was used to examine the relation of streamflow and progressively increasing forest cover and beaver populations during the period from 1912 to 1950. In November 1950, a severe storm, with 80 to 90 mph winds, heavy rain, and snow, caused extensive damage to the forest. Eschner and Satterlund (1966) were able to quantify the hydrologic effect of severe forest canopy damage by comparing streamflow in the period from 1951 to 1962 to the patterns that had been developing prior to the storm.

6.3.3.3 A Logging Experiment: Fernow Experimental Forest, Parsons, West Virginia

This "second generation" (USDA Forest Service) study was specifically designed to quantify the effects of four timber harvesting practices in the 1950s. The study watersheds were located on the Fernow Experimental Forest and ranged in size from 15 to 39 ha (37 to 96 acres). Streamflow, meteorological, and water quality data collection began in 1951. The calibration period lasted 6 years. The four treatment watersheds were cut between May 1957 and December 1958; the methods are summarized below and in Table 6.5. Fernow Watershed 4 was the control (undisturbed reference) for the experiment. Streamflow and suspended particulate matter (SPM) data were collected for 3 years after the treatments began. The results of this seminal study were dramatic and instructive (Hornbeck and Reinhart, 1964; Reinhart et al., 1963), and are summarized later in this chapter.

6.3.3.3.1 Commercial Clear-Cut (Circa 1957) (Fernow Watershed 1)

This treatment is more accurately described as a liquidation cut or high-grading operation. All merchantable wood products (sawtimber, mine timbers, pulpwood) were cut. The minimum size was about 15 cm (6 inches) dbh (diameter breast height); low value or undesirable "cull" trees were left. There were no restrictions on the location, construction, maintenance, or restoration of skid trails, stream crossings, or roads. This seemingly reckless treatment was imposed to quantify the effects of "the typical liquidation cutting all too commonly practiced throughout the mountain hardwood country" in the 1950s. Detailed pre- and postharvest timber inventories showed that this treatment removed or deadened 84% of the original stand volume. A recovery period of 60 to 80 years would be required before timber harvesting could be repeated on this site.

6.3.3.3.2   Diameter Limit (Fernow Watershed 2)

This treatment was another form of high grading in which all long-lived merchantable trees (e.g., oaks, black cherry, etc.) larger than 43 cm (17 in.) dbh and all short-lived merchantable species (e.g., black locust, sassafras, and mountain magnolia) larger than 18 cm (7 in.) were cut. (Typically this cutting practice was repeated at intervals of about 20 years.) There were no restrictions on skidding and road construction. After logging, water bars were installed at 40 m (132 ft) intervals on the roads. This treatment removed or deadened 42% of the original stand volume.

6.3.3.3.3   Extensive Selection (Fernow Watershed 5)

The term "extensive" refers to the management effort per unit area, not the spatial distribution of the silvicultural practice. In this treatment, all merchantable trees larger than 28 cm (11 in.) dbh were harvested; undesirable or "cull" trees were girdled and left standing. Skidding was excluded from stream channels and roads were limited to grades of less than 20%. Water bars to divert stormwater and sediment off roads were installed wherever they were needed immediately after the logging operations were completed. (Typically this cutting practice was repeated at intervals of 10 years.) This treatment removed or deadened 24% of the original stand volume.

6.3.3.3.4   Intensive Selection (Fernow Watershed 3)

The term "intensive" again refers to the management effort per unit area. Restated in contemporary silvicultural terms (Smith et al., 1997), this treatment was a combination of individual tree selection, small group selection, and improvement cutting (typically at 5 year intervals). Care was taken with road and skid trail location (e.g., 8 m [25 ft] buffer strips for streams), design (e.g., limiting grades to ≤10%), construction, and postharvest restoration (water bars, seeding exposed soil with grass, etc.). Most state forest cutting practices regulations now routinely exceed these standards. This treatment removed or deadened 16% of the original stand volume.

6.4   CHANGES IN STREAMFLOW, SEDIMENT, CHANNEL FORM, AND TEMPERATURE FOLLOWING FOREST CANOPY DISTURBANCE

6.4.1   STREAMFLOW

Hibbert (1967) summarized the short-term (5 years following treatment) effects of forest cutting on streamflow (water yield) in the United States, Africa, and Japan in a review of research completed before 1965. Seventeen of the 39 studies reviewed by Hibbert were paired watershed experiments at Fernow and Coweeta. Among these, the largest increases in water yield occurred on clearcut, north-facing slopes at Coweeta. The greatest increase (408 mm or 52%) in annual water yield occurred in the first year following a 100% clearcut on Coweeta Watershed 17. As noted earlier, Coweeta receives considerably more precipitation than other research forests in the eastern United States. Not surprisingly, it has generally been observed that the largest increases in water yield after cutting occur either in the geographic areas where precipitation is highest or in other areas during periods of unusually high precipitation (Hornbeck et al., 1993). North-facing slopes that receive the least direct radiation lose less water to evapotranspiration. From this 1967 review, Hibbert concluded that a "reduction of forest cover increased water yield," but the "response to treatment is highly variable and, for the most part, unpredictable." It also was clear that increases in streamflow diminished over time as vegetation regrew.

In a later review paper, Bosch and Hewlett (1982) analyzed the results reported by Hibbert's (1967) paper along with an additional 55 studies from around the world. They concluded that the aggregate results allowed forest managers to reliably predict that increased streamflow would be caused by a reduction in forest cover and that there was a positive correlation between the increase and the proportion of the forest overstory removed. Cutting coniferous trees produced a greater increase in streamflow than cutting deciduous trees. The many densely layered needles of conifers intercept more water than the broad leaves of deciduous trees and interception occurs throughout the year, not only during the growing season. Conifers also are actively photosynthesizing for a greater portion of the year and thus have greater evapotranspiration requirements (Swank et al., 1988). Streamflow increases were, as expected, greatest in those watersheds with the highest measurements of mean annual precipitation. Again, streamflow increases were relatively short-lived, as large quantities of available water fostered rapid regeneration of new forest vegetation (see also Lima et al., 1978). The Bosch and Hewlett (1982) review was generally limited to short-term results.

By the 1980s and 1990s, scientists at Coweeta, Marcell, Fernow, Leading Ridge, and Hubbard Brook were able to examine streamflow records on experimental watersheds that extended for 20 years or more. At Coweeta, streamflow increases following clearcutting lasted from 10 to 14 years. In one watershed, replacing hardwoods with white pine (*Pinus strobes* L.) caused a decline in streamflow below pretreatment levels after 8 to 10 years. These declines continued as the pine forest matured during the next 20 years. Commercial logging operations that involved road building and tractor skidding increased peak flows by about 15%. Logging with minimal forest floor disturbance produced an increase in stormflow response of about 7% (Swank et al., 1988). Increases in peak flow usually result from increased overland flow on logging roads; increased overland flow also leads to accelerated soil erosion.

Two review papers published in the 1990s summarized long-term results from the four northeastern research forests. The first (Hornbeck et al., 1993) compared results from Marcell, Fernow, Leading Ridge, and Hubbard Brook. The second (Hornbeck et al., 1997b) examined the streamflow effects of several different cutting treatments at Hubbard Brook. Regression analysis of the results from the four northeastern forests led Hornbeck and his coauthors to the following conclusions (Figure 6.6):

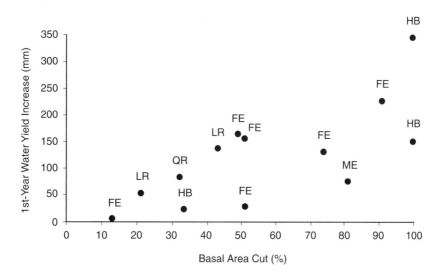

FIGURE 6.6 First-year water yield increases versus cutting: FE, Fernow Experimental Forest; LR, Leading Ridge Experimental Watershed; QR, Quabbin Reservoir; ME, Marcell Experimental Forest.

1. Initial increases in water yield occur promptly after forest cutting, with the magnitude being roughly proportional to the percentage reduction in basal area.
2. A 20% to 30% reduction in basal area was necessary to produce measurable increases in annual water yield.
3. The increases can be prolonged for an indeterminate length of time by controlling natural regrowth; otherwise they diminish rapidly, usually within 3 to 10 years (Hornbeck et al., 1993, 1997b).

Complete clearcutting with herbicide application to prevent regrowth—eliminating the influence of trees and other forest vegetation—resulted, predictably, in maximum annual streamflow increases—between 50 and 165 mm (10% to 25%) at Leading Ridge and Fernow and 347 mm (41%) at Hubbard Brook (Hornbeck et al., 1993). Streamflow increases diminished, with revegetation following the cessation of herbicide treatments. The maximum streamflow increase following clearcutting of upland aspen forests at Marcell Experimental Forest was 114 mm (58%) during the second year following treatment (Verry, 1972). Marcell is the area with the lowest precipitation and flattest topography. Harvesting on the peatland component of the Marcell watersheds did not produce an increase in annual streamflow, although it did cause a greater fluctuation in the water table. Reducing or eliminating the tree canopy over the peat bogs increased net precipitation and decreased transpiration. However, the increase in available energy at the wetland surface, coupled with increased wind speeds, led to more evaporation from the shallow saturated zone. The net effect on the volume of water available for streamflow was small (Verry, 1986). There also were distinct seasonal and flow distribution patterns. At all four sites, most streamflow increases occurred during the growing season, and the additional water increased baseflow while storm or peak flows changed little. This indicates that the increase in water yield was caused by the decrease in evapotranspiration and subsequent increase in soil water content and rates of subsurface flow.

At Hubbard Brook there were substantial differences in water yield among different types of clearcutting treatments (Table 6.2). Three treatments were used on three different watersheds: (1) Watershed 2—clearcutting with no product removal followed by 3 years of herbicide treatments (described earlier); (2) Watershed 4—progressive strip-cutting—the stand was clearcut in strips over a period of 3 years, leaving uncut strips during the first and second years; and (3) Watershed 5—whole-tree removal—clearcutting and removal of all harvested tree parts—boles, bark, and branches. The progressive strip-cut and whole-tree removal treatments were realistic silvicultural practices and did not include herbicide treatments. There was a maximum increase of 114 mm (8%) in the third year on the progressive strip-cut site, when two-thirds of the forest overstory was harvested, compared to a maximum of 347 mm (41%) increase on Watershed 2 in the first year following treatment. Whole-tree harvesting produced a maximum increase of 152 mm (23%) in the first year following harvesting, while increases in the other 2 years of the 3 year posttreatment period were not significantly different from predicted pretreatment values.

There were reductions in water yield from expected values from year 13 following treatment to year 30 on Watershed 2 and from year 8 to year 25 on Hubbard Brook Watershed 4 (progressive strip-cuts). As in the white pine stands at Coweeta, the reduction in streamflow was attributed to a change in species composition following cutting of the overstory. An overstory of sugar maple (*Acer saccharum* Marsh), American beech (*Fagus grandifolia* Ehrh.), and yellow birch (*Betula alleghaniensis* Britton) was replaced with early successional stands of rapidly growing pin cherry (*Prunus pensylvanica* L.) and paper birch (*Betula papyrifera* Marsh). The leaf area index (LAI) of these early successional stands was estimated to be approximately equal to that of mature forest stands in 3 to 4 years following stand initiation. In addition, stomatal resistance in leaves of paper birch and pin cherry is lower than that in sugar maple and beech leaves. This leads to higher rates of transpiration, reducing the amount of water available for streamflow (Federer and Lash, 1978; Hornbeck et al., 1997b).

TABLE 6.2
Changes in Annual Water Yield for Treated Watersheds at Hubbard Brook Experimental Forest, New Hampshire

Year after initial treatment	Clear felling with herbicide treatments years 1,2,3			Strip-cuts			Whole-tree removal		
	Estimated untreated streamflow mm	Change due to treatment mm	%	Estimated untreated streamflow mm	Change due to treatment mm	%	Estimated untreated streamflow mm	Change due to treatment mm	%
1	851	347*	41	777	22	3	648	152*	23
2	956	278*	29	1032	46*	4	876	47	5
3	919	240*	26	1417	114*	8	806	-15	-2
4	902	200*	22	819	67*	8	742	-11	-1
5	840	146*	17	1263	55*	4	681	4	1
6	787	44*	6	868	82*	9	1020	46	5
7	1059	12	1	1042	69*	7	1086	51*	5
8	1469	52	4	870	-14	-2	835	66*	8
9	832	67*	8	725	-30*	-4	860	47	6
10	1305	3	0	769	-26*	-3	879	20	2
11	884	48	5	1144	-18	-2	605	21	4
12	996	64*	6	869	-45*	-5	818	46	4
13	902	-13	-1	1069	-33*	-3			
14	764	-13	-2	709	-21*	-3			
15	807	-34	-4	927	-44*	-5			
16	1179	-41*	-3	856	-67*	-8			
17	885	-70*	-8	779	-29	-4			
18	1099	-62*	-6	702	-59*	-8			
19	715	-64*	-9	1079	-42*	-4			
20	958	-44*	-5	1158	-63*	-5			
21	872	-80*	-9	904	-42*	-5			
22	790	-82*	-10	943	-60*	-6			
23	708	-56*	-8	918	-30*	-3			
24	1109	-34	-3	623	-36*	-6			
25	1194	-48*	-4	1362	-23	-2			
26	923	-36	-4						
27	964	-80*	-8						
28	938	-71*	-8						
29	624	-54*	-9						
30	1410	-26	-2						

Hornbeck et al. (1997). Reprinted with the permission of the National Research Council, Canada.

* Statistically significant increase or decrease in water yield.

The variability in streamflow, among watersheds and over time, following overstory cutting can be attributed to a number of factors. The size of the cut, watershed aspect, annual variation in precipitation, soil hydraulic properties (upland mineral soils versus wetland organic soils), and changes in species composition have been discussed. In addition, the pattern and placement of cutting may have an effect—large clearcut areas near streams (the saturated source area) will produce a greater effect than the same total area divided into smaller cuts located farther from the stream. Herbicide treatments on some watersheds may destroy the remaining root systems of harvested trees that would have fostered rapid regrowth and use of newly available water. New trees growing from seeds require a longer period of time to become established than sprouts from mature root systems (Hornbeck et al., 1993).

In a natural experiment in a large watershed (Eschner and Satterlund, 1966), reforestation (an estimated increase in basal area of approximately 17 m²/ha to 30 m²/ha) from 1912 to 1950 and growth of the beaver population was associated with a 23% decrease in annual streamflow (196 mm). By constructing dams and ponds, beavers increase water storage and evaporation rates. Most of this decrease was attributed to a reduction in dormant season flow (132 mm; 67% of the total decrease). The storm that destroyed much of the forest canopy in November 1950 caused a statistically significant increase in mean annual streamflow of 45 mm from 1951 to 1962. The increase occurred in the dormant season only. The decrease in growing season streamflow was not significant. The increased dormant season streamflow was attributed to the reduced loss, via evaporation and sublimation, of intercepted snow after the predominantly white pine forest canopy had been damaged. In addition, less effective shading resulted in earlier and more rapid melting of the snowpack, leading to large streamflow increases in April.

Box 6.2   Forest Roads

Logging roads are one of the main causes of poor water quality in forested areas. On even the best of these roads, culverts and dips discharge sediment. But if the water that runs off the road can be filtered through the forest floor before it reaches the stream, the sediment will remain on the litter and there will be no impairment of water quality (Trimble and Sartz, 1957).

The construction of logging roads and landings exposes soil, providing a source of sediment that can be carried to streams with stormflow. While the greatest sediment input from roads generally comes during road construction and the active timber harvesting phase of management operations, roads can continue to be a source of sediment if they are not seeded with grass or stabilized with gravel (Swank et al., 2001). Soil compaction on unpaved forest road surfaces reduces infiltration and can channel overland flow to streams (Wemple et al., 2001). Roads also may intercept subsurface flow in forested watersheds and direct it to drainage ditches. Thus the presence of logging roads can increase the magnitude and frequency of high-frequency stormflow in forest streams. This also can increase stream channel erosion (Wemple and Jones, 2003). Three primary factors determine the degree to which forest roads affect streams: (1) the distance between the road and the nearest stream, (2) the slope of the road, and (3) the condition of road stream crossings.

Because logging roads and landings are the primary source of stream channel disturbance in logging operations, many of the regulations and recommendations prescribed by BMPs address the design, construction, and maintenance of these areas. The U.S. Department of the Interior National Park Service and the USDA Forest Service have funded programs to remove and restore old forest roads, particularly in California and the Pacific Northwest, as a means to reduce sediment loads and restore hydrologic processes (Madej, 2001; Switalski et al., 2004). A study in California (Madej, 2001) measured the effects of these restoration measures in Redwood Park following a 12 year recurrence interval storm. Treatments included decompacting road surfaces, removing culverts, and excavating sediment from stream channels. Following the storm, 10 m³, 135 m³, and 550 m³ of sediment per kilometer of stream were delivered from treated streams in the upper, middle, and lower slopes, respectively. In contrast, untreated roads generated 1500 to 4700 m³ of sediment per kilometer of stream. There was no detectable erosion on almost 80% of the treated roads.

6.4.2 SEDIMENT

The rate of natural or geologic erosion in forested areas of the Northeast is low compared to other climatic regions and land uses (Dunne and Leopold, 1978; Wolman and Schick, 1967). Patric (1976) analyzed data from forested watersheds in Arkansas, Kentucky, Maryland, Mississippi, New Hampshire, North Carolina, Ohio, Oklahoma, Pennsylvania, Tennessee, and Wisconsin and found that mean annual erosion rates ranged from 112 to 224 kg/ha. The mean annual sediment yield from a mature forest ecosystem at Hubbard Brook was estimated to be about 25 kg/ha (Bormann and Likens, 1994; Bormann et al., 1974). Erosion can vary substantially, however, even among adjacent watersheds that are similar in size and physical characteristics such as slope, soil type, and overstory vegetation. Stream sediment variability appears to be associated with the number and intensity of storm events rather than annual precipitation or watershed size. For example, sediment yields from three physically similar, undisturbed watersheds in the Hubbard Brook system were 18 kg/ha, 37 kg/ha, and 131 kg/ha in the same year—1978 (Martin and Hornbeck, 1994) (Table 6.3). Soil and channel erosion generally occurred during large dormant season rain events or high-intensity summer storms.

TABLE 6.3
Sediment Yields from Three Undisturbed Forested Watersheds at the Hubbard Brook Experimental Forest, New Hampshire

Year	WS-1 (11.8 ha) kg/ha	WS-3 (42.4 ha) kg/ha	WS-6 (13.2 ha) kg/ha	Precipitation (mm)
1975	68	35	18	1308
1976	20	10	15	1769
1977	62	29	79	1402
1978	131	37	18	1532
1979	61	47	25	1362
1980	141	64	32	1194
1981	41	25	34	1355
1982	47	28	10	1585
1983	10	11	3	1410
1984	84	47	35	1638
1985	7	5	4	1200
1986	52	45	17	1425
1987	108	71	52	1311
1988	6	4	1	1290
1989	18	19	7	1234
1990	13	24	13	1553
Mean	54	31	23	1411
SD	44	20	20	165
CV (%)	81	65	87	12

All are drained by first- or second-order streams. The last logging was prior to 1920 (Martin and Hornbeck, 1994).

Reprinted with the permission of the Society of American Foresters.

TABLE 6.4
Sediment Yields from Two Undisturbed Watersheds (WS-1 and WS-6)
and from Watershed 2 at Hubbard Brook

| | Sediment yields in kg/ha | | |
Year	WS-1 (undisturbed)	WS-2 (clearcut and herbicides applied)	WS-6 (undisturbed)
1966	—	13	4
1967	—	67	31
1968	—	92	10
1969	—	195	13
1970	—	365	42
1971	—	97	5
1972	—	22	6
1973	—	150	95
1974	—	—	25
1975	68	—	18
1976	20	13	15
1977	62	54	79
1978	131	51	18
1979	61	16	25
1980	141	19	32
1981	41	54	34
1982	47	18	10
1983	10	3	3
1984	84	39	35
1985	7	6	4
1986	52	50	17
1987	108	111	52
1988	6	7	1
1989	18	13	7
1990	13	19	13

Watershed 2 was clearcut in 1965 and herbicides were applied to prevent vegetative regrowth in 1966, 1967, and 1968. No wood products were removed and there were no skid trails or roads on the watershed (Martin and Hornbeck, 1994).

Reprinted with the permission of the Society of American Foresters.

Higher loads of suspended sediment in streams, following forest disturbance, can be the result of increased erosion from the watershed soils or erosion and suspension of sediment particles in the channel. Martin and Hornbeck (1994) stressed that forest cutting need not lead to elevated levels of sediment in streams if stream channels and riparian forests are protected. They concluded that the higher sediment load seen in Hubbard Brook Watershed 2 was primarily the result of stream channel erosion caused by the 3 years of herbicide treatments and the destruction of streamside trees (Table 6.4):

Streams at Hubbard Brook are characterized by a series of pools and cascades created by coarse woody debris. Since WS [watershed] 2 had been clearfelled and then sprayed with herbicides for 3 years, there was no material to replenish these dams. By 1969, the dams had decomposed and washed away. The greater velocities and greater base flows coupled with root system mortality from the herbicide caused the streambanks to fail and the stream to widen and contribute to sedimentation (Martin and Hornbeck, 1994). (Reprinted with permission of the Society of American Foresters.)

Studies conducted at Fernow during the 1950s and 1960s (Hornbeck and Reinhart, 1964; Reinhart et al., 1963) and described earlier in this chapter ("A Logging Experiment: Fernow Experimental Forest, Parsons, West Virginia") examined the difference in stream water sediment concentration brought about by exploitative logging versus carefully controlled timber harvesting. Experimental treatments took place between May 1957 and December 1958 (Table 6.5). The streams were sampled during and following the treatments from December 1957 to April 1960.

During the sampling period, SPM concentrations at Watershed 1 (commercial clearcut) averaged 490 mg/L versus 15 mg/L from the undisturbed reference or control watershed. The maximum suspended particulate matter (SPM) concentration after logging on Watershed 1 was 56,000 mg/L. The diameter limit cut (Watershed 2) resulted in mean SPM concentrations of 897 mg/L, with maximum values of 5200 mg/L. These very high values took place during the logging operations and were the result of unplanned skid roads (Box 6.2) and unprotected streams. SPM concentrations were considerably lower (maximum 210 mg/L) on the extensive selection cut (Watershed 5) and lower still (25 mg/L) on the intensive selection site (Watershed 3), where the harvest was light, skid roads were carefully planned, and the highest level of protection was afforded to streams. Water quality returned to pretreatment conditions on all sites, including the clearcut (Watershed 1), 1 to 2 years following the completion of logging operations. This was attributed to the rapid regrowth of vegetation (Edwards et al., 1999).

Because it was an unplanned, unsupervised, commercial clearcut, the Fernow study left unanswered the question regarding whether stream water quality could be maintained at high levels during a clearcut that employed careful planning of skid roads and restrictions on silvicultural operations designed to protect forest streams. This issue was addressed by a later Hubbard Brook study (Martin and Hornbeck, 1994). Watershed 5 at Hubbard Brook was subjected to a whole-tree clearcut in which 96% of the aboveground biomass was removed. Restrictions on the steepness of truck roads and skid trails, buffer strips along streamsides, and other restrictions and requirements for harvesting operations resulted in very low sediment concentrations, even though 70% of the forest floor was disturbed. Sediment yields during the year of logging were the same as they had been before the treatment. Sediment yields in the 3 years following the harvest were greater than those predicted had the watershed remained uncut, but the increases were modest. The pretreatment (1975–1983) maximum

TABLE 6.5
Watershed Treatments at Fernow Experimental Forest

Watershed no. and treatment	Timber cut	Percent reduction in basal area	Bulldozed skid roads		
			Maximum grade	Water bars	Other requirements
1, commercial clearcut	Everything merchantable (approximately 15 cm [6 in.] dbh or more)	84% removed (to 6 m²/ha [26 ft²/acre])	No limit	None	None
2, diameter limit	All trees over 43 cm (17 in.) dbh	42% removed	No limit	At 40 m intervals	None
5, extensive selection	Selected trees over 28 cm (11.0 in.) dbh	24% removed	20%	As needed	No skidding in stream channels
3, intensive selection	Selected trees over 13 cm (5 in.) dbh	16% removed	10%	As needed	No skidding in stream channels; skid roads located away from streams; grass seeding for soil stabilization where needed

dbh, diameter at breast height, 4.5 ft (1.4 m).

of 134 kg/ha/yr was exceeded only in the third year following cutting, when the soil loss through erosion reached 208 kg/ha/yr. In the next year, sediment yield decreased to 6 kg/ha/yr.

6.4.3 STREAM CHANNEL FORM

There is less documentation of changes in channel form associated with forest management than of the more obvious and dramatic changes caused by agriculture and urbanization (Chapters 7 and 8). As discussed in Chapters 2 and 5, stream channels change in response to increased streamflow. The shear stress or tractive force of increased streamflow causes bank erosion and the movement of sediment in the channel (Martin and Hornbeck, 1994). To the extent that changes in the forest canopy, either from timber harvesting or natural events, cause an increase in streamflow, stream channels will be altered. Although the sediment concentrations on the commercial clearcut watershed at Fernow returned to the pretreatment range, alterations of the stream channel—widening and incision—were still evident 40 years later (Edwards et al., 1999).

6.4.4 TEMPERATURE

Forest streams usually emerge from springs or via subsurface flow; hence water temperature is moderated by passing through soil. Groundwater is cooler than surface water in the summer and generally warmer than surface water in the winter.

Water temperature near a groundwater source is relatively stable. Water temperature is more strongly influenced by changes in air temperature during and shortly after stormflow events and as the distance from a groundwater source increases (Macan, 1958). Thus a disturbance that changes the air temperature in the vicinity of forest streams can change the temperature of the water in the stream channel. This is a concern because of the potentially damaging effects on fish and other aquatic organisms that are often adapted to a relatively narrow temperature range (Allan, 1995).

Many studies have shown that removing or damaging streamside vegetation increases stream water temperature in the spring and summer (Binkley and Brown, 1993). Stream water temperatures at Hubbard Brook Watershed 2 were generally higher in both summer and winter and showed greater fluctuations during the day than those of undisturbed streams that were shaded by streamside vegetation. A heavier winter snowpack in the cleared stream may have provided additional insulation, resulting in a net increase in winter water temperatures (Likens et al., 1970). Cutting all forest trees and understory vegetation in a watershed at Coweeta Experimental Forest caused maximum summer stream temperatures to increase from the normal mean temperature of 18.8°C to 22.7°C or higher in the first year following treatment. Diurnal temperature fluctuations also increased. Temperature increases at Coweeta exceeded optimum levels for trout habitat (Swift and Messer, 1971). At the Fernow Experimental Forest, researchers left a lightly cut riparian forest buffer strip 20 m wide on either side of a stream channel while clearcutting the remaining 31.7 ha of the watershed. There was no detectable increase in water temperature for the first 2 years following cutting. Cutting the protective strip caused a temperature increase of up to 7.8°C, to a maximum temperature of 25.3°C (Patric, 1980). A review of 20 stream temperature studies (Brown and Binkley, 1994) that included data from western states showed that removal of forest canopies over or adjacent to streams can lead to maximum stream temperature increases of 5°C or more.

6.5 WATER CHEMISTRY AFTER FOREST CANOPY DISTURBANCE

6.5.1 NUTRIENTS (NITROGEN, PHOSPHORUS, AND MINERAL NUTRIENT CATIONS)

In general, forests are sinks for nutrients. As discussed in Chapter 3, forests have an important function in protecting streams and downstream communities from nitrogen inputs in atmospheric deposition.

There are two related concerns about the loss of nutrients following timber harvesting: (1) excessive nutrient input to streams may adversely affect stream water quality and accelerate eutrophication both locally and downstream, and (2) the depletion of nutrients from forest soils may reduce future site productivity (Johnson et al., 1997; Likens et al., 1970). Changes in nutrient cycling are not limited to anthropogenic causes. Measurements by Swank (1988) showed an increase in nitrate export was correlated with forest defoliation by canker worms.

Forest practices do not appear to degrade water quality to phosphorus loading (Binkley and Brown, 1993). Because phosphorus is generally tightly bound to organic and mineral soil compounds, the loss of phosphorus from managed forests with natural regeneration is generally quite low. Paired watershed studies of undisturbed and clearcut and herbicided watersheds showed that the export of fine particulate phosphorus (sediment-associated phosphorus) primarily occurs during periods of high stream discharge or stormflow (Hobbie and Likens, 1973; Johnson et al., 1976; Meyer and Likens, 1979). Forest practices that minimize increases in streamflow (partial and patch cutting as opposed to clearcutting) and erosion also minimize phosphorus export. Binkley (2001) reported a mean concentration of 0.035 mg/L (median concentration 0.015 mg/L) for inorganic phosphate in forest streams in the Northeast.

The ecological experiment at Watershed 2 at Hubbard Brook in the White Mountains of New Hampshire produced dramatic losses of nitrates and nutrient mineral cations (Chapter 3, Box 3.2). Losses of this magnitude have not been caused by conventional types of forest management (i.e., with forest regeneration immediately after cutting) in the same region (Martin et al., 1986), in other areas of New England (Martin et al., 1984), and especially in more southerly regions with clay-rich, unglaciated soils (Corbett et al., 1978; DeWalle and Pionke, 1994; Reinhart, 1973) (Table 6.6).

A study of nine clearcut watersheds (2 to 24 ha) located within larger clearcuts (10 to 270 ha) in the White Mountains of New Hampshire showed that management caused disruption of the forest nutrient cycle, but that these changes were not long lasting (Martin et al., 1986). Following cutting, nitrate concentrations increased, with some streams reaching maximum levels of 23 to 28 mg/L between January and April of the second year after cutting (the U.S. Environmental Protection Agency [EPA] drinking water standard for nitrogen is 10 mg/L). Stream water nitrate concentrations in reference (untreated) watersheds in the same area ranged from 0.1 to 4.0 mg/L—generally lower in the summer and higher in the dormant season. Nitrate concentrations began to decline after the second year and approached reference values by the middle of the fourth year after cutting. In the winter of the second year following treatment, calcium concentrations had risen to between 6.5 and 8.1 mg/L compared to concentrations between 1.0 and 3.0 mg/L in reference and pretreatment streams.

Net losses of calcium from treated watersheds were estimated to be 29.2 kg/ha and 31.7 kg/ha in the first two years following treatment, compared with 12.5 kg/ha on the reference sites. Potassium (K^+) concentrations in stream water were three to four times higher than reference concentrations following cutting and remained higher throughout the study. The maximum stream water concentration for K^+ was 1.5 mg/L. Magnesium concentrations were stable.

Stream water pH decreased immediately after clearcutting two watersheds (average pH 5.5 versus 6.4) and remained lower during the first two years. Stream water pH remained similar to reference levels in the other seven clearcuts during this same time period and pH in all of the streams from harvested areas rose above reference levels in the third and fourth years. This may represent the combined effects of forest regeneration and calcium addition to streams. As noted in Chapter 4, an increase in stream water acidity, even for a brief period, is a concern because of the potentially damaging effects on fish and other stream biota. In another study (Martin et al., 1984) it was found that nutrient loss (nitrogen, potassium, calcium, magnesium) resulting from clearcuts in forests in Maine, Vermont, and Connecticut were "minor" compared to those in the White Mountains of New Hampshire and that losses that occurred were within the range of natural variation for uncut watersheds.

Although much of the research concerning nutrient loss has involved clearcuts, this is no longer the typical harvesting technique in much of the Northeast. Other silvicultural methods, shelterwood cuts, for example, in which the overstory is removed in two stages, allowing time for regeneration between cuts, patch retention, patch cuts, small group selection, and thinning are much more common. As would be expected, cuts that remove less than the entire overstory and leave forested

TABLE 6.6
Nitrate Concentrations in Stream Water from Treated Experimental Watersheds

Location	Soil type	Treatment	Treatment stream water nitrate concentrations (mg/L)	Control nitrate concentrations (mg/L)	Years following treatment in which maximum concentration occurred	References
Hubbard Brook, New Hampshire	Spodosols formed on glacial till and outwash	100% clearcut (no buffer) (9 watersheds, 2–24 ha) 70% clearcut (buffer)	17.6* (28) 8.8 (14.8)	0.2	2	Martin and Pierce (1980), Martin et al. (1986)
Fernow	Silt loams	13.8 ha 100% clearcut buffer strips	0.18 (growing season) 0.49 (dormant season) (1.42)	0–0.25	Second dormant season after treatment (1.5 years)	Aubertin and Patric (1974), Corbett et al. (1978)
Leading Ridge	Silt loams and stony loams	44.5 ha of 104 ha watershed clearcut, buffer strips	0.28 (3.5)	0.06 (1.5)	2 2	Lynch and Corbett (1990, 1994)
Coweeta	Gravelly and sandy loams	59 ha commercial clearcut/cable logged	(0.15)	0.005	3	Swank (1988)
Marcell	Clay loams	25.3 ha Clearcut	0.16	0.12	Measured one year after treatment	Verry (1972)

Mean annual values for the year of greatest concentration following treatment; maximum values are given in parentheses.

* mean of 9 treated watersheds.

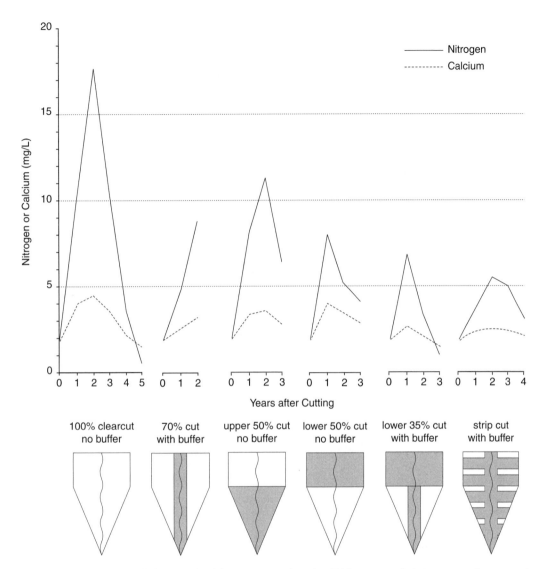

FIGURE 6.7 Mean annual nitrate and calcium concentrations (mg/L) in streams draining uncut, clearcut, and partially cut watersheds. Inserts show configurations of partially clearcut watersheds. In all cases water was sampled at the base of the watershed. (Martin and Pierce, 1980; reprinted with permission of the Society of American Foresters.)

riparian buffers cause smaller increases in stream water concentrations of nitrate and calcium (Martin and Pierce, 1980) (Figure 6.7).

Studies conducted at the Leading Ridge Experimental Forest in the headwaters of the Chesapeake Bay watershed examined nitrate export from managed forestland (DeWalle and Pionke, 1994; Lynch and Corbett, 1994). Limiting nutrient inflow to protect the Chesapeake Bay ecosystem and fisheries has been a major concern in this region for many years (Boesch et al., 2001; National Oceanic and Atmospheric Administration, 2005). An analysis of 20 years of streamflow and nitrate concentration data from the Leading Ridge Experimental Forest showed that a commercial clearcut (using BMPs) increased nitrate concentrations in streamflow for only 2 years following harvest due to rapid revegetation of the site. The highest losses occurred in the second year following clearcutting on Leading Ridge Watershed 3, when mean annual stream water concentration reached 0.28 mg/L compared to a pretreatment concentration of 0.04 mg/L. Nitrate export from the site never exceeded

0.54 kg/ha/yr. Researchers concluded that timber harvesting would not appreciably increase nitrate loading to Chesapeake Bay. Of far greater concern are inputs from excess fertilizer applied to agricultural lands, urban wastewater, and atmospheric deposition.

Similar experiments at Marcell Experimental Forest in north-central Minnesota found no detectable change in the concentration of any stream water constituents, including nitrate, phosphate, calcium, potassium, and magnesium, after clearcutting. Possible factors responsible for the lack of appreciable change in water quality following clearcutting in Minnesota include (1) climate—north-central Minnesota has a cooler climate and therefore lower decomposition rates than other sites in the East; (2) topography—leaching is far less likely on sites with minimal slopes; and (3) the availability of wetland storage sites—nutrients released from uplands may be assimilated by mosses and other plants in depressional wetlands (peat bogs) rather than being released to outlet streams (Verry, 1972).

In conclusion, the highest rates of nitrate loss following timber harvesting (clearcutting) in the Northeast appear to occur in the thin, sandy, podzolic soils of the White Mountains of New Hampshire. A recent comprehensive survey of nitrogen and phosphorus concentrations in forest streams throughout the United States confirms this finding (Binkley, 2001:35). Following a review of 43 harvesting experiments, Binkley observed that streams with high nitrate concentrations after logging either had high nitrate concentrations before logging (e.g., sites with nitrogen-fixing alders) or were located on the Hubbard Brook Experimental Forest. While increases in stream water nitrate concentrations occurred in 30 of the harvesting experiments, only 4 increased to more than 0.5 mg/L. Even in the White Mountains, when clearcutting is followed by natural revegetation, the rate of nutrient loss returns to preharvest levels within 3 to 4 years.

Nutrient losses are far lower at sites where soils contain more clay or organic matter and have greater cation exchange capacity. Clay soils are primarily found at floodplain and glacial lake bottom sites and south of the glacial boundary (Barnes et al., 1998; Corbett et al., 1978; Martin et al., 1984; Reinhart, 1973). Swank (1988) attributed low nutrient losses following forest harvesting to the "resistance and resilience" of the mixed hardwood forests at Coweeta. Within 3 years of the clearcutting on Coweeta Watershed 7, the net primary production (NPP) of regenerating trees and shrubs on that site was equal to 60% of the NPP of the original mature hardwood forest. Nutrients released by clearcutting and decomposition of the tops, branches, and foliage left onsite were quickly taken up and stored in the tissue of these rapidly growing early successional species. It was estimated that the amount of nitrogen, phosphorus, potassium, magnesium, and calcium stored in new vegetation was 29% to 44% of that of the mature forest within the first year following cutting.

6.6   EFFECTS OF FOREST CANOPY DISTURBANCE ON STREAM ECOSYSTEMS

As discussed in Chapters 4 and 5, the environment for aquatic biota (algae, plants, macroinvertebrates, and fish) is a product of streamflow, suspended and deposited sediment, availability of coarse woody debris, water temperature, and water chemistry. Forest management can cause changes in stream ecosystems when management activities substantially alter one or more of these parameters.

Irland (1999) noted the environmental effects of 19th century logging on channel habitat in Maine that persist to this day:

> Not only did the drives simply occupy the riverway for part of the year, the logs damaged riparian vegetation. The waterway modifications disrupted fish habitat and migrations. The bark and fiber, plus the "sinkers," covered the bottoms of streams and piled up in deadwaters and behind dams. Some of this material is still there, stifling habitat for fish, aquatic insects, and plants.

Current timber harvesting practices, while still causing changes in forest streams, cause far less disturbance. Binkley and Brown (1993) concluded that timber harvesting generally increases stream ecosystem productivity, while species diversity tends to decline. Cutting riparian trees increases stream temperature and solar radiation reaching the stream channel. If nutrient export increases as well, these conditions can fuel increased growth of periphyton (algae and diatoms) and plants, providing increased food supplies for the rest of the food web, macroinvertebrates, and fish. Because the nutrients are consumed within the stream channel (in-stream processing), nutrient concentrations measured at downstream stations may not reflect the total nutrient loss from the harvest site. These changes will benefit some species at the expense of others, altering the competitive dynamics in the aquatic community.

Hawkins et al. (1982) found that removing riparian canopies caused an increase in the abundance of invertebrates in all trophic guilds—collectors/gatherers, filter feeders, shredders, and predators. Swank et al. (2001) sampled and measured stream invertebrates following a commercial clearcut in the forests at the Coweeta Hydrologic Laboratory. Invertebrate diversity increased immediately following the clearcut, accompanied by a shift in functional feeding groups (increased grazers and reduced shredders). Taxa abundance increased in steep gradient, bedrock, and moss habitats and decreased in low gradient, sand, and pebble habitats. Sixteen years after the clearcut, invertebrate abundance was three times higher than that in a control stream. Current trends in taxa abundance indicate that the stream community is gradually returning to pretreatment conditions.

Culp and Davis (1983) conducted a study of the effects of logging on macroinvertebrate stream communities in Vancouver, British Columbia, Canada. This experiment compared the effects of two logging treatments. At one site, riparian vegetation was totally removed. Woody debris in the streams was removed as well. At the other site only merchantable timber was removed from the riparian buffer; large woody debris in the stream channel was not disturbed. Macroinvertebrate sampling was conducted from 1974 to 1976 to establish prelogging densities and community compositions at both sites as well as at an unlogged control site. Following timber harvesting, macroinvertebrates were sampled from 1977 to 1980. Significant declines in macroinvertebrate densities occurred where the riparian buffer was removed (Table 6.7).

Culp and Davis (1983) concluded that declines in macroinvertebrate populations were largely caused by increases in stream sediment. The removal of streamside vegetation did not cause an increase in periphyton biomass. Periphyton production at this site was phosphorus limited and phosphorus input and concentrations changed little following logging. Thus, while light increased at this site, a lack of nutrients prevented an increase in the periphyton food supply. There was a significant increase in fine (less than 9 mm) sediment particles in the stream channel at both sites. Streambanks remained stable at the site where the partial 10 m wide riparian buffer was maintained and in-channel large woody debris dams remained.

TABLE 6.7
The Influence of Logging and the Presence or Absence of Riparian Buffers on Benthic Macroinvertebrates in Three Forest Streams, British Columbia, Canada

Treatment	Logged (riparian buffer protection)		Logged (riparian buffer removed)		(Control, not logged)	
	Pre-logging	Post-logging	Pre-logging	Post-logging	Pre-logging	Post-logging
Macroinvertebrate densities (no./0.1 m^2)	899	707	1060	529	497	503

Culp and Davis (1983).

6.7 BEST MANAGEMENT PRACTICES

6.7.1 DEVELOPMENT OF BMPs

Best management practices are recommended methods designed to protect forest sites from erosion and forest streams from contamination by sediment and nutrients. Many of these management techniques were first developed during the 1950s, with implementation beginning during the 1960s (Ice and Stuart, 2001). The legislative mandate began in 1972 with the passage of the Federal Water Pollution Control Act Amendments (PL 92-500), also known as the Clean Water Act (CWA). The CWA identified forestry as a nonpoint source of pollution and required states to "develop area wide (watershed or regional) water quality management plans" (Ice et al., 1997). The Water Quality Act of 1987 cited examples of specific BMPs used in the management of nonpoint source pollution (Lynch and Corbett, 1990). The USDA Forest Service has promoted BMP programs that emphasize the development of principles and practices suited to local conditions. At the present time, all states advocate the use of BMPs (Ice and Stuart, 2001). State BMP programs in the Northeast and throughout the country differ both in the specific practices recommended and in the methods of enforcement (Irland, 2001).

In general, BMP protocols include guidelines for the following components or activities (Martin and Hornbeck, 1994. Reprinted with permission of the Society of American Foresters.):

1. Steepness of truck roads and skid trails.
2. Water control devices, including the construction and spacing of waterbars and broad-based dips.
3. Culverts, including type, size, spacing, and installation recommendations, such as ditch construction when seeps are involved.
4. Buffer strips of trees along stream channels to provide shade and a source of woody debris.
5. Filter strips of undisturbed land between disturbed sites and streams to trap sediment.
6. Filter devices, such as hay bales, to prevent sediment from flowing from a road or skid trail directly into a stream.
7. Minimizing the addition of logging slash and tree tops to streams and removal of this material when necessary.
8. Roads or skid trail stream crossings using culverts or bridges.
9. Location of landings and control of petroleum products and human waste.
10. Closing of logging operations during unfavorable weather.
11. Closing and rehabilitation of roads, landings, and skid trails, which often involves grooming, seeding, and mulching.

There are two primary questions in relation to BMPs: (1) Are they effective in protecting water quality, and (2) do loggers routinely follow BMP regulations and recommendations, in other words, what are typical rates of compliance?

6.7.2 EFFECTIVENESS OF BMPs

A substantial number of research studies have concluded that management practices are available to mitigate the impacts of timber harvesting on streamflow and water quality (these studies include the Fernow timber harvesting study [Reinhart et al., 1963] and Martin and Hornbeck [1994] discussed earlier in this chapter). A paired watershed study at the Leading Ridge Experimental Forest was used to evaluate the effectiveness of BMPs for the ridge and valley region in central Pennsylvania (Lynch and Corbett, 1990; Lynch et al., 1985). Leading Ridge Watershed 3 (LR3) was clearcut using BMPs. Data on streamflow, turbidity/sediment, temperature, and nutrient concentrations were compared to similar measurements in an undisturbed control watershed (LR1) over a 15 year period following the timber harvesting operation.

There was a significant increase in streamflow during the growing season in the first year following the clearcut. However, 4 years after the cut, no statistically significant increase (or decrease) was evident in either annual or seasonal streamflow. Turbidity at LR3 was slightly higher than predicted relative to pretreatment conditions throughout the data collection period. One possible cause of increased turbidity may have been gaps in the riparian buffer along intermittent streams. Increases in streamflow caused intermittent streams to become perennial streams. Trees that remained following cutting in these areas were exposed to higher wind velocities and fell into the stream channels, forming debris dams. Debris dams can interfere with the direction of flow, forcing water into new areas that are more vulnerable to erosion than established channels.

In response to these observations, the state of Pennsylvania modified the BMP regulations to require continuous buffers along perennial and intermittent stream channels (Lynch and Corbett, 1990). Total sediment yield based on water samples was estimated at 43 to 56 kg/ha/yr during the first 2 years following harvest, less than Patric's earlier (1976) estimates for undisturbed and well-managed forest land in the East. There was a slight increase in average daily stream water temperature ($\leq 1.6°C$ [$3°F$]), again attributed to gaps in the riparian buffers along formerly intermittent streams. The temperature returned to normal as vegetation became established along new stream channels. Changes in water chemistry were statistically significant, but still small and within water quality standards. Stream pH remained between 6.5 and 7.1 on the treated watershed. Stream water nitrate concentration increases were statistically significant. The highest mean annual concentration was 0.4 mg/L in the first year after cutting (well below the EPA drinking water standard of 10 mg/L).

Kochenderfer et al. (1997) examined the effectiveness of West Virginia's BMP guidelines developed between 1979 and 1989. The study compared streamflow and water quality data following harvesting in one watershed (the Haddix site) with data from a control watershed in the Fernow Experimental Forest. Pretreatment data was collected for 3 years and the study analyzed 8 years of posttreatment data.

Slopes on the Haddix site averaged 40%, so accelerated erosion was a distinct possibility. Special care was taken to plan forest roads in order to minimize environmental impacts. The entire road system was completely laid out on the ground before the harvesting operation began (Figure 6.8 [bottom]). Despite the steep slopes, roads were planned so that most truck roads had grades of less than 5%, with a maximum of 8%. Most of the skid roads had grades of less than 10%, with a maximum of 20% in a few short stretches. Landings were located on dry sites. A slope distance of 30 m was maintained between the roads and the streams except at stream crossings, where metal culverts were installed. Landings were located 46 m from any stream. Other protective measures included seeding critical areas, scattering slash on roadfills immediately following truck road construction, and constructing waterbars on skid roads during the logging operation. The roads were limed, fertilized, and seeded following completion of the logging operation. All trees larger than 36.5 cm in diameter were cut on 96% of the 39 ha watershed. Although harvesting was permitted in riparian areas, many residual trees and understory shrubs remained.

There were measurable changes in streamflow quantity and quality following the timber harvest. Individual peak flows and total stormflow increased significantly during the growing season. There was, however, no change in annual peak flows or total annual stormflow volumes. Again, the authors concluded that the growing season increase was due to a reduction in evapotranspiration, not soil compaction or disturbance. The largest increase in sediment export, relative to the control watershed, occurred during the active logging period. The estimated export (123.3 kg/ha) during that time was within Patric's (1976) estimated norms for erosion in the East. There were no statistically significant increases in the second year postharvest. Water chemistry data indicated that there was no significant change in pH after the logging operation; pH measurements ranged from 6.4 to 6.7 throughout the period of the study, both pre- and post-treatment. Nitrate concentrations in streamflow increased after the harvest. The maximum annual concentration (2.5 mg/L) occurred in the first year following

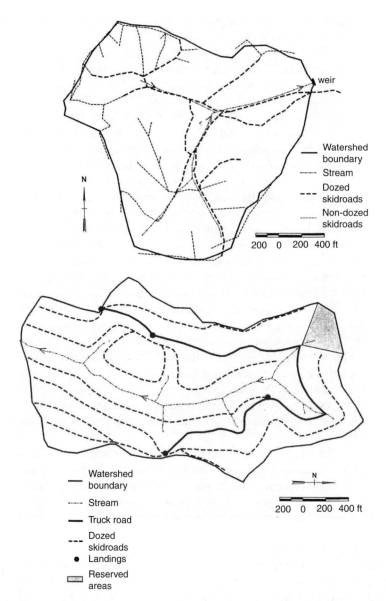

FIGURE 6.8 (Top) Unplanned road system in Watershed 1, Fernow Experimental Forest, West Virginia. No measures were taken to protect soil and water resources. Watershed 1 was commercially clearcut between May 1957 and September 1958. (Kochenderfer and Hornbeck, 1999; courtesy of the USDA Forest Service.) (Bottom) Road system in the Haddix Watershed, located 4 miles from the Fernow Experimental Forest, West Virginia. The road system conformed to West Virginia's best management practice standards. (Kochenderfer and Hornbeck, 1999; courtesy of the USDA Forest Service.)

the harvest then declined. Nitrate concentrations in streams from the control watershed were consistently higher than those on the treated watershed, evidence of the natural variation noted earlier. There was no significant increase in stream water temperature.

In a subsequent article (Kochenderfer and Hornbeck, 1999) compared the logging procedures and streamflow and water quality data from the Haddix watershed to the commercial clearcut completed in 1957 on Fernow Watershed 1 (Reinhart et al., 1963) discussed earlier (Figure 6.8 [top]). Road grades and the proximity of roads to streams were considered to be the primary factors that determined the different outcomes from these two timber harvesting operations. Forty-six percent of the bulldozed skid roads on Fernow Watershed 1 were built with grades of 21% or greater, compared to 2% of comparable roads in the Haddix watershed. Sixty-six percent of the skid roads on Fernow Watershed 1 were within 30 m of a stream, compared to 3% of the roads in the Haddix watershed. Sediment yield results are summarized in Table 6.8. Information from other BMP studies is summarized in Table 6.9.

6.7.3 BMP IMPLEMENTATION AND COMPLIANCE

Based on data collected through 1997 from 12 northeastern states (Pennsylvania, West Virginia, Maryland, Delaware, New York, New Jersey, Connecticut, Rhode Island, Massachusetts, Vermont, New Hampshire, and Maine), Ice and Stuart (2001) concluded that most state government agencies consider forest management to be a "modest" or "insignificant" source of water pollution relative to other nonpoint sources (i.e., agricultural and urban land use). Because of this, relatively few staff and little funding are devoted to the enforcement and monitoring of BMPs. Some states (e.g., Delaware, New Hampshire, New Jersey, Minnesota, Missouri, and Vermont) support logger education and rely on voluntary compliance or a combination of voluntary compliance and enforcement to promote the use of BMPs. Violations that involve wetlands are dealt with under wetland protection legislation. Other states, such as Massachusetts and Connecticut, have comprehensive forest cutting practice regulations, forester licensing, and logger certification enforced by state agencies and state-employed foresters (Table 6.10). In all, this report suggests that there is little research regarding compliance with and the effectiveness of BMP implementation. Studies that have been done seem to show widespread acceptance of some basic principles such as stream buffer zones, as well as changes in attitudes and on-the-ground performance that have resulted in general improvements.

A study in Vermont (Brynn and Clausen, 1991) evaluated timber harvesting operations to determine the degree of compliance with Vermont's acceptable management practices (AMPs) and erosion and stream sedimentation at various sites. Foresters completed 1 day evaluations of 78 sites selected randomly from 754 logging operations between August 1, 1987, and August 1, 1988. Typical of BMP programs nationwide, maintaining riparian buffers is a high priority of Vermont's AMPs, and more than 90% of the harvesting operations had complied with these guidelines. Increased stream sedimentation was evident at 46% of the operations with streams and this was attributed to widespread failure to meet AMP recommendations for truck roads and skid trails: "Stream crossings appeared to be the primary source of sedimentation from Vermont timber harvesting operations." Despite this, Brynn and Clausen (1991) concluded that soil erosion and stream water sedimentation were minimal overall and that it was important to avoid regulations that might seem overly restrictive and thereby discourage compliance.

TABLE 6.8
Annual Suspended Sediment Yields with and without BMPs

	Sediment yields	
	Fernow WS-1 (1957) (commercial clearcut)	Haddix (1986–1987) (with BMPs)
During logging operation	3228 kg/ha	123 kg/ha
First year after logging	323 kg/ha	77 kg/ha
Second year after logging*	8 kg/ha	58 kg/ha

* Modern sampling techniques would probably have produced values for WS-1 more comparable to Haddix.

Kochenderfer and Hornbeck (1999).

TABLE 6.9
Summary of Stream Water Changes Following Timber Harvesting Using BMPs

Location	Type of cut	Streamflow	pH	Temperature	Annual maximum sediment loss[a]	Nitrate (maximum mean annual concentration)	References
Pennsylvania: Leading Ridge Experimental Forest – LR3	Commercial clearcut	Statistically significant increases years 1–3 (growing season)	NA	<3°C	56 kg/ha* (second year after logging)	0.40 mg/L* (first year after harvest)	Lynch and Corbett (1990)
West Virginia: Haddix Watershed	Clearcut to 35.5 cm stump diameter	No statistically significant increases	No change	No change	123 kg/ha during logging; 77 kg/ha first year after logging	2.52 mg/L (first year after harvest)	Kochenderfer et al. (1997), Kochenderfer and Hornbeck (1999)
New Hampshire: Hubbard Brook Experimental Forest – Watershed 5	Whole-tree clearcut, 96% of biomass removed	NA	NA	NA	208 kg/ha* third year after logging	NA	Martin and Hornbeck (1994)

* Statistically significant increases.
[a] Compare to Patric's estimate of annual sediment loss from eastern forests (112 to 224 kg/ha).

TABLE 6.10
Summary of BMP Implementation in Northeast States

State	Estimated number of logging operations per year	Regulatory type	Compliance (%)	Monitoring and research
Connecticut	600	Regulatory		University of Connecticut, Storrs
Delaware	150	Voluntary plus enforcement		
Illinois	NA	Voluntary		
Indiana	NA	Voluntary plus enforcement	83	
Maine	~8,000	Combination	67	Briggs et al. (1998) (compliance)
Maryland	2,000	Regulatory	82	Koehn and Grizzel (1995) (compliance)
Massachusetts	800	Regulatory	100	Thompson and Kyker-Snowman (1989)
Michigan		Voluntary plus enforcement	82	
Minnesota		Voluntary plus enforcement	75	
Missouri		Voluntary		
New Hampshire	5,000	Voluntary plus enforcement		
New Jersey	240	Voluntary plus enforcement		
New York	6,000	Combination	Field evaluation found compliance was low	Neversink Watershed Research Program
Ohio		Voluntary plus enforcement		
Pennsylvania	12,000	Regulatory		Lynch and Corbett (1990); Leading Ridge
Rhode Island	200	Combination	95	
Vermont	2,000	Voluntary plus enforcement	90 for stream avoidance and stream buffers; soil erosion practices on truck roads and skid roads commonly failed BMP requirements	Brynn and Clausen (1991), Wilmot and Scherbatskoy (1997)
West Virginia	3,000	Regulatory	Compliance varies based on specific management recommendation; highest (70–90) for grades and construction of roads and landings; lowest (19–70) for seeding and mulching	Kochenderfer et al. (1997), Fernow
Wisconsin		Voluntary plus enforcement	85	

Ice and Stuart (2001:Appendix D) and Irland (2001).

Briggs et al. (1998) evaluated BMP compliance in Maine from an inspection of 120 randomly selected sites that were logged between January 1, 1993, and December 31, 1994. They found that compliance varied in relation to the type of BMPs. The environmental and economic benefits of carefully planning the location of roads and skid trails were widely recognized. Compliance for BMPs associated with planning new roads was greater than 90%. Some roads with stormwater control and sediment problems were established years or decades ago and it would be expensive and sometimes create a greater disturbance to bring them up to current standards. Increases in stream sediment that could be attributed to not using one or more BMPs were found at 34 of 120 sites.

6.8 SUMMARY AND CONCLUSIONS

6.8.1 Logging History, Conservation, and Forest Research

Logging in the 1800s and early 1900s cleared the landscape of "virgin timber" and damaged the rivers used to transport logs from forest to market. The abuses of the early timber and pulpwood industries inspired public support for forest conservation. In the early 20th century, U.S. Forest Service scientists and administrators moved to establish forest reserves and research stations. One of the primary tasks of these experimental forests has been to investigate the link between forest management and streamflow and water quality. The longest continually operating forest research site in the United States is the Coweeta Hydrologic Laboratory and NSF-LTER site in western North Carolina. In the northeastern area, prominent research forests include the Hubbard Brook Experimental Forest and NSF-LTER in the White Mountains of New Hampshire, the Fernow Experimental Forest in West Virginia, the Leading Ridge Watershed Research Unit in Pennsylvania, and the Marcell Experimental Forest in Minnesota (Ice and Stednick, 2004).

6.8.2 Changes in Streamflow, Sediment, Channel Form, and Temperature Following Forest Canopy Disturbance

Timber harvesting increases the magnitude of streamflow in forest streams because evapotranspiration is reduced when forest vegetation is removed. Increases in streamflow are measurable when more than 20 to 30% of the forest canopy is removed. Streamflow returns to reference conditions as the forest regenerates, usually within 3 to 10 years after cutting. Streamflow may decrease in relation to reference (preharvest) conditions when actively growing early successional tree species replace a mature forest on the site (Hornbeck et al., 1997b).

Suspended particulate matter concentrations may increase in streams during and after timber harvesting. Erosion and sediment delivery is largely related to the planning, design, and management of forest roads and the management of riparian buffers. Sediment loading can be minimized if roads are designed to avoid steep slopes, riparian buffers are protected, and road-stream crossings are carefully designed and maintained. Increased stream sediment may also be the result of channel instability and erosion caused by prolonged increases in streamflow (Martin and Hornbeck, 1994).

Water temperature increases (typically 4°C to 8°C) when trees and other vegetation that shade the stream are removed or damaged. Maintaining a riparian forest buffer virtually eliminates this effect (Brown and Binkley, 1994).

6.8.3 Changes in Water Chemistry Following Forest Canopy Disturbance

6.8.3.1 Nutrients (Nitrogen, Phosphorus, and Mineral Nutrient Cations)

Phosphorus is tightly held in forest soils. Increases in phosphorus loading following timber harvesting are rare (Brown and Binkley, 1994). Nitrate and mineral cation losses directly attributable to timber harvesting are usually small, or in some cases not detectable, when forest vegetation is allowed to regrow immediately following the cut. The greatest increases have occurred in research studies involving clearcutting in the White Mountains of New Hampshire. Stream water nutrient concentrations typically return to preharvest levels within 4 years after cutting (Binkley, 2001; Martin et al., 1986).

6.8.4 Effects of Forest Canopy Disturbance on Aquatic Biota

Timber harvesting has the potential to alter streamflow, sediment, nutrients, temperature, and light resources within the stream channel. The supply of these resources defines stream habitat for algae, aquatic plants, macroinvertebrates, and fish. The response of the stream community is variable, depending on the extent of the disturbance and the physical properties (soils, topography) and nutrient resources of the site. In some cases (Hawkins et al., 1982; Swank et al., 2001), increased resources following riparian removal brought about an increase in production and macroinvertebrate abundance. In others, removal of riparian trees and woody debris, and the associated increases in sediment, reduced macroinvertebrate and fish abundance and diversity (Culp and Davis, 1983; Jones et al., 1999; Chap. 4, this volume). It is likely that any change in the aquatic environment will favor some species over others and result in changes in community composition (Swank et al., 2001).

6.8.5 Best Management Practices

Best management practices are forest management practices designed to prevent the degradation of forest lands and water quality during and after timber harvesting. In general, BMPs address the location and management of forest roads and landings with special attention to road-stream crossings and the management and protection of riparian buffers. A number of studies have shown that BMPs are effective at minimizing increases in streamflow magnitudes, sediment concentrations, and temperature associated with timber harvesting (Lynch and Corbett, 1990; Kochenderfer et al., 1997). BMP implementation and enforcement programs vary from state to state. While Massachusetts and Connecticut have regulatory programs, most states rely on programs of logger education and voluntary enforcement (Ice and Stuart, 2001).

6.8.6 Forests, Water, and 21st Century Challenges

The watershed studies presented in this chapter were designed to quantify the effects of natural and anthropogenic disturbance on forests and water. In the early 1900s, many people still needed to be convinced that forests were an essential and renewable natural resource rather than an industrial raw material that had to be sacrificed in the name of progress. The paired watershed studies conducted by the U.S. Forest Service provided the empirical evidence needed to explain and justify investments in forest conservation and resource management throughout the 20th century. The early studies (1930s to 1950s) at Coweeta and Fernow, for example, clearly demonstrated the differences between exploitive logging and "scientific" forest management. The application of scientific forestry and sound professional judgment led to enormous decreases in ecological impact and modest increases in economic return. Obviously there was no valid excuse or incentive—economic or otherwise—for "cut and run" logging. Later studies (1960s to 1990s) showed the minimal short-term impact of silviculture when (1) roads and skid trails are carefully designed, (2) harvesting operations are controlled and supervised, and (3) post-harvest stabilization is undertaken. Other watershed studies and regional surveys showed that the use of BMPs such as riparian forest buffers, stormwater and soil erosion controls, and stable road-stream crossings could avoid or minimize adverse effects on streamflow and water quality. In other words, sustainable forest management can (and should) be fully compatible with aquatic ecosystem and source water protection (Barten et al., 1998). At a minimum, this approach enables landowners to derive an income from the land while maintaining forest cover—the most desirable watershed condition. In other cases, because our forests are the product of centuries of land use, forest management is sometimes needed to restore and enhance the diversity of tree species, age classes, and vertical structure. Silvicultural methods such as patch cuts, small group selection, and seed tree cuts in conifer plantations can be used to guide homogeneous, even-aged stands back to a more resistant and resilient (with respect to disturbance) mixed species, uneven-aged condition. A recent National Research Council study (2000) endorsed the idea of a "working landscape" or a mosaic of protected areas, carefully managed forests, farms, and other open space as the ecological counterweight to sprawl and unplanned development. This review of New York City's watershed management approach emphasized the critical function of forests as the "first barrier" in the time-tested multiple barrier approach to source water protection.

The knowledge and experience generated by the forest hydrology and watershed management research summarized in this chapter and by Ice and Stednick (2004) will be needed to address the threat posed by the conversion of forests to residential and commercial land use. The conversion of forest land to other uses is not as obvious as the blighted landscape that inspired the Conservation Movement a century ago, yet as Chapters 8 and 9 will show, the hydrologic effects can be just as serious and pervasive. Furthermore, they are more likely to be permanent. A century ago, forests grew back on abandoned marginal farmland and worn out pastures. The water balance and the biotic integrity of large areas were gradually restored. It is also noteworthy that during the 20th century, the population of the United States has increased from about 90 million to nearly 300 million people.

A recent study entitled "Forests on the Edge: Housing Development on America's Private Forests" quantified conditions in 2000, then examined the population and development projections for 2030 across the 48 conterminous states (Stein et al., 2005). The authors noted that between 1982 and 1997, approximately 4,200,000 ha of forest—an area almost twice the size of Adirondack Park in northern New York—was lost to development nationwide. If current patterns and trends continue, another 17,900,000 ha of forest will be converted to residential and commercial land uses by 2030. The 20 state Northeastern Area had 69,000,000 ha of forest in 1997 (Alig et al., 2003); 80% was privately owned and potentially vulnerable to conversion. Stein et al. (2005) used a geographic information system (GIS) overlay process on 1026 large watersheds (eight-digit hydrologic unit code [HUC] delineated by the U.S. Geological Survey) to characterize the effects of forest conversion during the next 30 years. Six of the 15 watersheds with the greatest proportion of forest conversion are located in the Northeast; all 15 are along the East Coast (Table 6.11). (Three of them are in Maine—the Androscoggin, Kennebec, and Penobscot Rivers.) This outcome would threaten the survival of endangered Atlantic salmon (National Research Council, 2002, 2004). This rather ominous result, and the research described in this and later chapters, should forewarn us about the serious and costly results of business as usual. Chapter 10, "Watershed Management Implications," presents alternatives and highlights approaches that could prevent a repetition of the circa 1900 floods, droughts, sedimentation, and aquatic ecosystem degradation that demonstrated what Hugh Hammond Bennett (1939) termed our "astonishing improvidence."

TABLE 6.11
Northeastern Watersheds with Projected Major Increases in Housing Density

Watershed	Projected forest loss (ha)	State	National rank ($n = 1026$)
Lower Penobscot	125,500	Maine	1
Little Kanawha	91,400	West Virginia	10
Middle Hudson	89,600	New York and Massachusetts	11
Lower Androscoggin	86,500	Maine and New Hampshire	13
Lower Kennebec	85,000	Maine	14
North Branch Potomac	84,700	Maryland, Pennsylvania, and West Virginia	15

Modified from Stein et al. (2005).

REFERENCES

Aber, J.D., Ollinger, S.V., Federer, C.A., Reich, P.B., Goulden, M.I., Kicklighter, D.W., Melillo, J.M., and Lathrop, R.G., Predicting the effects of climate change on water yield and forest production in the northeastern United States, *Climate Res.*, 5, 207–222, 1995.

Adams, M.B., Edwards, P.B., Wood, F., and Kochenderfer, J.N., Artificial watershed acidification on the Fernow Experimental Forest, USA, *J. Hydrol.*, 150, 505–519, 1993.

Alig, R.J., Plantinga, A.J., Ahn, S., and Kline, J.D., Land use changes involving forestry in the United States: 1952 to 1997, with projections to 2050, General Technical Report PNW-GTR-587, U.S. Department of Agriculture, Forest Service, Pacific Northwest Research Station, Portland, OR, 2003, http://www.fs.fed.us/pnw/publications/gtrs2003.shtml; accessed July 2006.

Allan, J.D., *Stream Ecology: Structure and Function of Running Waters*, Chapman & Hall, London, 1995.

Anderson, H.W., Hoover, M.D., and Reinhart, K.G., Forests and water: effects of forest management on floods, sedimentation, and water supply, General Technical Report PSW-18, USDA Forest Service, Washington, D.C., 1976.

Aubertin, G.M. and Patric, J.H., Water quality after clearcutting a small watershed in West Virginia, *J. Environ. Qual.*, 3, 243–249, 1974.

Baker, M.B., Jr., Effects of ponderosa pine treatments on water yield in Arizona, *Water Resourc. Res.*, 22, 67–73, 1986.

Barnes, B.V., Zak, D.R., Denton, S.R., and Spurr, S.H., *Forest Ecology*, 4th ed., John Wiley & Sons, New York, 1998.

Barten, P.K., Modeling streamflow from headwater catchments in the northern Lake States, Ph.D. dissertation, University of Minnesota, Minneapolis, 1988.

Barten, P.K., Kyker-Snowman, T., Lyons, P.J., Mahlstedt, T., O'Connor, R., and Spencer, B.A., Massachusetts: managing a watershed protection forest, *J. For.*, 96(8), 10–15, 1998.

Beaver Creek Watershed, http://ag.arizona.edu/OALS/watershed/beaver/; accessed July 2006.

Bennett, H.H., *Soil Conservation*, McGraw-Hill, New York, 1939.

Beschta, R.L., Forest hydrology in the Pacific Northwest: additional research needs, *J. Am. Water Resourc. Assoc.*, 34, 729–741, 1998.

Binkley, D., Patterns and processes of variation in nitrogen and phosphorus concentrations in forested streams, Technical Bulletin 836, National Council for Air and Stream Improvement, Research Triangle Park, NC, 2001, http://www.ncasi.org//Publications/Detail.aspx?id=46; accessed July 2006.

Binkley, D. and Brown, T.C., Forest practices as nonpoint sources of pollution in North America, *Water Resourc. Bull.*, 29, 729–740, 1993.

Binkley, D., Ice, G.G., Kaye, J., and Williams, C.A., Nitrogen and phosphorus concentrations in forest streams of the United States, *J. Am. Water Resourc. Assoc.*, 40, 1277–1292, 2004.

Boesch, D.F., Brinsfeld, R.B., and Magnien, R.E., Chesapeake Bay eutrophication: scientific understanding, ecosystem restoration, and challenges for agriculture, *J. Environ. Qual.*, 30, 303–320, 2001.

Bormann, F.H. and Likens, G.E., *Pattern and Process in a Forested Ecosystem: Disturbance, Development and the Steady State, Based on the Hubbard Brook Ecosystem Study*, Springer-Verlag, New York, 1994 [first published in 1979].

Bormann, F.H., Likens, G.E., Siccama, T.E., Pierce, R.S., and Eaton, J.S., The export of nutrients and recovery of stable conditions following deforestation at Hubbard Brook, *Ecol. Monogr.*, 44, 255–277, 1974.

Bosch, J.M. and Hewlett, J.D., A review of catchment experiments to determine the effect of vegetative changes on water yield and evapotranspiration, *J. Hydrol.*, 55, 3–23, 1982.

Briggs, R.D., Cormier, J., and Kimball, A.J., Compliance with forest BMPs in Maine, *Northern J. Appl. For.*, 15(2), 57–68, 1998.

Brooks, K.N., Ffolliott, P.F., Gregerson, H.M., and DeBano, L.F., *Hydrology and the Management of Watersheds*, Iowa State University Press, Ames, IA, 2003.

Brown, T.C. and Binkley, D., Effect of management on water quality in North American forests, General Technical Report RM-248, USDA Forest Service, Washington, D.C., 1994.

Brynn, D.J. and Clausen, J.C., Postharvest assessment of Vermont's acceptable silvicultural management practices and water quality impacts, *Northern J. Appl. For.*, 8, 140–144, 1991.

Carlson, M. and Ober, R., The Weeks Act: how the White Mountain National Forest came to be, in *People and Place: Society for the Protection of New Hampshire Forests, The First 100 Years*, Conroy, R.G. and Ober, R. (eds.), Society for the Protection of New Hampshire Forests, Concord, NH, 2001.

Corbett, E.S., Lynch, J.A., and Sopper, W.E., Timber harvesting practices and water quality in the eastern United States, *J. For.*, 76, 484–488, 1978.

Coweeta Hydrologic Laboratory, http://coweeta.ecology.uga.edu (July, 2006).

Culp, J.M. and Davis, R.W., An assessment of the effects of streambank clear-cutting on macroinvertebrate communities in a managed watershed, *Can. Tech. Rep. Fish. Aquat. Sci.*, 1208, 1983.

DeWalle, D.R. and Pionke, H.B., Nitrogen export from forest land in the Chesapeake Bay region, in *Towards a Sustainable Coastal Watershed: The Chesapeake Experiment*, Hill, P. and Nelson, S. (eds.), Chesapeake Research Consortium, Edgewater, MD, 1994, pp. 649–655.

Dissmeyer, G.E. (ed.), Drinking water from forests and grasslands: a synthesis of the scientific literature, General Technical Report SRS-039, U.S. Department of Agriculture, Southern Research Station, Asheville, NC, 2000.

Douglass, J.E., The potential for water yield augmentation from forest management in the eastern United States, *Water Resourc. Bull.*, 19, 351–358, 1983.

Douglass, J.E. and Hoover, M.D., History of Coweeta, in *Forest Hydrology and Ecology at Coweeta*, Swank, W.T. and Crossley, D.A. (eds.), Springer-Verlag, New York, 1988, pp. 17–31.

Dunne, T. and Leopold, L.B., *Water in Environmental Planning*, W.H. Freeman, New York, 1978.

Edwards, P.J., Carnahan, D.L., and Henderson, Z., Channel cross-section and substrate comparisons among four small watersheds with different land-disturbance histories, in AWRA Symposium Proceedings Specialty Conference Wildland Hydrology, TPS-99-3, American Water Resources Association, Herndon, VA, 1999, pp. 217–218.

Eschner, A.R. and Satterlund, D.R., Forest protection and streamflow from an Adirondack watershed, *Water Resourc. Res.*, 2, 765–783, 1966.

Federer, C.A. and Lash, D., Simulated streamflow response to possible differences in transpiration among species of hardwood trees, *Water Resourc. Res.*, 14, 1089–1097, 1978.

Fernow Experimental Forest, http://www.fs.fed.us/ne/parsons/; accessed July 2006.

Gove, B., *Log Drives on the Connecticut River*, Bondcliff Books, Littleton, NH, 2003.

Graves, H., *Purchase of Land Under the Weeks Law in the Southern Appalachian and White Mountains*, USDA Forest Service, Washington, D.C., 1911.

Greeley, W.B., The relation of geography to timber supply, *Econ. Geogr.*, 1, 1–14, 1925.

Hawkins, C.P., Murphy, M.L., and Anderson, N.H., Effects of canopy, substrate composition, and gradient on the structure of macroinvertebrate communities in Cascade Range streams of Oregon, *Ecology*, 63, 1840–1856, 1982.

Henderson, F.M., *Open Channel Flow*, Macmillan, New York, 1966.

H. J. Andrews Experimental Forest, http://www.fsl.orst.edu/lter/; accessed July 2006.

Hibbert, A.R., Forest treatment effects on water yield, in *International Symposium on Forest Hydrology, Proceedings of a National Science Foundation Advanced Science Seminar held at The Pennsylvania State University, University Park, Pennsylvania, Aug. 29–Sept. 10, 1965*, Sopper, W.E. and Lull, H.W. (eds.), Pergamon Press, Oxford, 1967, pp. 527–564.

Hobbie, J.E. and Likens, G.E., Output of phosphorus, dissolved organic carbon, and fine particulate carbon from Hubbard Brook watersheds, *Limnol. Oceanogr.*, 18, 734–742, 1973.

Holbrook, S.H., *Holy Old Mackinaw: A Natural History of the American Lumberjack*, Macmillan, New York, 1938.

Hornbeck, J.W., Stormflow from hardwood forested and cleared watersheds in New Hampshire, *Water Resourc. Res.*, 9, 346–354, 1973.

Hornbeck, J.W. and Reinhart, K.G., Water quality and soil erosion as affected by logging in steep terrain, *J. Soil Water Conserv.*, 19, 23–27, 1964.

Hornbeck, J.W., Adams, M.B., Corbett, E.S., Verry, E.S., and Lynch, J.A., Long-term impacts of forest treatments on water yield: a summary for northeastern USA, *J. Hydrol.*, 150, 323–344, 1993.

Hornbeck, J.W., Bailey, S.W., Buso, D.C., and Shanley, J.B., Stream water chemistry and nutrient budgets for forested watersheds in New England: variability and management implications, *For. Ecol. Manage.*, 93, 73–89, 1997a.

Hornbeck, J.W., Martin, C.W., and Eager, C., Summary of water yield experiments at Hubbard Brook Experimental Forest, New Hampshire, *Can. J. For. Res.*, 27, 2043–2052, 1997b.

Hubbard Brook Experimental Forest, http://www.hubbardbrook.org/; accessed July 2006.

Ice, G.G. and Stednick, J.D., *A Century of Forest and Wildland Watershed Lessons*, Society of American Foresters, Bethesda, MD, 2004.

Ice, G.G. and Stuart, G.W., *State Nonpoint Source Pollution Control Programs for Silviculture Sustained Success, 2000 Progress Report*, National Association of State Foresters, Washington, D.C., 2001, http://www.stateforesters.org/reports.html#2003; accessed July 2006.

Ice, G.G., Stuart, G.W., Wade, J.B., Irland, L.C., and Ellefson, P.V., 25 years of the Clean Water Act: how clean are forest practices, *J. For.*, 95, 9–13, 1997.

Irland, L., *The Northeast's Changing Forest*, Harvard University Press, Cambridge, MA, 1999.

Irland, L., Forestry operations and water quality in the northeastern states: overview of impacts and assessment of state implementation of nonpoint source programs under the Federal Clean Water Act, Technical Bulletin 820, National Council for Air and Stream Improvement, Research Triangle Park, NC, 2001, http://www.ncasi.org//Publications/Detail.aspx?id=30; accessed July 2006.

Johnson, A.H., Bouldin, D.R., Goyette, E.A., and Hedges, A.M., Phosphorus loss by stream transport from a rural watershed: quantities, processes, and sources, *J. Environ. Qual.*, 5, 148–157, 1976.

Johnson, C.E., Romanowicz, R.B., and Siccama, T.G., Conservation of exchangeable cations after clear-cutting of a northern hardwood forest, *Can. J. For. Res.*, 27, 859–868, 1997.

Jones, E.B.D., III, Helfman, G.S., Harper, J.O., and Bolstad, P.V., Effects of riparian forest removal on fish assemblages in southern Appalachian streams, *Conserv. Biol.*, 13, 1454–1465, 1999.

Kittredge, J., *Forest Influences: The Effects of Woody Vegetation on Climate, Water, and Soil, with Applications to the Conservation of Water and the Control of Floods and Erosion*, McGraw-Hill, New York, 1948.

Kochenderfer, J.N. and Hornbeck, J.W., Contrasting timber harvesting operations illustrate the value of BMPs, in *Proceedings, 12th Central Hardwood Forest Conference, February 28–March 2, 1999, Lexington, KY*, Stringer, J.W. and Loftis, D.L. (eds.), General Technical Report SRS-24, U.S. Department of Agriculture, Forest Service, Southern Research Station, Asheville, NC, 1999, pp. 128–135.

Kochenderfer, J.N., Edwards, P.J., and Helvey, J.D., Land management and water yield in the Appalachians, in *Watershed Planning and Analysis in Action*, Riggins, R.E., Jones, E.B., Singh, R., and Rechard, P.A. (eds.), American Society of Civil Engineers, New York, 1990, pp. 522–532.

Kochenderfer, J.N., Edwards, P.J., and Wood, F., Hydrologic impacts of logging an Appalachian watershed using West Virginia's best management practices, *Northern J. Appl. For.*, 14, 207–218, 1997.

Koehn, S.W. and Grizzel, J.D., *Forestry Best Management Practices: Managing to Save the Bay. Assessment and Analysis Report on Forestry BMP Implementation in Maryland*, Maryland Department of Natural Resources, Annapolis, MD, 1995.

Leopold, L.B. and Maddock, T., Jr., *The Flood Control Controversy: Big Dams, Little Dams, and Land Management*, Ronald Press, New York, 1954.

Likens, G.E., Bormann, F.H., Pierce, R.S., Eaton, J.S., and Johnson, N.M., *Biogeochemistry of a Forested Ecosystem*, Springer-Verlag, New York, 1977 [2nd ed. published in 1995].

Likens, G.E., Bormann, F.H., Johnson, N.M., Fisher, D.W., and Pierce, R.S., Effects of forest cutting and herbicide treatment on nutrient budgets in the Hubbard Brook watershed-ecosystem, *Ecol. Monogr.*, 40, 23–47, 1970.

Lima, W.P., Patric, J.H., and Holowaychuk, N., Natural reforestation reclaims a watershed: a case history from West Virginia, Research Paper NE-392, USDA Forest Service, Washington, D.C., 1978.

Long-Term Ecological Research (LTER) Network, http://www.lternet.edu/; accessed July 2006.

Lopes, V.L. and Ffolliott, P.F., Sediment rating curves for a clearcut ponderosa pine watershed in northern Arizona, *Water Resour. Bull.*, 29, 369–382, 1993.

Lynch, J.A. and Corbett, E.S., Evaluation of best management practices for controlling nonpoint pollution from silvicultural operations, *Water Resour. Bull.*, 26, 41–52, 1990.

Lynch, J.A. and Corbett, E.S., Nitrate export from managed and unmanaged forested watersheds in the Chesapeake Bay watershed, in *Towards a Sustainable Coastal Watershed: The Chesapeake Experiment*, Hill, P. and Nelson, S. (eds.), Chesapeake Research Consortium, Edgewater, MD, 1994, pp. 656–664.

Lynch, J.A., Corbett, E.S., and Mussallem, K., Best management practices for controlling nonpoint-source pollution on forested watersheds, *J. Soil Water Conserv.*, 40, 164–167, 1985.

Lynch, J.A., Sopper, W.E., and Partridge, D.B., Changes in streamflow following partial clearcutting on a forested watershed, in *Watersheds in Transition*, Csallany, S.C., McLaughlin, T.G., and Striffler, W.D. (eds.), American Water Resources Association, Middleburg, VA, 1972, pp. 313–320.

Macan, T.T., The temperature of a small stony stream, *Hydrobiologia*, 12, 89–106, 1958.

Madej, M.A., Erosion and sediment delivery following removal of forest roads, *Earth Surf. Processes Land-forms*, 26, 175–190, 2001.

Marcell Experimental Forest, http://www.ncrs.fs.fed.us/EF/marcell/; accessed July 2006.

Martin, C.W. and Hornbeck, J.W., Logging in New England need not cause sedimentation of streams, *Northern J. Appl. For.*, 11, 17–23, 1994.

Martin, C.W. and Pierce, R.S., Clearcutting patterns affect nitrate and calcium in streams of New Hampshire, *J. For.*, 78, 268–272, 1980.

Martin, C.W., Noel, D.S., and Federer, C.A., Effects of forest clearcutting in New England on stream chemistry, *J. Environ. Qual.*, 13, 204–210, 1984.

Martin, C.W., Pierce, R.S., Likens, G.E., and Bormann, F.H., Clearcutting affects stream chemistry in the White Mountains of New Hampshire, Research Paper NE-579, USDA Forest Service, Washington, D.C., 1986.

McCullough, J.S.G. and Robinson, M., History of forest hydrology, *J. Hydrol.*, 150, 189–216, 1993.

Meyer, J.L. and Likens, G.E., Transport and transformation of phosphorus in a forest stream ecosystem, *Ecology*, 60, 1255–1269, 1979.

National Oceanic and Atmospheric Administration, Chesapeake Bay Office, 2005, http://noaa.chesapeakebay.net; accessed July 2006.

National Research Council, *Watershed Management for Potable Water Supply: Assessing the New York City Strategy,* National Acadamies Press, Washington, D.C., 2000.

National Research Council, *Atlantic Salmon in Maine*, National Academies Press, Washington, D.C., 2004, http://www.nap.edu; accessed July 2006.

National Research Council, *Genetic Status of Atlantic Salmon in Maine: Interim Report*, National Academies Press, Washington, D.C., 2002, http://www.nap.edu; accessed July 2006.

O'Meara, W., *We Made it Through the Winter: A Memoir of a Northern Minnesota Boyhood*, Minnesota Historical Society Press, St. Paul, MN, 1974.

Patric, J.H., Soil erosion in the eastern forests, *J. For.*, 74, 671–677, 1976.

Patric, J.H., Effects of wood products harvest on forest soil and water relations, *J. Environ. Qual.*, 9, 73–80, 1980.

Pinchot, G., *Breaking New Ground*, estate of Gifford Pinchot, 1947; republished by Island Press, Washington, D.C., 1998.

Reinhart, K.G., Timber-harvest clearcutting and nutrients in the northeastern United States, Research Note NE-17, USDA Forest Service, Washington, D.C., 1973.

Reinhart, K.G., Eschner, A.R., and Trimble, G.R., Effect on streamflow of four forest practices in the mountains of West Virginia, Research Paper NE-1, U.S. Forest Service, Washington, D.C., 1963.

Russell, C.E., *A-Rafting on the Mississip'*, University of Minnesota Press, Minneapolis, 1928.

Smith, D.M., Larson, B.C., Kelty, M.J., and Ashton, P.M.S., *The Practice of Silviculture, Applied Forest Ecology*, 9th ed., John Wiley & Sons, New York, 1997.

Sopper, W.E. and Lull, H.E. (eds.), *Forest Hydrology: Proceedings of a National Science Foundation Advanced Science Seminar*, Pergamon Press, New York, 1967.

Stein, S.M., McRoberts, R.E., Alig, R.J., Nelson, M.D., Theobald, D.M., Eley, M., Dechter, M., and Carr, M., Forests on the edge: housing developments on America's private forests, General Technical Report PNW-GTR-636, USDA Forest Service, Pacific Northwest Research Station, Portland, OR, 2005, http://www.fs.fed.us/projects/fote/reports/fote-6-9-05.pdf; accessed July 2006.

Swank, W.T., Stream chemistry responses to disturbance, in *Forest Hydrology and Ecology at Coweeta*, Swank, W.T. and Crossley, D.A. (eds.), Springer-Verlag, New York, 1988, pp. 339–358.

Swank, W.T. and Crossley, D.A. (eds.), *Forest Hydrology and Ecology at Coweeta*, Springer-Verlag, New York, 1988.

Swank, W.T., Swift, L.W., Jr., and Douglass, J.E., Streamflow changes associated with forest cutting, species conversions, and natural disturbances, in *Forest Hydrology and Ecology at Coweeta*, Swank, W.T. and Crossley, D.A. (eds.), Springer-Verlag, New York, 1988, pp. 297–312.

Swank, W.T., Vose, J.M., and Elliott, K.J., Long-term hydrologic and water quality responses following commercial clearcutting of mixed hardwoods on a southern Appalachian catchment, *For. Ecol. Manage.*, 143, 163–178, 2001.

Swift, L.L. and Messer, J.B., Forest cuttings raise the temperatures of small streams in the southern Appalachians, *J. Soil Water Conserv.*, 26, 111–116, 1971.

Switalski, T.A., Bissonette, J.A., DeLuca, T.H., Luce, C.H., and Madej, M.A., Benefits and impacts of road removal, *Frontiers Ecol. Environ.*, 2, 21–28, 2004.

Thompson, C.H. and Kyker-Snowman, T., Evaluation of nonpoint source pollution problems from crossing streams with logging equipment and off-road vehicles in Massachusetts, Research and Demonstration Project 88-03, Massachusetts Department of Environmental Quality Engineering, Boston, MA, 1989.

Trimble, G.R. and Sartz, R.S., How far from a stream should a logging road be located? *J. For.*, 55, 339–341, 1957.

Troendle, C.A. and King, R.M., The effect of timber harvest on the Fool Creek Watershed, 30 years later, *Water Resourc. Res.*, 21, 1915–1922, 1985.

Verry, E.S., Effect of an aspen clearcutting on water yield and quality in northern Minnesota, in *Watersheds in Transition*, Csallany, S.C., McLaughlin, T.G., and Striffler, W.D. (eds.), American Water Resources Association, Middleburg, VA, 1972, pp. 276–284.

Verry, E.S., Forest harvesting and water: the Lake States experience, *Water Resourc. Bull.*, 22, 1039–1047, 1986.

Verry, E.S., The effect of aspen harvest and growth on water yield in Minnesota, in *Forest Hydrology and Watershed Management*, Swanson, R.H., Bernier, P.Y., and Woodward, P.D. (eds.), Publication 167, International Association of Hydrologic Sciences, Wallingford, UK, 1987, pp. 553–562.

Wemple, B.C. and Jones, J.A., Runoff production on forest roads in a steep mountain catchment, *Water Resourc. Res.*, 39(8), 2–17, 2003.

Wemple, B.C., Swanson, F.J., and Jones, J.A., Forest roads and geomorphic process interactions, Cascade Range, Oregon, *Earth Surf. Processes Landforms*, 26, 191–204, 2001.

West, T.L., *Centennial Mini-Histories of the Forest Service*, FS-518, USDA Forest Service, Washington, D.C., 1992.

Whitney, G.G., *From Coastal Wilderness to Fruited Plain: A History of Environmental Change in Temperate North America 1500 to the Present*, Cambridge University Press, 1994.

Williams, R.L., *The Loggers*, Time-Life Books, Alexandria, VA, 1976.

Wilm, H.G., Statistical control of hydrologic data from experimental watersheds, *Trans. Am. Geophys. Union*, Part 2, 616–622, 1944.

Wilm, H.G., How long should experimental watersheds be calibrated?, *Trans. Am. Geophys. Union*, 30, 272–278, 1949.

Wilmot, S. and Scherbatskoy, T. (eds.), *Vermont Forest Ecosystem Monitoring: 1995 Annual Report*, Vermont Department of Forests, Parks, and Recreation, Waterbury, VT, 1997.

Wolman, M.G. and Schick, A.P., Effects of construction on fluvial sediment, urban and suburban areas of Maryland, *Water Resourc. Res.*, 2, 451–464, 1967.

Zon, R., Forests and water in the light of scientific investigation, USDA Forest Service, Washington, D.C., 1927 [reprinted with revised bibliography from Appendix V of the Final Report of the National Waterways Commission 1912, Senate Document 469, 62nd Congress, 2nd Session].

7 Forest Conversion to Agriculture

In fifteen decades, Americans have transformed a wilderness into a mighty nation. In all the history of the world, no people ever built so fast and yet so well. This will be a land of liberty, they said in the beginning, and as they hacked the forest, drove the ploughshares deep into the earth and spread their herds across the ranges, they sang of the land of the free they were making. All that they finally built upon this continent is founded on that faith—that there would be opportunity and independence and security for man.

Hugh Hammond Bennett, 1939, *Soil Conservation*

7.1 INTRODUCTION

European settlement of the Northeast initiated large-scale forest clearing and conversion of forests to cultivated fields and pastureland. The conversion of the precolonial forest began in the 1600s in Massachusetts. By the 1870s, forests and grasslands in Illinois, Wisconsin, and Minnesota had been replaced by cultivated fields of corn, oats, wheat, and hay (Knox, 2001; Whitney, 1994). This process also involved the extensive clearing and draining of forested and prairie wetlands and coastal marshes (Box 7.1) (Carlson and Fowler, 1979; McCorvie and Lant, 1993). "Few areas on the Earth's surface have experienced as extensive and dramatic a change as the mid-latitude forests and grasslands of eastern North America" [in such a short period of time] (Whitney, 1994). Clearing for agriculture constituted the greatest disturbance since the retreat of the North American glacial ice sheets approximately 12,000 to 15,000 years ago.

While it is certain that this vast unplanned experiment changed streamflow, channel form, and sedimentation patterns in streams and rivers in the Northeast, the greatest changes occurred 100 to more than 200 years ago. Recent studies attempt to discern the extent of historical disturbance caused by agriculture and describe the influence, or legacy, of past land use on current watershed and stream conditions. Paired watershed experiments of clearcutting in experimental forests give some indication of the initial effects of clearing on streamflow (Hibbert, 1967; Swank and Crossley, 1988). Historical records reveal the extent and pattern of agricultural development and offer contemporary impressions of the environmental changes that accompanied settlement (Knox, 1977; Trimble, 1982; Whitney, 1994). In addition, several current studies have quantified historical patterns of soil erosion by examining the geologic record of sediment deposition in pre- and post-European settlement periods (Brugam, 1978; Davis, 1976; Knox, 1987).

In the last 50 years, nonpoint source pollution of waterways caused by agriculture has become a primary environmental, policy, and management concern (U.S. Environmental Protection Agency, 2002a). Excess nitrogen and phosphorus from fertilizer and animal waste enter streams in agricultural areas and cause the eutrophication of surface waters (Foy and Withers, 1995; Jordan et al., 1997a,b,c). Excess nitrogen that reaches household wells is a potential threat to public health (Pionke and Urban, 1985). Pesticides used in farming are another source of groundwater and surface water contamination (Gallagher et al. 1996; Pionke and Glotfelty, 1989; Schottler et al., 1994). Waterborne pathogens from livestock manure are also a public health concern (Mitchell, 2002).

In this chapter we will review and discuss the historical changes in streamflow, soil erosion, sedimentation, and channel form caused by forest conversion, agricultural development, and contemporary problems of pollution associated with agriculture in the northeastern United States.

Box 7.1   Draining and Filling of East Coast Tidal Salt Marsh

Tidal estuaries are among the most productive ecosystems on Earth (Figure 7.1). They are critical spawning, nesting, and rearing habitat for many species of finfish, shellfish, and birds. Many early settlements in New England and the Mid-Atlantic colonies were located near the marshlands that form a major part of the estuarine environment. This was partly due to the need to harvest salt marsh hay, which included native grasses such as black grass (*Juncus gerardi*) and *Spartina patens*. Although these grasses were low in nutrients, they were sufficient to keep livestock alive during the winter months (Whitney, 1994). Later uses of salt hay included insulation, mulch for crops, packing material, curing concrete, and paper production. In an effort to produce more salt hay for harvesting and more pasture land for grazing animals, dikes were built in and around the marshes to lower the water table. These actions wreaked havoc with the salt marsh ecosystems, preventing tidal inundation, destroying habitat, and creating an environment that encouraged mosquito breeding (Carlson and Fowler, 1979).

Destruction of salt marsh continued during the 19th and 20th centuries. Estuaries were ditched, drained, and filled first for agriculture and then increasingly for development. Because the marsh lands were seen as worthless, they were also used as garbage dumps. The Civilian Conservation Corps conducted widespread ditching outside developed areas in the 1930s. This effort actually exacerbated the mosquito problem because the ditches were not maintained and areas of stagnant water increased. By 1938, it was estimated that 90% of the Atlantic Coast salt marsh from Maine to Virginia had been ditched (Carlson and Fowler, 1979; Rosza, 1995).

Currently the condition of coastal marshlands remains precarious (U.S. Environmental Protection Agency, 1998, 2002b). There are multiple threats to marshlands all along the East Coast because excess nutrients, sediments, and contaminants from human land use in heavily populated upland areas continue to flow into these ecosystems. Restrictions on agricultural conversion and development in the marshes are fairly recent (U.S. Environmental Protection Agency, 1998). An average of 200 ha of marsh wetlands were drained each year between 1956 and 1979 in the Chesapeake Bay region. This declined to 28 ha/yr between 1982 and 1989 following the enactment of state and

FIGURE 7.1 East Coast salt marsh (courtesy of Ethan Nedeau).

federal conservation plans. In northern New Jersey, a master plan enacted in February of 2004 rezoned 3500 ha of the Hackensack Meadowlands for conservation and designation as a nature reserve. The Meadowlands, which originally covered more than 10,000 ha, are located in northern New Jersey, within sight of the New York City skyline. They served as a regional garbage dump for many years. Lately, escalating property values have made the area increasingly attractive to developers (Hackensack Riverkeeper, 2005).

Since the 1970s, there has been increased interest in salt marsh restoration and creation, and research in this field is expanding (Grant, 2002). Results in some instances have been rapid, if not remarkable. Ten years after the removal of impoundments and restoration of tidal flushing *Spartina alterniflora* cover increased from 1% to 45% in a tidal marsh in Connecticut. Although fauna were not measured systematically, invertebrate, fish, and bird communities typical of saltmarsh habitat were present following restoration as well (Sinicrope et al., 1990). The time required for complete restoration of habitat and biogeochemical functioning in these marshes may be considerably longer (Mitsch and Wilson, 1996; Rosza, 1995).

7.2 DEVELOPMENT OF THE AGRICULTURAL LANDSCAPE (1620 TO 1930)

Clearing for agriculture produced a particular suite of hydrologic and erosion and sediment responses throughout the Northeast. The magnitude and extent of those responses depended on the ecological characteristics (local climate, geology, soils, and topography) of a particular region and on the population density, economic development, and technology available at the time of settlement. (For a comprehensive examination of the environmental effects of the settlement of the Northeast, the interested reader is referred to Gordon G. Whitney's extensively referenced and eminently readable book, *From Coastal Wilderness to Fruited Plain: A History of Environmental Change in Temperate North America, 1500 to the Present*, published by Cambridge University Press, 1994.)

7.2.1 THE EAST COAST

Clearing progressed relatively slowly during the colonial period. It took 170 years (1620 to 1790) to clear the land and establish farming communities throughout New England and the Mid-Atlantic states. Settlers in most parts of New England and New York were confronted with dense forests and thin, rocky, glaciated soils that were poorly suited for farming. While settlers in the Mid-Atlantic states, south of the glacial boundary, were spared the backbreaking work of clearing stones and boulders, they shared the limitations imposed by a shortage of labor and relatively primitive methods and tools. Before 1790, farm tools were handmade and primarily constructed of wood with small iron parts, such as plow tips, manufactured individually on the farm or by local blacksmiths (Danhof, 1972). These factors all combined to limit rates of forest clearing on the East Coast during the colonial period.

Most colonial New England farmers lived in relatively isolated communities and practiced subsistence agriculture. The population and road network required to engage the markets necessary for larger scale commercial farming did not exist. Hence there was no economic incentive to clear more land than was needed to provide food for an individual family, and possibly the common needs of a small community (Raup, 1966). Concord, Massachusetts, for example, retained more than half its forested land in the mid-1700s, more than 100 years after it was first settled (O'Keefe and Foster, 1998). In general, land clearing proceeded at rates between 0.4% and 0.8% per year in colonial communities in Massachusetts, Rhode Island, and southeastern Pennsylvania prior to 1750 (Whitney, 1994:153).

The number of people of European descent on the East Coast increased dramatically in the latter half of the 18th century. In the areas surrounding the Chesapeake Bay, for example, the emigrant population grew from 700 in 1650, to 19,500 in 1700, to 144,000 in 1750, and 1,150,000 in 1800 (Percy, 1992). This increase in population density fueled the development of regional markets along with an expanding network of roads. Farmers all along the East Coast began clearing more land for commercial crops (Foster and O'Keefe, 2000).

Tobacco was the primary commercial crop in the Chesapeake Bay region. Tobacco is a demanding crop and the nutrient capital and productivity of land used for growing tobacco is reduced after only two or three harvests. During the first century of settlement, farmers cleared small plots of land and planted tobacco among the tree stumps. As soil fertility declined, they replaced tobacco with corn, then followed corn with wheat. The soil was then allowed to lie fallow for several years to regain its fertility while new areas were cleared. As long as the population remained small, the total effect on the landscape was not severe. As the population grew and economic opportunities expanded, larger and more fragile and marginal (clay soils on steeper slopes) areas were cleared for farming. Plows could now be used in areas where tree stumps and roots had rotted away or been removed. Population density and a relative shortage of arable land compelled farmers to reduce or eliminate the time allowed for fields to recover. All these practices led to decreased soil fertility and increased soil erosion (Gottschalk, 1945; Percy, 1992). There was some recognition at the time that these practices were wasteful but, as is often the case, immediate needs superseded long-term goals; soils eroded and streams filled with sediment.

> The circumstances alone of a new country, almost boundless, in ... which, fresh lands of great fertility, have been for sale during many years, at very low nominal prices, has greatly contributed to accelerate among our land killers, the exhaustion of our soil; and to prevent attempts at improvement. ... (James E. Garnett, 1818, *Defects in Agriculture: Memoirs of the Virginia Society for Promoting Agriculture: Containing Communications on Various Subjects in Husbandry and Rural Affairs*, quoted in Percy [1992]).

7.2.2 THE MIDWEST

By the late 18th and early 19th centuries, settlers were moving westward into the Ohio Valley. Here the soils were deep, fertile, free of stones, and more easily tilled. While the potential for crop production was greater here than in the East, there was, at first, no transportation system linking the Midwest to eastern markets. Hence the first settlers in the Midwest also lived in small, relatively isolated communities and practiced subsistence farming. This changed with the opening of the Erie Canal in 1830, later augmented by railroad lines that connected the farmlands of the Midwest with the burgeoning cities of the eastern seaboard (Raup, 1966). The economic opportunities offered by newly available markets and efficient transportation provided the incentive for the rapid clearing and agricultural development of the Midwestern "Corn Belt"—Ohio, Indiana, Illinois, Iowa, Missouri, and parts of Michigan and Minnesota.

Agricultural expansion in the Midwest was accelerated by technological improvements that allowed landowners to cultivate larger areas quickly and economically. Tools such as axes, sickles, and scythes were standardized and mass produced by specialized manufacturers; the cast iron plow replaced wooden plows (Danhof, 1972; Hurt, 1994). The McCormick reaper was patented in 1833 and 1834 by Obed Hussey and Cyrus McCormick. McCormick established a factory for the production of these machines in Chicago in 1854. By continuing to technically improve the machine, advertising, and offering easy credit, McCormick steadily increased sales. By 1860 the McCormick reaper was viewed as a necessity for farming operations (Cronon, 1991). Three men were required to operate the mechanical thresher patented by Hiram and John Pitts in 1836. Although large and expensive, it processed three to four times as much wheat as these same workers might have processed by hand in the same amount of time (Danhof, 1972). Tile drainage systems also were introduced during this period (Colman, 1968). The development of this new technology coincided with the settlement of the treeless, fertile prairie areas of Illinois, Wisconsin, Iowa, and Minnesota. Large areas of open prairie were converted to farmland in a much shorter period of time than had been possible in the East. Government land policy made it possible for individual farmers to secure large areas of land with very little money. As settlement progressed, more successful commercial farmers bought out neighbors who could not afford the capital investment that farming now required, further increasing the size of individual farms (Gates, 1932).

Settlers in the Midwest drained freshwater wetlands to a far greater extent than had been the case in the New England and Mid-Atlantic states. The cumulative effects on the hydrological and ecological characteristics of the region have been severe and long lasting and continue to affect flooding patterns and wildlife habitat. Wetland drainage was promoted by state and federal legislation—the Swamp Land Acts of 1849, 1850, 1860, and the Graduation Act of 1854—and by the formation of rural drainage districts throughout the area. Draining wetlands was seen as a means of reclaiming "worthless" land for agriculture and as a way to reduce or eliminate diseases, such as malaria and cholera, associated with swampy areas.

> They [swamps] are evils common to all countries, . . . rendering portions of the earth not only desolate and unsusceptible of cultivation, but fruitful promoters of disease and death. . . . [labor and capital] when properly employed must redeem portions of the land from sterility, and make it valuable and useful, instead of the generator of disease (*Congressional Globe*, 31 Congress 1, Session, 2: 1191–1192, 1826–1827, 1848–1850, quoted in McCorvie and Lant [1993]).

Wetland drainage was aided and accelerated by the increasing availability of manufactured drainage tiles. Handmade tiles had been used on a limited scale by a few farmers in upstate New York in the 1830s (Colman, 1968). By the 1870s, drainage tiles were being mass produced and were affordable and widely available throughout the Midwest. Federal funding supporting wetland drainage continued into the 1950s. It is now estimated that 71% of the wetlands (19 to 20 million ha) in the Corn Belt states have been drained since 1850. The states of Illinois, Indiana, Iowa, and Ohio have lost between 90% and 99% of wetland areas that existed before 1850 (McCorvie and Lant, 1993). Beginning in the 1960s and 1970s, research in hydrology and ecology demonstrated the value of wetlands in flood mitigation and as wildlife habitat, leading to a reassessment of the last century of land use policies:

> Viewed from the perspective of current attitudes and policies toward wetlands, the widespread drainage of marshes . . . throughout the Corn Belt seems less than an accomplishment and rather more of a massive environmental insult (McCorvie and Lant, 1993).

7.2.3 REFORESTATION

Agricultural expansion in the Midwest and the influx of large amounts of inexpensive commodities eventually caused many farmers in the East to abandon land that was no longer economically viable. At first many New Englanders turned to sheep farming (Wilson, 1935). This caused the most extensive forest clearing in the history of the region. Sixty to 80% of the New England landscape was cleared in all but the most remote and mountainous areas. Most of this land was used for pasture—less than 10% was tilled for crops (Foster and O'Keefe, 2000). In 1836, virtually every town in Vermont had more than 1000 sheep. The towns of Walpole, Lebanon, and Hanover, in the Connecticut Valley Region of New Hampshire, each claimed to have more than 10,000. More prosperous landowners bought out their neighbors and converted farms to sheep pasture. After 1840, increasing amounts of wool, produced at lower cost in western states, were transported into the region. In order to remain profitable, some farms in New York State and New England were converted to dairy farms, producing butter and cheese and later milk for local markets (Brunger, 1955; Wilson, 1935). Many farms were simply abandoned, as people moved into towns and cities and found jobs in manufacturing. Thus, while the population of New England is 11 times greater today than it was in 1800 (U.S. Census Bureau), the landscape is once again predominantly forested.

Reforestation of abandoned farms also occurred in northern regions of the Lake States—Michigan, Wisconsin, and Minnesota. Timber companies sold cleared land to settlers following the white pine logging boom of the late 19th and early 20th centuries. As in the northern regions of New England and New York State, clearing land that supported impressive forests did not necessarily create productive farmland. The growing season was too short, the soils were poor, and markets were too distant. Many of the farmers, who had tried to settle the "cutover" regions, soon abandoned their unproductive farms. These lands then reverted to state ownership, often because of delinquent taxes, and became the core of state and national forests. Most of the forests regenerated naturally, with fast-growing aspen and birch rather than white pine. During the 1930s the Civilian Conservation Corps (CCC) was very active in reforestation in this region, planting almost 500 million trees in Michigan alone (Cook, 2005; Enger, 1998; Minnesota Historical Society, 2002; Stearns, 1997). Forests now cover close to half of Michigan (19.3 million acres) and are primarily located in the northern two-thirds of the state (Cook, 2005). Minnesota now has approximately 7 million ha of forest, down from the original forest area of 12.7 million ha, but greater than that in the early 1900s (Minnesota Historical Society, 2002). There are 6.4 million ha of forested land in Wisconsin (Wisconsin Department of Natural Resources, 2005). Native prairie here and in other Midwestern states (Illinois, Missouri, Iowa) has been almost entirely converted to corn, wheat, or soybean production. Less than 1% of the original prairie remains in Illinois, Indiana, Iowa, and Missouri (Whitney, 1994).

7.2.4 Current Patterns of Agricultural Land Use

At present, agricultural land use in the Northeast tends to be specialized by region (Figure 7.2). Ohio, Indiana, Illinois, Iowa, and parts of Wisconsin and Minnesota continue to be the predominant agricultural areas in the Northeast. Extensive farming activity also occurs in the Mid-Atlantic states

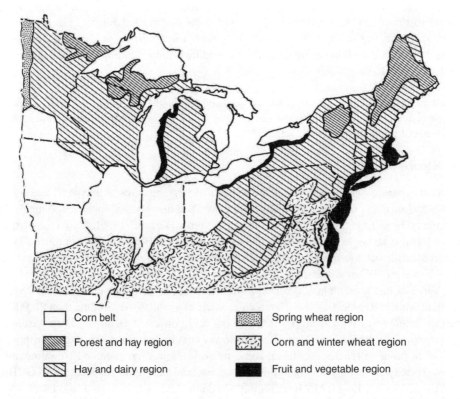

	Corn belt		Spring wheat region
	Forest and hay region		Corn and winter wheat region
	Hay and dairy region		Fruit and vegetable region

FIGURE 7.2 Major agricultural regions of the Northeast and Midwest United States in 1920. Modified slightly from Meyer (1987). (Whitney, 1994; reprinted with the permission of Cambridge University Press.)

in the Chesapeake Bay and Delaware River watersheds (U.S. Environmental Protection Agency, 1998). Industrial forest management is concentrated in New England, New York, Pennsylvania, West Virginia, Michigan, Wisconsin, and Minnesota. Fruit, vegetable, and dairy farming are common in the lowlands of New England, New York, and Pennsylvania. Corn and soybean production is concentrated in western Ohio, Indiana, Illinois, Iowa, and Missouri (Whitney, 1994).

7.3 AGRICULTURAL RESEARCH: SOIL AND WATER

7.3.1 DEVELOPMENT OF RESEARCH PROGRAMS

7.3.1.1 The Natural Resources Conservation Service

Government-sponsored research involving the effects of agricultural practices on erosion, stream sediment, and water quality began in the 1930s, after the cumulative effects of poor agricultural practices, drought, and the economic devastation of the Great Depression combined to destroy the livelihoods of many Midwestern farmers. The U.S. Department of Agriculture (USDA) has sponsored research programs through the Soil Conservation Service (SCS), renamed the Natural Resources Conservation Service (NRCS), and the Agricultural Research Service (ARS), which was organized in the early 1950s. The first project designed to test the effectiveness of soil conservation measures was located in the Coon Valley watershed (Box 7.2). Throughout the 1940s and 1950s, projects also involved flood control through dams and channel modifications. Currently the NRCS conducts research related to water quality through the National Water and Climate Center and NRCS National Technology Support Centers (NTSCs) located in North Carolina, Texas, and Oregon. The activities of the NRCS Watershed Science Institute were transferred to the NTSCs in 2004 following an NRCS reorganization. Research topics include the movement of pesticides to ground and surface waters, and nutrient management (Natural Resources Conservation Service, 2005).

7.3.1.2 The Agricultural Research Service

The ARS investigates a wide range of topics pertaining to agricultural science and watershed management. Individual ARS units focus on particular research topics. In the Northeast and Midwest, there are several that specialize in agricultural impacts on streamflow and water quality. Most are affiliated with land grant universities.

Established in the 1930s, the North Appalachian Experimental Watershed (NAEW), in Coshocton, Ohio, was originally part of the SCS and was transferred to the ARS in 1954. There are 22 instrumented watersheds at the NAEW site, which covers 425 ha of farmland and forest in an area of rolling hills and rich soils developed on sandstone, shale, and limestone. Like the Coweeta Experimental Forest and Hydrologic Laboratory, the NAEW now has close to 70 years of continuous data collected from small watersheds and groundwater lysimeters. Early work at the NAEW led to the development of the curve number method for estimating overland flow potential from different land surfaces. NAEW scientists have also investigated no-till or conservation tillage practices, nutrient and pesticide movement through macropores, and the effects of different pasture management schemes on water quality. Current research areas include agricultural management practices, carbon sequestration, and evaluation of agricultural best management practices (BMPs).

The Pasture Systems and Watershed Management Research Unit (PSWMRU) is affiliated with Pennsylvania State University. The PSWMRU was formed in 1992, the result of a merger between the Northeast Watershed Research Center (NWRC) and the U.S. Regional Pasture Laboratory (USRPL). The research focus of this group has evolved over time. Initially the major concern was the generation of overland flow and flood prediction in small watersheds. In the late 1970s and 1980s, research was conducted on subsurface hydrology and flow pathways from the soil surface and the root zone to groundwater and streams. Current research focuses on the sources, transformation, transport, and fate of phosphorus, nitrogen, and pesticides in groundwater and surface water (Agricultural Research Service, 2004).

Additional research stations with programs related to water quality include the National Soil Erosion Research Laboratory (NSERL), West Lafayette, Indiana; the Soil and Water Management Research Unit at the University of Minnesota, St. Paul; the National Soil Tilth Laboratory, Ames, Iowa; and the Cropping Systems and Water Quality Research Unit, Columbia, Missouri. The USDA-ARS Southeast Watershed Research Laboratory, located in Tifton, Georgia, has investigated land use effects on watersheds in the Coastal Plain ecosystem. Much of the research on nutrient processing in riparian buffers has come from this research unit (Lowrance et al., 1995, 1997, 2000).

7.3.1.3 Other Research Programs

The National Oceanic and Atmospheric Administration (NOAA) of the U.S. Department of Commerce supports research involving agriculture and urban land use effects on aquatic systems through the Sea Grant program (http://www.nsgo.seagrant.org). The Sea Grant program works with 32 colleges and universities located in the coastal and Great Lakes states. Finally, the Smithsonian Environmental Research Center (SERC) (http://www.serc.si.edu) has supported research on coastal zone environmental issues for nearly 40 years. Much of this research focuses on agricultural land use effects.

7.4 CHANGES IN STREAMFLOW, SEDIMENT, AND CHANNEL FORM FOLLOWING AGRICULTURAL CLEARING AND DEVELOPMENT

7.4.1 Streamflow

If agricultural clearing were the subject of a paired-watershed experiment, it would involve clearcutting, removal of all trees and brush, and the prevention of any forest regrowth. The soil would either be tilled and planted with crops or planted to grass and used as pasture for domestic grazing animals (Whitney, 1994). Clearcutting experiments on paired watersheds can suggest the range of the increase in streamflow that might have followed. Whole-tree harvesting in New Hampshire resulted in a maximum streamflow increase of 152 mm (23%) in the first year, while clearcutting with suppression of regrowth by herbicides produced a 347 mm (41%) increase (Hornbeck et al., 1997). One paired watershed study at Coweeta Hydrologic Laboratory in North Carolina was specifically designed to test the effect of conversion to agriculture in that region. In 1940, all the vegetation was burned or removed from Coweeta Watershed 3. This was followed by a 12 year period of unregulated agriculture (farming and grazing) on a 6 ha area (Swank and Crossley, 1988). This treatment resulted in a 127 mm increase (above the pretreatment mean annual streamflow of 607 mm) in mean annual streamflow during the first year. Increases of 95 mm, 59 mm, 113 mm, and 80 mm continued in years 2 through 5, respectively (Hibbert, 1967). Of course, once established, crops or pasture grasses would increase evapotranspiration and thereby reduce water yield. Nevertheless, shallow rooted annual plants or grasses would not intercept or transpire as much water as deeply rooted woody vegetation.

Unlike forest management, which can produce short-term, small-scale changes in streamflow, conversion to agriculture causes large-scale and persistent alterations in hydrologic pathways (Verry, 1986). In forests, rain and snowmelt infiltrate through the litter layer to the organic (O), mixed (A), and mineral (B and C) horizons of the soil. As discussed in Chapter 2, water may flow rapidly through macropores—channels created by worms and other soil fauna and by the growth and decay of tree roots. The large, stable aggregates increase soil porosity and permeability. The formation of soil aggregates is enhanced by soil organic matter, derived from annual leaf fall, and by diverse populations of microbes and invertebrates. Decomposers, especially fungi, bind soil particles together into larger aggregates while processing organic matter (Bardgett et al., 2001; Brady and Weil, 2002; Wischmeier and Mannering, 1965).

Tilling breaks up the soil and incorporates surface organic matter into the mineral soil, forming an AP horizon, or plow layer. By design, this provides an enriched and aerated seedbed and favorable seed-soil contact in newly plowed fields. Repeated tilling destroys macropores and accelerates the decomposition of soil organic matter by increasing the oxygen supply to fungi and bacteria. Once the forest is removed, the annual supply of organic matter from falling leaves and branches is eliminated. Soil aggregates that are low in organic matter (decomposition greater than the input rate) become unstable. They break down and form compacted areas of fine particles when subjected to wetting and drying. This often results in the formation of a compacted layer of soil just beneath the AP horizon—a plow pan—that has low permeability and hinders vertical flow of soil water (Figure 7.3) (Brady and Weil, 2002). Tilling and other farming practices can also cause substantial changes in the biotic community—loss of biodiversity and relative increases or decreases in populations of different species. Declines in earthworm populations, for example, have been attributed to mechanical damage from tilling and decreased organic matter (Edwards and Lofty, 1982). The loss of earthworms and other organisms can adversely affect soil processes that enhance infiltration (Anderson, 1988). Soil compaction is exacerbated by the repeated trampling of grazing animals (Barnes et al., 1998; Sartz and Tolsted, 1974). The net effects of converting forest to farmland are (1) an increase in the portion of precipitation available for stormflow (by reducing evapotranspiration) and (2) the reduced ability of the soil to absorb and store water (Knox, 2001; Sartz, 1978; Sartz and Tolsted, 1974; Verry, 1986; Whitney, 1994).

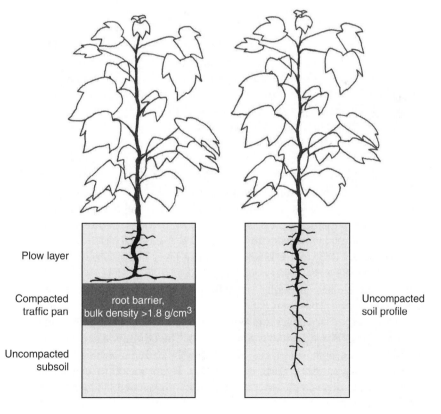

Plow layer

Compacted traffic pan — root barrier, bulk density >1.8 g/cm^3

Uncompacted subsoil

Uncompacted soil profile

FIGURE 7.3 Traffic (plow) pan beneath the AP horizon, caused by tilling and compaction from heavy equipment, can restrict root growth in agricultural fields.

Forest conversion to row crops or pasture typically leads to overland flow in areas where it was formerly infrequent or rare (Sartz, 1969). John Bartram, a Pennsylvania naturalist, made the following observation in 1760:

> About 20 years past, when the woods was not pastured & full of high weeds and the ground light, then the rain sunk much more into ye earth and did not wash and tear up the surface (as now). The rivers and brooks in floods would be black with mud but now the rain runs most of it off on the surface, is collected into the hollows which it wears to the sand and clay which it bears away with the swift current down to brooks and rivers whose banks it overflows (Bartram, 1760; in Cronon, 1983).

The increase in overland flow reduces the proportion of precipitation directed to soil storage and groundwater recharge. There is an obligate increase in the magnitude and frequency of peak flows. At the same time, baseflow from groundwater discharge is typically lower during dry periods and the level of the saturated zone may decline as well (Sartz, 1969). Several early writers observed that small streams and springs dried up in the years following settlement (Fitzpatrick et al., 1999; Knox, 1977, 2001; Whitney, 1994). For example, the earliest map of New Haven, Connecticut (1641), shows two streams, one of "substantial" size. On subsequent maps, the stream reaches were shortened, and the smaller stream had disappeared by 1802. Neither stream currently exists (Cronon, 1983).

Holford wrote these observations concerning streamflow in southwestern Wisconsin in 1900:

> Fifty years ago, this was one of the best watered regions on earth. Springs of ample volume bubbled up at the head of every ravine and in hundreds of places along every stream, and trickled from many a stony hillside. Today the large majority of these springs are dry and the rising generation knows nothing of them unless by tradition (Holford, 1900; in Knox, 1977).

Without accurate precipitation records from the period, it is difficult to quantify the actual extent and severity of this phenomenon. These early, anecdotal reports of changes in streamflow are consistent with more recent research showing similar reductions in baseflow following the soil compaction and increase in the area of impervious surface (roads, parking lots, and roofs) associated with urban development (Simmons and Reynolds, 1982).

Verry (1986) showed that the annual flood peak in the Red River of the North in Minnesota increased by 43% in 1908, following extensive forest clearing from 1869 to 1910 (Figure 7.4). Fitzpatrick et al. (1999) determined, by studying floodplain sediment deposits, that the magnitude of floods and sediment loads in the North Fish Creek watershed in northern Wisconsin increased after the conversion of forests to cropland and pasture. Their modeling work suggested that between the mid-1920s and mid-1930s, the time when the greatest portion of watershed land was devoted to agriculture, flood peaks were three times larger than in the presettlement era. A reduction in the proportion of the watershed used for growing crops and a corresponding increase in grazed and abandoned pastures has reduced this effect somewhat, so that current floods are estimated to be two times larger than what would be generated from similar precipitation events in a forested watershed.

Many noteworthy studies involving the effects of clearing and agriculture on streamflow, erosion, and sedimentation have been conducted in the Driftless Area, a region of about 25,900 km^2 that includes parts of southwestern Wisconsin, northwestern Illinois, northern Iowa, and southeastern Minnesota (Bowman, 1911; Sartz, 1978). The Driftless Area is of particular interest because of its unique geologic and glacial history and the extreme sensitivity of the landscape to changes in land use (Knox, 2001; Sartz et al., 1977). It is a region of unglaciated, hilly terrain with deep loessal (wind-deposited) soils surrounded by a glaciated landscape. The natural vegetation is a combination of prairie and forest with treeless grasslands on drier upland ridges and plateaus. Forests of oak, hickory, and some evergreen species occupy the slopes between the uplands and the river valleys. The uplands are about 120 m higher than the river valleys (Sartz, 1970; Sartz et al., 1977).

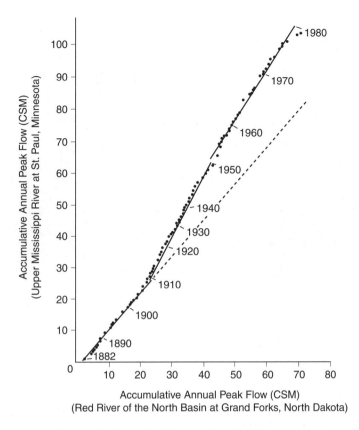

FIGURE 7.4 Annual peak flow on the Mississippi River at St. Paul, Minnesota, increased approximately 43% in 1908, coincident with the conversion of native forests to agricultural or young forest land (stands less than 16 years old). (Verry, 1986; reprinted with the permission of the American Water Resources Association.)

Sartz (1970) analyzed the effect of land use on stormflow in this region. The measurements were taken on several small (less than 10 ha), homogeneous watersheds within the Coulee Experimental Forest in the Driftless Area of southwestern Wisconsin and compared stormflow from five major storms in 1965. Sartz found that peak discharge from tilled (essentially bare) land averaged 37.4 mm/hour, while peak discharge in alfalfa meadows averaged 15.1 mm/hour. Peak discharge from forested pasture averaged 3.1 mm/hour (even though the forested area had been heavily grazed), while peak discharge from heavily grazed open pasture averaged 25.2 mm/hour. The peak discharge from tilled land was 12 times greater than that from the forested pasture, even though the forested pasture watershed was the largest of the five watersheds and had the greatest slope (35% versus 15% to 25% for the other watersheds). The unusual hydrogeologic setting of the Driftless Area highlights the influence of soil surface condition on flow patterns and peak flows.

Box 7.2   Restoration in the Coon Valley Watershed

Coon Valley in the Wisconsin Driftless Area is the site of the first soil conservation demonstration project in the United States (Helms, 1992; Leopold, 1935; Roldán, 2002). This 37,000 ha watershed was cleared and settled beginning in the 1850s. As the number of farms and population of grazing animals (mostly dairy cows and beef cattle) increased, the steep slopes between the upland plateaus and river valleys were cleared and used for pasture and crops. Erosion from the uplands and sediment accumulation in floodplains were severe (Figure 7.5)

FIGURE 7.5 Erosion gullies in the Coon Creek Watershed. Note the people standing in the gully (courtesy of NRCS).

(Leopold, 1935; Trimble, 1983). Measurements taken in the late 1930s showed that the elevation of the floodplain had increased 0.75 m between 1853 (presettlement) and 1904 (1.5 cm/y), 1.5 m between 1904 and 1930 (0.6 cm/y), and 1.2 m between 1930 and 1938 (15 cm/yr) (Trimble, 1983).

In 1934, farmers in this area were invited to participate in a conservation program initiated by the U.S. Department of the Interior, Soil Erosion Service (SES), a new government program that was renamed the Soil Conservation Service (SCS) in 1935 and is currently known as the Natural Resource Conservation Service (NRCS) (Roldán, 2002). The renowned conservationist Aldo Leopold described this first government sponsored soil conservation effort in the following manner:

> ... after selecting certain demonstration areas on which to concentrate its work, it [the SES] offered to each farmer on each area the cooperation of the government in installing on his farm a reorganized system of land-use, in which not only soil conservation and agriculture, but also forestry, game, fish, fur, flood-control, scenery, songbirds, or any other pertinent interest were to be duly integrated. It will probably be another decade before the public appreciates either the novelty of such an attitude by a bureau, or the courage needed to undertake so complex and difficult a task (Leopold, *American Forester Magazine,* 1935, with permission).

More than half of the 800 farmers in the Coon Valley signed cooperative agreements with the SES between 1933 and 1935 (Gaumnitz, 2002; Helms, 1992). By doing so, they agreed to participate in a 5 year demonstration project to test new concepts in land management.

Under this new system, land use was determined primarily by slope. First, CCC workers fenced areas with greater than 40% slopes to exclude grazing animals. In forested areas, the understory was allowed to regrow while open land was planted to conifers. Regulated grazing was permitted in previously cleared fields where slopes ranged between 25% and 40% (no new clearing was permitted); permanent pastures were established in areas of 15% to 25% slopes. Areas of less than 15% slope could be farmed, but the crops were planted in contour strips with terraces used for corn (Figure 7.6). In addition, farmers were encouraged to improve

FIGURE 7.6 Contour plowing, Coon Creek watershed (courtesy of NRCS).

wildlife habitat by planting trees and hedges, and providing feeding stations for birds (Helms, 1992). At the end of the first 5 year period, the productivity of pastureland and cropland in the program had improved markedly. The original participants in the government program continued these conservation efforts and farmers not enrolled in the demonstration project began to imitate their neighbors' farming practices. These general land use guidelines are now accepted practice throughout the project area (Gaumnitz, 2002).

Fifty to 60 years after the first farmers enrolled in the program, the effects of conservation oriented farming practices can be seen in the current status of sediment accumulation, streamflow, and stream habitat. A geographic information system (GIS) analysis of the watershed (Roldán, 2002) shows that forest cover in the watershed increased from 37% in 1939 to 50% in 1993. Improved soil conditions have changed flow patterns in the watershed. Precipitation is now more likely to infiltrate into the soil and recharge groundwater reserves; less of the water is directed to overland flow. Gebert and Krug (1996) analyzed long-term trends in annual peak discharge and minimum flows over the last 50 years. Flood peaks have diminished while low flows have increased significantly. A comparison of Driftless Area streams with a forested watershed (Wolf River, near Shawano, Wisconsin, 1916 to 1992) and other data showed that long-term changes in climate were not a causal factor.

The change in flow path, from overland to subsurface flow, has meant a decrease in erosional soil loss from more than 90,000 kg/ha/yr to about 9,000 kg/ha/yr. Sediment accumulation in the floodplain between 1938 and 1976 was 0.6 m (1.6 cm/yr). Since that time, the rate has decreased even more. Recent measurements have shown a sediment accumulation rate of only 0.5 cm/yr (Figure 7.7) (Trimble, 1983, 1999).

Increased groundwater input helps to lower and stabilize stream temperatures, improving conditions for coldwater fish such as trout. Non-native brown trout were found in area streams in the 1970s. These are being replaced by native brook trout as stream temperatures continue to decrease, dissolved oxygen levels increase, and sediment levels decrease. Recent increases in the number of streams that support self-sustaining native brook trout populations has reduced reliance on fish stocking in this area (Gaumnitz, 2002).

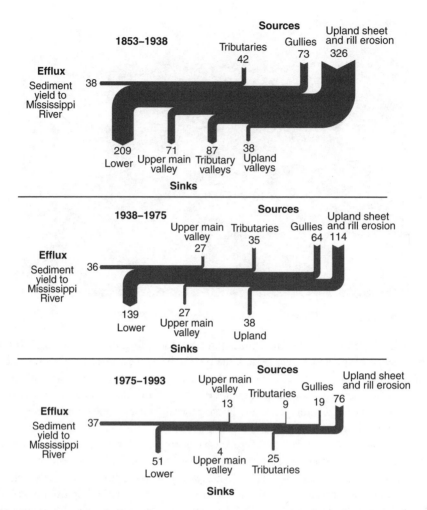

FIGURE 7.7 Sediment budgets for Coon Creek, Wisconsin, 1853–1993. This agricultural basin is approximately 25 km southeast of LaCrosse, Wisconsin, and has an area of 360 km². Numbers are annual averages for the periods in 10^3 Mg (tonnes)/yr. All values are direct measurements except "Net upland sheet and rill erosion," which is the sum of all sinks and the efflux minus the measured sources. The lower main valley and tributaries are sediment sinks, whereas the upper main valley is a sediment source. (Redrawn and modified slightly from Trimble [1999]; reprinted with permission from Trimble, S.W., Decreased rates of alluvial sediment storage in the Coon Creek Basin, Wisconsin, 1975–93, *Science*, 285, 1244–1246, 1999, copyright © 1999 American Association for the Advancement of Science.)

7.4.2 Sediment

As discussed in Chapter 6, natural erosion losses from undisturbed and carefully managed eastern forests are quite low. Estimates from various regions range from 25 kg/ha/yr to 224 kg/ha/yr (Bormann et al., 1974; Patric, 1976; Patric et al., 1984). Elevated levels of sediment in streams flowing through managed forests are usually only evident in the first few years following timber harvesting and decrease as the regenerating forest restores the water balance. Furthermore, as described in Chapter 6, most erosion occurs on logging roads and skid trails and this can be minimized with best management practices (Martin and Hornbeck, 1994; Patric, 1980).

Accelerated erosion associated with agricultural development is a large-scale phenomenon that affects entire regions (Verry, 1986). Soil particles from upland farm fields are dislodged by raindrops falling on soil no longer protected by forest vegetation and transported by overland flow (Gottschalk,

1945; McGuinness et al., 1971; Wilkin and Hebel, 1982). While much of the eroded soil is redeposited in rills and gullies within the watershed, a portion is delivered to streams, where smaller suspended particles increase stream water turbidity and larger particles reshape stream channels. Large quantities of eroded sediment may be deposited in streams and rivers, on floodplains, and in bays and coastal estuaries (Evans et al., 2000; Fitzpatrick et al., 1999; Gottschalk, 1945; Knox, 1977, 2001).

Regional variations in climate, geology, glacial history, soils, and topography (ecoregion characteristics) influence the amount of soil lost to erosion (Whitney, 1994; Wischmeier and Smith, 1978). Erosion is relatively low in areas of poorly sorted glacial till or in sandy glacial outwash areas with high infiltration. Generally these soils also are the least amenable to crop production. Farming of unglaciated, loess-derived, silt-loam soils in hilly terrain leads to some of the highest erosion rates (Knox, 1987, 2001; Sartz, 1978). Patterns of soil erosion have also been affected by the types of farming practices common to certain regions and influenced by the cultural heritage of the farmers who settled in particular areas (Trimble, 1982). Improved farming practices and soil conservation measures implemented since the 1940s have reduced, but not eliminated erosion losses (Box 7.2) (Knox, 1987, 2001). In some cases, sediment concentrations remain high despite improved farming practices, partly because sediment deposited in channels and on floodplains during earlier decades of high erosion is available for transport during flood events (Evans et al., 2000).

Erosion and sedimentation caused by clearing for agriculture are less problematic in the sandy, rocky, glacial till soils of New England than in other areas (Brady and Weil, 2002; O'Keefe and Foster, 1998; Whitney, 1994). Still, historical records contain anecdotal evidence of landscape-scale problems in this region. The city of New Haven, Connecticut, was forced to extend its main wharf more than 1200 m (3900 ft) between 1765 and 1821 (a period of widespread forest clearing) due to the accumulation of sediment from rivers flowing into the harbor. Similar actions were necessary in Boston, Barnstable Bay, and Nauset (Cronon, 1983).

A study in New Hampshire (Likens and Davis, 1975) and another in Connecticut (Brugam, 1978) examined changes in the rates of sediment deposition before and after European settlement of New England. A sediment core taken from Mirror Lake in New Hampshire indicated an increase in organic matter input to the lake from the surrounding watershed in the last 200 years. There also were associated increases in phosphorus, sulfur, and metals during this period. This coincided with a period of logging, clearing, plowing, and settlement (Likens and Davis, 1975). Sediment deposition in Linsley Pond, a shallow (maximum depth 14.8 m) kettle hole pond near New Haven, Connecticut, was examined by Brugam (1978). The area surrounding the pond was settled in 1700 and became the site of a small farm with dairy cattle. Sediment was dated using carbon 14 (^{14}C) and radioactive lead isotopes, as well as by pollen analysis. Here, as in other pollen analysis studies, there is a marked change in pollen types at the time of settlement. Older, deeper sediments are dominated by the pollen of trees and other forest species. At an elevation in the sediment core corresponding to the time of settlement, the pollen composition changes abruptly. Pollen from agricultural crops and weeds, especially ragweed, increases dramatically, while tree pollen declines sharply. Thus the change in sedimentation rates before and after settlement can be measured with reasonable accuracy. Sediment deposition averaged 0.23 ± 0.05 cm/yr from 1113 to 1700, prior to settlement, and rose to 0.33 ± 0.1 cm/yr during the period from 1700 to 1938, a 40% increase. Between 1938 and 1974 there were substantially higher rates of sediment influx (0.40 ± 0.05 cm/yr from 1938 to 1960 and 0.94 ± 0.24 cm/yr after 1960) associated with suburban development.

In the coastal and Piedmont regions of Maryland, presettlement erosion is estimated to have been less than 350 kg/ha/yr. Measurements of sediment yield from completely forested watersheds ranged from 17 to 52 kg/ha/yr. However, rich clay soils in this region are quite vulnerable to erosion when cultivated on slopes greater than 2% or 3% (Wolman, 1967). Gottschalk (1945) describes numerous examples of colonial towns that lost their harbors and navigable rivers because of sedimentation. Repeated dredging was required to maintain open waterways and harbors (Figure 7.8). Wolman (1967) estimated that sediment yield from farmed land in the Piedmont region of

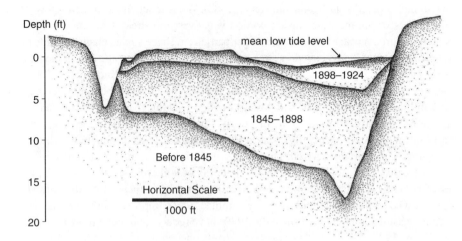

FIGURE 7.8 Sedimentation of the Patapsco River arm of Baltimore Harbor near the Hanover Street Bridge. (Gottschalk, 1945; reprinted with the permission of the American Geographical Society.)

Maryland averaged 2100 kg/ha/yr from the early 1800s to the 1940s, compared to approximately 35 kg/ha/yr in completely forested watersheds in the same area.

In the Chagrin Valley of Ohio, Evans et al. (2000) examined the sedimentary profile in a reservoir that had been emptied in 1994 after the dam failed. The highest rates of sedimentation occurred between the 1840s and 1870s, when settlement led to almost complete deforestation of the watershed. Dairy farming and sheep raising were the dominant activities during this period.

One of the most carefully designed studies of historical changes in sedimentation following European settlement involved sediment sampling in two lakes in a glaciated landscape in southern Michigan (Davis, 1976). Sediment cores were collected along east-west and north-south transects to account for differential rates of deposition. Segments of the cores were dated using pollen analysis and radiocarbon dating. The abrupt increase in ragweed pollen concentration in the sediment cores coincided with a change in the sediment from material that was high in organic matter to clay. This, again, is a typical consequence of deforestation, increased overland flow and soil erosion. Sediment accumulation ranged from 0.3 to 2.5 cm/yr in different parts of the lake basin between 1830, the time of settlement, and the 1970s. An estimated 2240 tonnes of sediment were deposited in the lake during this 146 year period. This represents a post-agricultural sediment yield of 900 kg/ha/yr from the watershed compared to a pre-settlement rate of 90 kg/ha/yr. More detailed measurements showed that the sedimentation rates were much larger during a short period at the time of forest clearing, approximately 40 times greater than pre-settlement rates.

Studies in the Driftless Area document extreme examples of erosion and sedimentation caused by clearing and farming. In addition to the streamflow research discussed earlier, Sartz (1970) also compared levels of suspended sediment in stormflow from small homogeneous watersheds in the Coulee Experimental Forest (LaCrosse County, Wisconsin). The high-intensity convective storms (thunderstorms) that are common in this area in the spring and summer can substantially accelerate soil erosion. Rainfall intensity quickly exceeds the infiltration capacity of the soil and causes overland flow and surface erosion. In addition, high-velocity raindrops from these storms strike the ground with considerable force and detach more soil than low-intensity, long duration storms (Knox, 2001; Sartz et al. 1977). Maximum concentrations of suspended sediment in stormwater from tilled land were 238,000 mg/L, compared to 19,800 mg/L from an alfalfa meadow, 3600 mg/L from a logged forest, and 100 mg/L from an undisturbed forest.

Knox (1987) studied sediment deposition in river valley floodplains of the lead-zinc mining district, located in the southern part of the Driftless Area. He used measurements of concentrations of trace metals from mining operations (dates of mining activities in the area are well documented) and radiocarbon dating of wood fragments in sediment cores to reconstruct the pattern of erosion and sedimentation back to presettlement times. The depths of sediment deposits in a floodplain are more variable over a larger area than those in lake bottoms in closed basins. A portion of eroded sediments is carried out of the watershed entirely as suspended particles in streamflow and sediment deposition may vary depending on the configuration of the river valley at different points along the course of the river (Magilligan, 1985). Thus sedimentation rates found in these studies are not directly comparable to the limnological studies in Connecticut and Michigan (Brugam, 1978; Davis, 1976). Similar rates of sedimentation in river valleys may indicate higher rates of erosion.

As noted earlier, the Driftless Area was first settled in the 1830s. Accelerated erosion began immediately following settlement. Presettlement sedimentation rates were estimated at 0.02 cm/yr. Sedimentation rates measured in floodplain sediment cores increased moderately (compared to levels after 1880) and were quite variable by location. In the lead-zinc district, the height of agricultural development in the region (1880 to 1920) was also a period of above average precipitation. Erosion created gullies 4 to 5 m deep and up to 8 m wide (Knox, 1987, 2001). Sedimentation occurred along the banks of tributaries and on main stem valley floodplains. By the 1920s, tributary channels had widened in response to the increased streamflow and sediment load and sedimentation on floodplains was less noticeable. However, sedimentation in the floodplains continued at rates of 4 to 5 cm/yr, reaching a maximum in the 1930s (Knox, 1987). Sedimentation has been greatly reduced during the second half of the 20th century with the widespread implementation of improved land management and soil conservation practices (Box 7.2) (Knox, 1987; Helms, 1992; Roldán, 2002).

Subsequent investigations have shown that much of the topsoil may have been lost from agricultural soils in this region during periods of accelerated erosion. Knox (2001) compared the soil profiles of an agricultural field and an undisturbed prairie. In the surface soil (A horizon) of the prairie, levels of organic matter were relatively high (5% to 6%), while clay content was relatively low (18% to 22%). The surface soil of the agricultural field had levels of organic matter (1.5% to 2.5%) and clay (32%) that were comparable to the subsoil (B horizon) of the prairie. Knox interpreted this to mean that the original, organically enriched topsoil had been lost from the agricultural fields through years of erosion, with severe consequences for the region as a whole:

> ...the historic stripping of porous, permeable surficial sediment and the exposure of less porous and less permeable soil at the ground surface in cultivated areas have lowered rates for soil infiltration capacities. In turn, reduced infiltration has produced a progressive positive feedback effect that favors additional surface runoff and more soil erosion (Knox, 2001, reprinted with permission from Elsevier).

7.4.3 CHANNEL FORM

Research designed to measure historic stream channel changes following clearing for agriculture has been conducted in the Platte River watershed of Wisconsin—within the Driftless Area (Knox, 1977). A federal land survey of the watershed (including both the Big Platte and Little Platte subwatersheds) conducted in 1832–1833 measured and recorded stream widths prior to settlement. At this time the watershed was 70% forested and 30% open prairie. These same streams were resurveyed in the 1970s, when woodlands occupied approximately 25% of the Big Platte watershed.

After 146 years of clearing and farming, channel form had changed considerably. Stream headwater channels (approximately the first 37 km) were shallower and more than twice the width of the conditions in the 1830s. The middle reaches had also increased in width, by up to 40%. Bank erosion had increased because the stream channels were conveying more water during high frequency flood events as a result of increased overland flow. In the lower reaches of the river, the influence of increased streamflow was less important than the vast quantities of sediment delivered both from upland soils and from eroding stream channels. Beyond the first 56 km, the streams were narrower and deeper. In general, channel top widths had decreased by about 25%. This was caused by the deposition of large amounts of sediment on the floodplain during overbank flow events. This was exacerbated by the construction of a dam on the Mississippi River, 10 km downstream at Dubuque, Iowa. The dam caused the Mississippi to deposit "extreme magnitudes of historical siltation on the floodplain and channel margins" (Knox, 1977).

An analysis of floodplain soils revealed that the original prairie topsoil was buried under as much as 4 m of sediment. The sediment consisted of silts and sandy silts. The rates of channel erosion and sediment deposition have varied with the intensity of agricultural land use as well as the type of farming. The greatest changes occurred between the 1880s and late 1930s, when the upland fields were cultivated to grow row crops, primarily corn, and soil conservation practices were uncommon.

The channel changes in the Platte River system were more severe than those identified in studies of other streams in the same region (southwestern and southern Wisconsin) — Kickapoo Creek, Otter Creek, and Turtle Creek. The Platte River is underlain by dolomitic limestone, which provides abundant coarse bedload. In contrast, the Kickapoo and Otter Creek watersheds are located in areas dominated by sandstone bedrock that does not provide coarse bedload. The Turtle Creek watershed is located in a glaciated region of southern Wisconsin, an area of low-lying rolling hills. Large areas of sandy glacial outwash soils are found in the watershed. These soils are far more permeable than the fine-textured silt and clay soils in the unglaciated Driftless Area. This shows how increased infiltration capacity and larger soil particles can, in some cases, mitigate the effect of land use change. Increases in overland flow and erosion in these areas would have been smaller and associated channel changes less marked (Knox, 1977).

It should be noted that even larger changes in stream characteristics and habitat in the Northeast were brought about by the construction of dams. During colonial times, settling an area and establishing a town commonly involved building a small dam to power a gristmill or a sawmill (Cronon, 1983). Larger dams were built during the 1900s to provide water and water power for industrial growth (Judd, 1997). Eventually even more massive structures were built for hydroelectric power, municipal reservoirs, and flood control. Writing in 1956, noted geologist and hydrologist Luna Leopold concluded that "The effect of these structures on changes in the channels greatly overshadows the effects due to varying proportions of sediment to water produced by man's use of the land" (Leopold, 1956). Examination of this topic is beyond the scope of this publication, but for a brief introduction and general discussion of the ecological effects of dams, the reader is referred to Leopold (1997). The topic is covered extensively in Postel and Richter (2003).

7.4.3.1 Water Temperature

As in forested systems, removal of riparian vegetation causes temperature increases in smaller headwater streams in agricultural landscapes (Younus et al., 2000). The mechanisms and effects are the same in forested and agricultural landscapes (see Chapters 5 and 6).

7.5 AGRICULTURAL LAND USE AND WATER CHEMISTRY

7.5.1 Nutrients

Large quantities of nitrogen and phosphorus are introduced to agricultural fields as artificial and manure-based fertilizers. When nutrient input exceeds the amount that can be used by plants and stored in the soil, excess nitrogen and phosphorus is transported by overland flow and subsurface flow to streams, lakes, and estuaries. Soluble compounds such as nitrate flow vertically and laterally through the soil and can accumulate in groundwater and water supply wells. Nutrient pollution from agriculture has contributed to the eutrophication of aquatic ecosystems on local and regional scales. On the East Coast, excess agricultural nutrients flow to coastal bays and estuaries (Boesch et al., 2001; U.S. Environmental Protection Agency, 1998). In the Midwest, nitrogen and phosphorus from agriculture drain into the Great Lakes (Richards et al., 2002) and through the Mississippi

River system to the Gulf of Mexico (Burkhart and James, 1999; Goolsby et al., 2000, 2001). These nutrient inputs contribute to eutrophication and hypoxia and anoxia in receiving waters, resulting in the deterioration of freshwater, estuarine, and marine ecosystems.

7.5.1.2 Nitrogen and Phosphorus in Streamflow

Beaulac and Reckhow (1982) analyzed 40 studies of nutrient export in stream water. Eighteen of these had been conducted in Maryland, Ohio, Wisconsin, and Minnesota. In general, they found that both nitrogen and phosphorus were exported in greater quantities from land in row crops than from forested land. Manure storage facilities released the most nitrogen and phosphorus.

Johnson et al. (1997) studied 62 watersheds within the Saginaw Bay watershed of central Michigan. They found that alkalinity, total dissolved solids, and nitrogen concentrations (in the form of nitrate [NO_3] and nitrite [NO_2]) were higher in streams flowing from watersheds occupied primarily by row crop agriculture than in watersheds where other land uses predominated. Other land uses included urban, non-row crop agriculture (orchards, pasture, and occasional feedlots), forests and rangeland, nonforested wetland, and forested wetlands. (Animal feedlots were not a large factor in this study.) The mean concentration of nitrogen (nitrates and nitrites) in nine watersheds in which 80% to 98% of the land area supported row crop agriculture was 3.6 mg/L. In 26 watersheds with 18% to 75% of the land in row crop agriculture, the mean concentration of nitrate and nitrite nitrogen was 1.0 mg/L. Categorization of the watersheds was based on GIS analysis of aerial photographs provided by the Michigan Department of Natural Resources. Land use, in particular row crop agriculture, was the most important variable explaining the highest nitrate concentrations; they occurred in the summer after the largest fertilizer application. Overall watershed land use, land cover patterns, and management activities did not correlate well with total phosphorus export. This was attributed to two factors: (1) water samples were collected twice a season, not necessarily during storm periods when sediment-associated phosphorus levels would be expected to be highest, and (2) the most intensively farmed areas in this region are associated with soils high in lacustrine clays that increase phosphorus retention. A similar study of 18 small watersheds and other sites in the Aroostook River basin in Maine (Cronan et al., 1999) found that the mean level of nitrate nitrogen in waters from agricultural watersheds was 20 times higher than that from forested watersheds (2 mg/L versus 0.10 mg/L) during periods of moderate flow.

Nutrient export in the Chesapeake Bay watershed has been studied intensively (Correll et al., 1999a,b; Gburek et al., 1991, 1994; Jordan et al., 1997a,b,c; Pionke et al., 1996; Vaithiyanathan and Correll, 1992). Higher concentrations of nitrogen in streams and groundwater have repeatedly been associated with cropland and dairy farms (Correll et al., 1992; Jordan et al., 1986, 1997a,b,c). Peterjohn and Correll (1984) found that a cornfield discharged 39 kg/ha/yr of total nitrogen (including nitrate, nitrite, ammonium, and organic nitrogen) in overland flow and groundwater discharge. This compared to a nitrogen discharge of 0.42 kg/ha/yr from forested land. Additional research (Correll et al., 1992; Jordan et al., 1986) found that excessive amounts of nitrogen and phosphorus from croplands entered streams, despite the presence of riparian forests. This study measured nitrogen and phosphorus flux through a 2286 ha watershed and its receiving waters from March 1981 to March 1982. Watershed land use was distributed among forests (62%), cropland (23%), pastures (12%), and wetlands (3%).

> Croplands discharged far more nitrogen per hectare in [overland and subsurface flow] than did forests and pastures. Most of the nitrogen released by the croplands was absorbed by adjacent riparian forests. However, nutrient discharges from these riparian forests still exceeded discharges from pastures and other forests (Jordan et al., 1986).

One study in the Chesapeake Bay area (Vaithiyanathan and Correll, 1992) compared a 6.3 ha forested watershed and a 16.4 ha watershed with four cultivated fields totaling 10.4 ha. The agricultural watershed exported more phosphorus (2.41 kg/ha/yr) than the forested watershed (0.31 kg/ha/yr). In the forest, phosphorus was primarily bound in organic form. Phosphorus released by the decomposition of leaves and other forest litter was taken up and held in new forest vegetation. The small amount of phosphorus lost in streamflow was replaced by phosphorus from weathering and precipitation. In the agricultural watershed, there was less organic input to the soil because of the lack of a forest canopy and the removal of crops. Furthermore, tilling promoted soil respiration and organic matter decomposition, and released organically bound phosphorus. In these cultivated fields, most of the phosphorus was in mineral (inorganic) form. Inorganic phosphorus is more soluble and mobile than organic phosphorus. Sediment-associated phosphorus is more easily lost through erosion from open fields than from forested land. The phosphorus load was also increased by fertilizer additions in cultivated fields.

Jordan et al. (1997a) sampled 17 watersheds with varying proportions of forest, cropland, and pasture. The watersheds were located in four clusters in the Coastal Plain region of the Chesapeake Bay watershed and ranged in size from 100 to 2000 ha. Over this large area it was found that phosphorus export correlated with the concentration of suspended sediment in stream water, but that this did not necessarily correlate with the proportion of agricultural land use within the watershed. Differences in soil erodibility and landform had a greater effect than land use on phosphorus export from these watersheds.

The Rhode River, in the Maryland Atlantic Coastal Plain, was one of the four watershed clusters included in the study of Jordan et al. (1997a). It consists of seven small (6.1 to 253 ha) contiguous watersheds. Correll et al. (1999a,b) studied nutrient fluxes within these watersheds over a 25-year period from 1971 to 1996. They examined the relationships between precipitation and land use, and total suspended sediment in stream water and the nutrient content of that sediment. Phosphorus loading from a cropland watershed ranged from 0.14 to 13.3 kg/ha/yr. In a forested watershed, phosphorus loading ranged from 0.013 to 1.18 kg/ha/yr. Phosphorus export correlated with rainfall (and stormflow) and this correlation was highly significant ($p < .001$). The volume of precipitation explained 54% to 73% of the variance in total phosphorus flux.

In sum, studies of a small number of proximate, predominantly single-use watersheds are able to find a clear correlation between agricultural land use and increased phosphorus input to streams (Correll et al., 1999a,b; Vaithiyanathan and Correll, 1992). Studies that encompass larger landscapes and mixed-use watersheds are unable to demonstrate clear associations between varying amounts of agricultural land use and the amount of phosphorus discharged to streams (Johnson et al., 1997; Jordan et al., 1997a,b,c). This is understandable because there are many factors in addition to land use (e.g., land use history, topography precipitation patterns, soil chemistry, and movement of sediment through the watershed) influencing phosphorus storage and transport at larger spatial and longer temporal scales.

7.5.1.3 Nitrogen in Groundwater

Lichtenberg and Shapiro (1997) analyzed water quality data from 810 wells in Maryland. Their regression analyses found a relation between elevated nitrate levels (greater than 3 mg/L) and poultry farms and cornfields. Poultry farms produce large quantities of manure. Chemical fertilizers are applied most heavily in cornfields. The hydrogeologic characteristics associated with high nitrate in groundwater include unconfined aquifers, limestone bedrock formations, and fractured bedrock. In general, nitrate pollution is found far more frequently in shallow groundwater than in deep wells (Natural Resources Conservation Service, 1997). A study in Pennsylvania (Pionke and Urban, 1985) tested 14 wells located in a 7.4 km^2 watershed. The watershed was 65% agricultural (about 57% cropland and 8% pasture) and 35% forested. Eleven of the wells were located in cropland and three in forested areas. The wells were sampled 10 times over a period of 10 years from 1973 to 1982. Each well was sampled five times during the growing season and five times during the dormant season. Mean concentrations of nitrate, phosphate, and chloride were five to seven times higher in groundwater beneath cropland than forests. The mean nitrate concentration in groundwater under forested lands was 0.71 mg/L, with a maximum of 4.1 mg/L, compared with a mean of 3.0 mg/L under cropland, with a maximum of 23.3 mg/L. The highest median nitrate concentration (13 mg/L) reported for shallow groundwater in the U.S. Geological Survey (USGS) National Water Quality Assessment Program (NAWQA) was found in an agricultural area in the Coastal Plain of southern New Jersey (Ayers et al., 2000). Nitrogen contamination of shallow groundwater is a continuing concern in the Coastal Plain region because of the predominance of sandy soils with high infiltration rates (Watt, 2000). Infertile sandy soils require large amounts of fertilizer to be productive for agriculture and excessive application rates almost immediately lead to contamination.

Case studies of adverse health effects related to nitrates in groundwater were reported in the 1950s (Dunne and Leopold, 1978). Nitrate is chemically reduced to nitrite (NO_2) in the human digestive tract. Nitrite combines with hemoglobin in the blood to form methemoglobin. This reduces the capacity of blood cells to transport oxygen. Infants less than 3 months old are most susceptible and may become cyanotic ("blue baby syndrome") when seriously affected. Nitrate in groundwater has also caused serious problems for domestic livestock—horses, pigs, and chickens. Symptoms include cyanosis, shortness of breath, rapid heartbeat, staggered gait, frequent urination, and collapse, in some cases leading to coma and death (Canter, 1996).

7.5.2 Pesticides

The publication of Rachel Carson's groundbreaking work *Silent Spring* in 1962 alerted the public to the dangers of persistent agricultural pesticides, especially organochlorine insecticides such as DDT. Over the last 40 years the government and the agricultural industry have been compelled to acknowledge and respond to the continuing public concern regarding agricultural pesticides in food and water (Lear, 1992). This led to bans on the most overtly dangerous and persistent chemicals, as well as the development of new shorter-lived herbicides and insecticides. Concerted efforts at integrated pest management (IPM) involving biological controls have also increased. Due to the persistence of the organochlorine pesticides, DDT, dieldrin, and chlordane are still found in streams and stream bed sediments in the United States. Twenty years or more after their use was restricted, sediment quality guidelines for these compounds were exceeded at more than 20% of agricultural sites tested (U.S. Geological Survey, 2001). The use of insecticides declined between 1964 and 1992, while the use of herbicides increased substantially. Today herbicides constitute approximately 70% of the pesticides used in agriculture (Figure 7.9).

Herbicides currently in use have half-lives ranging from 8 days (e.g., alachlor, an herbicide used for corn, soybeans, and peanuts) to more than 60 to 100 days (e.g., atrazine, the most commonly used herbicide, was applied to 64 million acres of cropland in the United States in 1990) (EXTOX-NET, 2002). In some cases, however, the intermediate metabolites formed during the breakdown of these chemicals (e.g., deethylatrazine from atrazine) also appear to have toxic effects on aquatic organisms (DeLorenzo et al., 1999).

If these chemicals remain in the soil for a sufficient period of time they break down to nontoxic compounds. (Chapter 3 contains more detailed information on pesticide soil processing and transport.) However, if pesticides are transported by overland flow or leached to groundwater systems before they break down, they can threaten aquatic ecosystems and present possible risks to human health. Groundwater and drinking water well studies typically identify low concentrations of pesticides, well below established maximum contaminant loads (MCLs) set as health standards by the EPA, in a small number of samples (Gallagher et al., 1996; Holden et al., 1992; Natural Resources Conservation Service, 1997; Novak et al., 1998; Pionke and Glotfelty, 1989). In contrast, pesticide surface water sample concentrations can be considerably greater. Analyses of 146 water samples taken from eight locations along the Mississippi, Ohio, and Missouri Rivers found atrazine present in all of the samples, with 27% at concentrations above the EPA's MCL of 0.003 mg/L

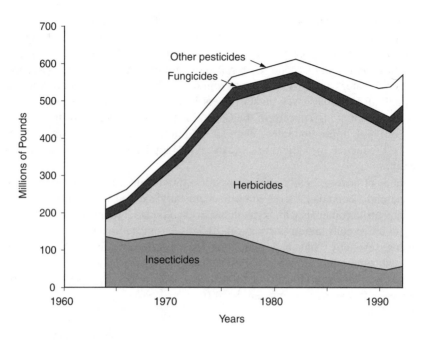

FIGURE 7.9 Pesticide use on selected crops by pesticide type, 1964 to 1992 (courtesy of USDA).

(USGS cited in EXTOXNET, 2002). Recent studies raise questions about whether MCLs currently determined by the EPA for drinking water are sufficient to protect aquatic organisms and, ultimately, human health (Box 7.3).

Box 7.3   Health Effects of Herbicides and Fungicides (Frogs and Humans)

Compared to earlier compounds such as DDT, modern pesticides have been considered by many to be relatively safe. These compounds decompose relatively quickly and often do not appear to negatively affect growth and survival rates for laboratory animals. Atrazine, the most persistent agricultural herbicide and the one most commonly found in waterways, is classified as being only slightly toxic to fish, although it does bioaccumulate in fish organs. Questions remain, however, about long-term effects of chronic low-level exposure (U.S. Geological Survey, 1999). Recent studies indicate that, although not immediately toxic, these chemical compounds may disrupt hormonal pathways in biological organisms in ways that can interfere with reproduction and accelerate the growth of tumors (Sanderson et al., 2000).

In laboratory studies at the University of California, African clawed frogs (*Xenopus laevis*) were immersed in water containing varying amounts of atrazine (Hayes et al., 2002). The EPA MCL for atrazine is 0.003 mg/L (or 0.003 ppm). Exposure to atrazine produced no changes in mortality, time to metamorphosis, or weight at metamorphosis. However, frogs developed gonadal abnormalities at doses as low as 0.0001 mg/L (0.1 ppb). Between 16% and 20% of exposed animals developed either multiple gonads or became hermaphroditic with multiple testes and ovaries. Testosterone levels in mature male frogs were reduced by 90% following exposure to solutions containing atrazine at 0.025 mg/L. Exposure to atrazine at 0.001 mg/L significantly reduced the laryngeal size of male frogs. Hayes et al. hypothesize that atrazine stimulates the production of aromatase, an enzyme that promotes the conversion of testosterone to estrogen. These experiments have been repeated 51 times at University of California–Berkeley with similar results.

Researchers in The Netherlands in collaboration with the Institute of Environmental Toxicology at Michigan State University, have confirmed that relatively low concentrations of the related 2-chloro-s-triazine herbicides (atrazine, simazine, and propazine) significantly increase the production of the aromatase enzyme in human cancer cells. This was linked to the conversion of androgens to estrogens and the possible acceleration of tumor growth (Sanderson et al., 2000).

A third study (Cavieres et al., 2002) measured the effect of adding different levels of a commonly available commercial herbicide (a mixture of 2-4-D, mecoprop, dicamba, and inactive ingredients) to the drinking water of pregnant mice. Litter size was significantly reduced in all dosing groups and the effect was greatest in animals exposed to the lowest herbicide concentration, less than the MCL of 0.07 ppm set by the EPA. The authors speculate that low dose herbicides are more likely to interfere with the endocrine balance governing the survival and implantation of embryos in female mice, possibly because low levels of pesticides tend to mimic hormonal activity.

Garry et al. (1996) examined the relationship between the rate of birth defects in infants born between 1989 and 1992 in Minnesota and parental pesticide exposure. There was a significantly higher rate of birth defects of all types among children born to licensed pesticide appliers. Within this group, the greatest number of birth defects was found among children of pesticide appliers conceived in the spring, when pesticides are applied in large amounts. The rate of birth defects also varied by region. The highest birth defect rate among children of pesticide appliers (30.0 per 1000 live births) and among the children in the general population (26.9 per 1000 live births) was found in western Minnesota, a major wheat, sugar beet, and potato growing area. A subsequent study (Garry et al., 2002) found that neurologic and neurobehavioral problems were most common among children born to parents exposed to the fumigant phosphine and the herbicide glyphosate.

U.S. Geological Survey (1999) scientists have voiced concerns about the interpretations of EPA limits for pesticides in both agricultural and urban environments:

> Current standards and guidelines do not completely eliminate risks because: (1) values are not established for many pesticides, (2) mixtures and breakdown products are not considered, (3) the effects of seasonal exposure to high concentrations have not been evaluated, and (4) some types of potential effects, such as endocrine disruption and unique responses of sensitive individuals, have not yet been assessed.

Results of a national study (Holden et al. 1992) conducted with the active participation of the EPA found that less than 1% of 1430 sampled drinking water wells in 89 counties in the Northeast, Southeast, and Midwest had detectable levels of alachlor, a rapidly decomposing herbicide. Atrazine, a more persistent chemical, was found in 12% of the wells. Using this sample, it was predicted that atrazine concentrations exceeded the 0.003 mg/L MCL in 1% of the estimated population of 6 million wells in the target area, or 5700 wells. A significant correlation was also found between nitrate concentrations of 10 mg/L or greater in well water and the presence of herbicides, particularly atrazine and simazine. A study in Pennsylvania (Pionke and Glotfelty, 1989) found that atrazine concentrations in wells correlated with the intensity of corn production within a 100 m radius semicircle upslope of each well. Eighteen wells and two free-flowing springs were sampled three times each (two wells were added during the second sampling and another was added during the third sampling). Corn production intensity (CPI), defined as "the average percentage of the near well area planted to corn over the crop rotation cycle," was rated for each well. Little or no atrazine was found in most wells with a CPI of less than 50%. Wells with a CPI of 75% to 100% contained atrazine at a concentration of 50 to 1100 ng/L (0.00005 to 0.001 mg/L). Water samples were also tested for other pesticides (alachlor, metolachlor, carbofuran, terbufos, chlorpyrifos, and fonofos).

None of these were found. Although Pionke and Glotfelty (1989) originally intended to sample groundwater during both recharge and discharge periods, all three sample sets occurred during recharge. It is possible therefore that atrazine concentrations were higher because downward flow during recharge reduces the residence time in the soil.

Gallagher et al. (1996) investigated the transport of pesticides (atrazine, cyanazine, alachlor, metolachlor, and carbofuran) and nitrate from unconfined aquifers to tidal surface waters in Virginia streams that are tributaries of the Chesapeake Bay. They collected samples every 4 to 6 weeks from May 1992 to February 1993 from upland soils and nearshore sediments at four agricultural sites. They collected water samples from three monitoring wells at each of the agricultural sites and from estuarine groundwater discharge sites (one at each site), also at 4 to 6 week intervals. Pesticides were found in upland soil (approximately 50%) and groundwater samples (approximately 40%). Groundwater concentrations were generally less than 0.0001 mg/L. The herbicides alachlor, atrazine, cyanazine, and metolachlor were found in 18% of the samples from nearshore groundwater discharge sites (concentrations of 0.05 to 0.5 µg/L). The authors concluded that subsurface flow transports agricultural chemicals from upland farms to nearshore environments and that, despite low concentrations, this pathway should be considered in management policies designed to protect tidal estuaries.

Aside from effects on individual organisms, herbicides such as atrazine may have broad-based, ecological consequences. DeLorenzo et al. (1999) showed that atrazine and deethylatrazine significantly reduce photosynthetic activity and dissolved oxygen concentrations in estuarine microbial communities, leading to a decrease in phytoplankton, an increase in bacterial abundance and productivity, and alteration of the microbial food web. Another chemical, endosulfan, reduced the number of cyanobacteria. Significant effects were only observed at relatively high concentrations (0.05 mg/L for atrazine and 0.25 mg/L for deethylatrazine), but the authors expressed concern about the possibility of synergistic effects from a variety of herbicides, all targeting photosynthetic pathways and accumulating in the same waterway.

7.5.3 Pathogens

Waterborne intestinal parasites such as *Cryptosporidium* and *Giardia* live and reproduce in the digestive tracts of wildlife, farm animals, and humans. Infection results in severe intestinal distress (diarrhea, nausea, vomiting) and dehydration and can be life threatening to infants, the elderly, and persons with weakened immune systems (e.g., AIDS patients and patients being treated for cancer). Infected animals (including humans) shed the eggs or oocysts of these parasites in their feces. *Cryptosporidium* and *Giardia* are transmitted from animals to humans when a previously uninfected individual consumes the oocysts either from direct contact with fecal matter or, more commonly, in contaminated drinking water (Centers for Disease Control and Prevention; Lawhorn, 1996; Mitchell, 2002). The presence of fecal coliform bacteria in water is frequently measured as an indicator of the probable presence of *Cryptosporidium* and *Giardia* (Fraser et al., 1998; National Research Council, 2000).

A study in northern Saskatchewan found that "concentrations of *Cryptosporidium* and *Giardia* were significantly higher in streams in watersheds with more intensive agricultural [livestock] production compared with non-agricultural areas" (Mitchell, 2002). Fecal coliform bacteria contamination is frequently associated with areas where livestock have unrestricted access to streams. Streams where riparian areas were protected from grazing animals even for intermittent periods had significantly lower fecal coliform concentrations in a study conducted in the Driftless Area of southeastern Minnesota (Sovell et al., 2000). A study in the lower Susquehanna River watershed in Pennsylvania found that water from nearly 70% of 146 household wells contained detectable levels of coliform bacteria (Lindsey et al., 1998). Bacteria included fecal coliform, fecal streptococcus, and *Escherichia coli*. Possible sources of contamination included inadequate protection of wellheads from overland flow and septic system failure, and manure application to agricultural fields. Not surprisingly, it was found that "bacteriological contamination was more likely to occur in water from wells in agricultural areas than in water from forested areas."

7.6 AGRICULTURAL EFFECTS ON AQUATIC ECOSYSTEMS

Agricultural land use typically changes the quantity and timing of streamflow, increases sediment and nutrient loading, and may affect water temperatures and water chemistry (Dance and Hynes, 1980)—all fundamental features of stream habitat. Roth et al. (1996) measured fish abundance and biodiversity at 23 sites in the River Raisin watershed in southeastern Michigan. They found that an index of biotic integrity (IBI) for that region decreased by more than 50% (from 46 to 22) as the proportion of agricultural land in subwatersheds increased from zero to 100%. Land use and the presence and extent of riparian forests in the watershed were the most important factors determining abundance and diversity of fish populations at the sampling sites. A study of agricultural and forested watersheds in Montana showed that predominantly agricultural watersheds had the lowest levels of aquatic macroinvertebrate abundance and diversity, including Ephemeroptera, Plecoptera, and Trichoptera (EPT) (Rothrock et al., 1998). In this case, agriculture included hayfields and pasture for beef cattle. In nutrient-poor environments, the addition of agricultural nutrients may enhance the growth and abundance of some fish species (Dance and Hynes, 1980; Lenat and Crawford, 1994).

Dance and Hynes (1980) compared macroinvertebrate communities in two streams in adjacent watersheds of comparable size in southwestern Ontario. Climatic and geologic conditions in the two watersheds were similar and historical records showed that both areas had similar forest cover in 1840. Since that time the western watershed has been more extensively developed, primarily for dairy farming. The western watershed retained only 6.9% forest cover, compared to 34.1% in the eastern watershed, and far more (64% versus 30%) of the streambanks in the eastern watershed were protected by riparian forest. Only two species of Plecoptera (stoneflies) were found in the stream draining the western watershed, compared to eight in the forested streams in the eastern watershed. There were more taxa (110 versus 98) in streams in the eastern watershed and more of these taxa represented intolerant species. In addition, most of the macroinvertebrates in the agricultural streams in the western watershed were grazers or detritivores, trophic behaviors typical of species that tolerate warm waters with higher nutrient and sediment concentrations (see Chapter 4).

7.7 AGRICULTURAL BMPs

As methods for protecting soil and water in agricultural areas, BMPs are well established. The development of these conservation practices began in the 1930s, during the Great Depression, with the establishment of the SCS. The NRCS *National Handbook of Conservation Practices* now contains 159 "baseline national conservation practice standards." Agricultural engineers and extension specialists select relevant BMPs and adapt them to local conditions. Broadly categorized, agricultural BMPs for water quality protection include methods for (1) source reduction methods and (2) sediment, nutrient, and pesticide transport control or mitigation. Practices designed to reduce sediment sources include erosion control through conservation tillage, contour farming, and strip cropping (e.g., planting alternating strips of corn and soybeans). Riparian forest buffers help to intercept sediment, nutrients, and pesticides. Riparian forest buffers also provide storage sites for sediment and phosphorus, denitrification, and the decomposition of pesticides. BMPs for grazing animals typically exclude livestock from riparian areas with fencing. This prevents the destruction of riparian vegetation, trampling of streambanks, and the direct deposit of manure in stream channels (Meals, 2001). Nutrient management BMPs focus on the amount and timing of fertilizer and manure applications. Properly designed manure storage facilities are an integral part of most nutrient management programs (Inamdar et al., 2002). Detailed information on agricultural BMPs is available from the NRCS website and state NRCS offices and websites.

7.7.1 EFFECTIVENESS AND COMPLIANCE

It is difficult to separate BMP effectiveness and compliance. Forestry BMPs can be evaluated on individual harvesting sites in otherwise undisturbed watersheds. In contrast, agricultural BMPs usually must be evaluated within agricultural landscapes. Studies designed to test the effectiveness of BMPs usually measure the results that can be achieved in watersheds with many individual farms and farmers who comply to varying degrees with multiple management recommendations. In some instances, a few noncompliant farms can add enough nitrogen or phosphorus to negate the water quality improvements produced by diligent BMP implementation across the rest of the watershed. Measurement of BMP effectiveness is also confounded by time lags between implementation and water quality improvement since the store of nitrogen and phosphorus may be the legacy effect of decades of land use. In addition, while the implementation of BMPs may bring about substantial improvements in water quality, they may not be sufficient to meet water quality standards (Clausen and Meals, 1989). The studies described below evaluate the implementation of agricultural BMPs in a variety of locations and environmental contexts.

7.7.2 CASE STUDIES

7.7.2.1 Maumee and Sandusky Watersheds, Northwestern Ohio

Baker and Richards (2002) evaluated the results of BMP implementation in the Sandusky and Maumee River watersheds in northwestern Ohio. The Sandusky and Maumee Rivers flow into Lake Erie. During the 1950s and 1960s, excessive phosphorus loading in the Lake Erie watershed caused eutrophication and associated deterioration in water quality from algal blooms and reductions in dissolved oxygen. Phosphorus sources included point sources (municipal wastewater treatment plants) and nonpoint sources (agricultural land use). Bans on phosphate detergents and improvements in wastewater treatment facilities produced major reductions in phosphorus loading. Loading studies indicated that an additional reduction of 2000 tonnes/yr would be needed to reach target levels required to reduce algal blooms and anoxia in Lake Erie. Counties in Ohio were assigned a phosphorus reduction target of 1390 tonnes/yr. Plans for reducing phosphorus loading focused on erosion control because 75% to 80% of phosphorus loading was from sediment-associated particulate phosphorus. Soil phosphorus levels were high and it seemed unlikely that reductions in phosphorus applications would have any immediate effect. Still, reductions in phosphorus application rates did occur, as farmers began to calculate phosphorus application rates based on the amount needed to replace phosphorus lost through crop removal rather than the amount required to increase phosphorus concentrations in soil tests.

Baker and Richards used a mass balance model to construct a phosphorus budget using county records of fertilizer sales and numbers of beef cattle, dairy cows, swine, and sheep to estimate commercial fertilizer and manure inputs. The amount of phosphorus in sewage sludge applied to fields as fertilizer was estimated from data from 77 municipal wastewater treatment plants in the two watersheds. Estimates showed that more than 75% of phosphorus inputs came from commercial fertilizer, 20% from manure, 2% from sludge, and 1% from precipitation. Phosphorus inputs for 1993 to 1995 were 30% to 40% lower than in 1979 to 1981 period because of decreased application rates for commercial fertilizer and a reduction in the number of farm animals in the Sandusky and Maumee watersheds. During the same time period, crop yields increased, resulting in increased phosphorus removal by crops. The concentration and loading of total suspended sediment (TSS), total phosphorus (TP), and soluble reactive phosphorus (SRP) in streams and rivers all decreased between 1975 and 1995. SRP decreased by 65% to 89% in the Maumee and Sandusky Rivers following corrections for seasonal and annual variation in discharge. In the Sandusky watershed, average annual SRP export (not corrected for variation in discharge) decreased from 92.8 tonnes/yr between 1975 and 1984 to 40.1 tonnes/yr between 1986 and 1995. Despite reductions in fertilizer and manure inputs, soil phosphorus concentrations increased during the same period. This indicated that phosphorus inputs still exceeded outputs and that phosphorus soil storage remained high.

The reduction in sediment and total phosphorus was expected after the widespread adoption of conservation (no-till) tillage throughout the Maumee and Sandusky watersheds, as well as the focused protection of highly erodible lands. The success of conservation tillage in reducing phosphorus transport in this area also may be related to soil type. Soils in northern Ohio are poorly or very poorly drained alfisols with high clay content and low infiltration rates. Overland flow and sediment transport (and associated phosphorus transport) are more likely to occur on these sites than in areas with well-drained, sandy soils. It would therefore be expected that erosion control measures would have a substantial effect in reducing phosphorus export in a system in which the predominant phosphorus pathway is sediment transport in overland flow. The reduction in SRP was not anticipated. In fact, plot studies indicated that SRP export would increase in no-till systems, as phosphorus storage in surficial soil layers increased. There was no clear reason for the decrease in SRP. Possible explanations include a reduction in winter application of fertilizer (the season in which sediment-associated losses tend to be highest) and a trend toward warmer, drier winters. This reduced overland flow and SRP export during spring snowmelt.

7.7.2.2 Big Spring, Iowa

Water quality in the 267 km^2 Big Spring basin, Clayton County, Iowa, has been monitored since 1981. Monitoring began in response to the concerns of local citizens after high nitrate concentrations were measured in domestic wells. The Big Spring Basin Demonstration Project, initiated in 1986, is a cooperative effort by the Iowa Department of Natural Resources, U.S. Geological Survey, Iowa State University, NRCS, and EPA. This program has supported public education and farm demonstration projects, in addition to an extensive groundwater and surface water monitoring network.

The substrate and hydrogeologic setting of the Big Spring basin is limestone bedrock. Groundwater flows through the carbonate Galena aquifer; about 85% of the groundwater is discharged through Big Spring. The basin is subject to moderate karst development and overland flow may be channeled directly to subsurface pathways through sinkholes during storm events. It is estimated that approximately 10% of the recharge of Big Spring comes from these sinkholes.

Corn and alfalfa fields comprise 50% and 35% to 40% of land cover, respectively. Many farms also include livestock (dairy cattle and hogs) and bring the total area of agricultural land use to 97% of the Big Spring basin. There are approximately 200 farms with an average size of 134 ha. There are no major urban point sources of pollution (Liu et al., 2002). Pollution concerns focus on nitrate and herbicide (primarily atrazine) contamination of the groundwater. The primary modes of transport reflect the differences in the state and solubility of the compounds. Dissolved nitrate enters the soil and is transported by subsurface flow. Less soluble herbicides are typically transported by overland flow. Ammonium and organic nitrogen concentrations increase in direct proportion to suspended sediment concentrations in overland flow during stormflow events.

Nitrate contamination of groundwater and surface water in the Big Spring basin is not only a local concern. Nitrate from this basin enters the Turkey River, where it is carried to the Mississippi River and ultimately to the Gulf of Mexico (Rowden et al., 2000). Nutrients flowing from the Big Spring basin and many similar agricultural watersheds in the upper Midwest contribute to eutrophication in the Gulf of Mexico. Between 1985 and 1992, the region of hypoxic bottom water in the Gulf averaged between 8000 and 9000 km^2. From 1993 to 2000, the hypoxic zone extended from 16,000 to 20,000 km^2 during the summer months (Goolsby and Battaglin, 2001; Goolsby et al., 2001; Rabalais et al., 2001). Eutrophication in the Gulf of Mexico is primarily attributed to increased use of nitrogen fertilizers in the upper Midwest from the mid-1960s to the early 1980s. In a pattern that was typical of the entire region, fertilizer use in the Big Spring basin tripled during this time period (Goolsby et al., 2001; Rowden et al., 2000).

Twenty-four BMP demonstration projects focused on nutrient management were established between 1987 and 1992 in the Big Spring basin. Monitoring showed that it was possible to reduce fertilizer and herbicide use while maintaining crop yields. In response to educational outreach and direct assistance programs, the input of both nitrogen fertilizers and herbicides has declined. The fertilizer application rate for corn decreased from an average of 195 kg/ha in 1981 to 129 kg/ha in 1993 (Miller and Brown, 1998; Rowden et al., 2000).

Despite the program's success in reducing nitrate and herbicide inputs, improvements in water quality have been difficult to measure because of interannual variation in climate. In general, an annual increase in precipitation leads to an increase in soil moisture and subsurface flow. The larger volume of water moving through the soil transports larger amounts of dissolved nitrate, leading to increased nitrate concentrations in groundwater discharge. Some of the nitrate moved during wet years may be nitrate that was stored in previous dry years. Reductions in groundwater nitrate loading from 396,000 to 88,400 kg and concentration from 8.7 to 5.7 mg/L were measured at Big Spring from 1982 to 1989, and some of this was probably due to reductions in fertilizer use. During 1990 and 1991, however, annual precipitation and groundwater discharge increased. During these wet years, nitrate concentrations in groundwater discharge increased from 8.2 to 12.5 mg/L and nitrate loading increased from 176,000 to 655,600 kg. Nitrate concentrations and loading continued to vary with annual precipitation throughout the 1990s.

Atrazine has been detected in 86% of the groundwater samples from Big Spring since monitoring began in 1982. Changes in atrazine concentrations in groundwater discharge are related to the timing and intensity of precipitation rather than the volume of groundwater flow. High-intensity storms, especially those that occur after herbicide applications, transport atrazine with overland flow and sediment to sinkholes and increase the proportion of groundwater recharge that bypasses the root zone. Storms in June 1996 and May 1999 were followed by several weeks of increased atrazine discharge. In addition, atrazine decomposition rates decline when the soil is dry. This phenomenon leads to the accumulation of atrazine and subsequent increases in concentration in stormwater at the end of extended droughts (Rowden et al., 2000). The findings of the Big Spring basin study demonstrate that improvement of water quality in agricultural landscapes requires a long-term commitment to BMPs and performance monitoring.

7.7.2.3 White Clay Lake, Wisconsin

White Clay Lake is a 95 ha, shallow lake (mean depth 4.2 m, maximum depth 13.5 m) located in an agricultural watershed in Wisconsin. Ninety percent of the shoreline consists of protected wetland marshes. In the 1950s and 1960s, dairy farmers in the White Clay Lake watershed increased the size of their herds and moved many of their animals from pastures to feedlots. Corn production in the watershed also increased. By 1978 there were 25 concentrated animal feeding operations (CAFOs) located in the 1215 ha watershed. At that time, the lake, although slightly eutrophic, was in relatively good condition. The phosphorus concentration in the lake water was 0.029 mg/L and water clarity was 2.7 m. Nonpoint pollution control BMPs were implemented in the early 1980s to protect the lake from future water quality deterioration. Watershed BMPs were installed to prevent pollution from animal waste. These included storage facilities for 6 to 12 months of manure, redesigning feedlots to prevent water from passing through, installing filter strips between feedlots and streams, and not spreading manure on frozen ground. Cropland management BMPs included using contour strip planting, building settling ponds, and changing crop rotations to grow more hay and less corn.

Garrison and Asplund (1993) evaluated the results of this program. Manure storage facilities or drainage controls had been installed at 19 of 25 CAFOs. Annual phosphorus loading was highly variable and strongly influenced by precipitation. Most of the phosphorus was transported during spring snowmelt and annual variations in snowfall and spring rains increased or decreased the amount of phosphorus transported during a particular year. Despite this, water quality in the lake deteriorated during the 1980s. Mean annual phosphorus concentrations in lake water had increased to 0.044 mg/L and water clarity had decreased to 2.1 m. Summer chlorophyll concentrations increased as well, from 0.009 mg/L in the late 1970s to 0.013 mg/L in the late 1990s.

Garrison and Asplund (1993) attributed the decline in water quality to extremely high concentrations of phosphorus in the soil and to two farms that had not participated in the BMP program. One of the nonparticipating farms was located on the lakeshore and animals had been observed in the water at this site. The barnyard closest to one of the major tributaries to the lake also did not install BMPs.

7.7.2.4 Lake Champlain, Vermont

The Lake Champlain watershed includes forest and agricultural land in Vermont, New York, and Quebec, Canada. Increases in phosphorus loading have accelerated eutrophication of the lake. As a result, a series of studies have been conducted to measure the effectiveness of agricultural BMPs to reduce phosphorus loading to the lake (Meals, 1996, 2001; Meals and Hopkins, 2002).

The first of these studies (Meals, 1996) describes BMP implementation in the LaPlatte River and St. Albans Bay watersheds (Table 7.1). Dairy farming is the primary agricultural activity in both watersheds. Manure and commercial fertilizers are applied to corn and hayfields. Land treatment programs to reduce nonpoint source pollution were established in 1985 and 1986. In the LaPlatte River watershed, these included 25 manure storage systems and implementation of BMPs to regulate waste disposal (no spreading on frozen ground and soil incorporation). In addition, 18 milk house waste BMP systems and 10 barnyard waste control BMPs were installed. Erosion control practices were implemented on 950 ha of cropland. In the St. Albans Bay watershed, 64 manure storage systems were constructed and animal waste management BMPs were applied on 3760 ha. Forty-three milk house waste systems were established, 61 barnyard waste control systems were constructed, and soil erosion controls were applied to 3760 ha of cropland. Riparian zone management and the exclusion of livestock from streams were not undertaken in either watershed.

Fecal streptococcus bacterial counts in major tributaries were reduced by 50% to 75% during the sampling period (1979 to 1989) and sediment loads were reduced, but phosphorus concentrations and loads did not decrease. Meals offers several possible reasons for this result. First, the lack of riparian zone management and livestock exclusion from streams may have been a particularly important omission from the BMP program. Second, while the participation rate was high, it is possible that a few nonparticipating farmers may continue to contribute a large portion of the phosphorus input to streams. Third, it is possible that there is a considerable time lag between the

TABLE 7.1
Subwatershed and BMP Information: Study of Agricultural BMPs, Lake Champlain Watershed, Vermont

	La Platte River	St. Albans Bay
Area (ha)	13,800	13,000
No. of farms	40	102
BMPs		
Manure storage facilities	25	64
Milk house waste systems	18	43
Barnyard waste	10	61
Erosion control (ha)	950	3,760
Percent of animals under BMP management	68	75
Percent of cropland under BMP management	80	75

Meals (1996).

implementation of BMP programs and noticeable effects, especially in areas where soils in a cool, wet climate have been enriched with phosphorus for decades.

In a second study (Meals, 2001; Meals and Hopkins, 2002), the Lake Champlain Basin Agricultural Watersheds National Monitoring Program (NMP) Project evaluated the effectiveness of "livestock exclusion, streambank protection, and riparian restoration practices in reducing concentrations and loads of nutrients, sediment, and bacteria in surface waters." This paired watershed study measured water quality responses to BMPs implemented in the 690 ha Samsonville Brook and 1422 ha Godin Brook watersheds. The 954 ha Berry Brook watershed served as a control. The three streams are tributaries of the Missiquoi River. Land use in the three watersheds is similar, approximately 60% forested and 30% agricultural, with 2% to 3% residential. The primary agricultural activity is dairy farming and stream water quality was impaired (high phosphorus, organic matter, and bacteria) as a result of the unrestricted activity of livestock.

Data from the calibration period demonstrated that nutrient loading and bacterial counts in the three watersheds were similar. Treatments in Godin Brook watershed included "fencing, watering systems, reducing the number and size of livestock crossing areas, and streambank erosion control." Two years of post-treatment water quality measurements showed significant reductions in total phosphorus concentration (21%) and export (21% to 41%) in the Samsonville and Godin Brook watersheds.

7.7.2.5 Virginia

The state of Virginia has funded a cost-sharing program to help farmers implement BMPs to improve water quality and reduce nonpoint source pollution to the Chesapeake Bay. Three studies have analyzed BMP performance in Virginia by measuring changes in the concentration and loading of (1) sediment and nutrients (Inamdar et al., 2001), (2) pesticides (Shukla et al, 2001), and (3) bacteria–fecal coliform (FC) and fecal streptococci (FS) (Inamdar et al., 2002) following BMP implementation.

7.7.2.5.1 Sediment and Nutrients

The 1463 ha Nomini Creek watershed, located predominantly in the upland region of the Virginia Coastal Plain, is approximately 47% forested and 49% agricultural with 4% of the land occupied by residential land use and roads. The primary agricultural activity is row crop production. The major crops are corn, soybeans, and small grains (wheat and barley). Water samples were collected and analyzed for sediment, nitrogen, and phosphorus during baseflow and stormflow events between 1986 and 1989, before the implementation of BMPs. Twenty-six farms adopted various BMPs in 1989 and 1990. These included strip cropping, conservation tillage, nutrient management, and IPM. Vegetated filter strips and erosion control structures also were put in place. This led to a 50% reduction in the area of corn being grown with traditional methods; there was a five-fold increase in the area of corn production using conservation tillage. Water samples were collected during baseflow periods and stormflow events between 1990 and 1997.

Measurements of sediment and nutrient export in the post-BMP period were likely affected by climate variations. There was a 13% increase in rainfall, primarily the result of an increase in high-intensity, short-duration summer storms. This produced a 26% increase in streamflow (184 mm versus 231 mm). These conditions would be expected to increase sediment and nutrient transport relative to the pre-BMP period. There were no significant changes in the concentration of suspended solids found in post-BMP sampling, nor were there peaks in sediment concentration during stormflow periods—an indication that BMPs were effective in reducing sediment transport.

Total nitrogen export for the entire watershed decreased by 26% (9.57 to 7.05 kg N/ha/yr) despite the increase in streamflow. Decreases were observed for dissolved ammonium nitrogen, soluble organic nitrogen, and particulate nitrogen. Nitrate export increased by 36% in the post-BMP period from 1.49 to 2.02 kg N/ha/yr and mean annual concentrations rose from 0.81 to 0.87 mg/L. Changes in phosphorus loading and concentrations were not significant in the post-BMP period. There was a slight decrease in the export of particulate phosphorus, but the export of dissolved phosphorus in organic and inorganic forms increased (Inamdar et al., 2001).

7.7.2.5.2 Pesticides

Shukla et al. (2001) measured decreases in the concentration of two pesticides—atrazine and metolachlor—following the implementation of BMPs at an agrichemical mixing and handling facility located in Calverton, Virginia, in the Piedmont region. BMPs included the installation of a pesticide mixing and loading concrete pad with a sump and pump fitting, structures designed to contain drainage from the facility and divert it from critical areas, and recycling of rinsate. Significant reductions in pesticide concentrations were recorded at the stream leaving the site and at the watershed outlet. The pre-BMP sampling period was from 1986 to 1988; post-BMP sampling was conducted from 1989 to 1996. The annual mean concentration of atrazine was reduced from 2.690 to 0.165 mg/L and metolachlor was reduced from 4.579 to 0.402 mg/L at the facility outlet. Mean annual atrazine concentrations at the watershed outlet decreased from 0.015 to 0.007 mg/L and metolachlor decreased from 0.015 mg/L to 0.003 µg/L. Both atrazine and metolachlor were detected in 100% of the samples during the 7 year sampling period.

7.7.2.5.3 Bacteria

Inamdar et al. (2002) studied the effectiveness of BMPs in reducing bacterial pollution in the 1163 ha Owl Run watershed in the Piedmont area of Virginia. Watershed land use includes cropland (31%), pasture (18%), and woodlands, streams, and wetlands (36%). The watershed is also home to five major dairies, one replacement heifer, and three beef cattle operations. Total agricultural land use is 60%. Only one dairy farm had a waste management program when the study began. The other farmers disposed of manure with daily or weekly field applications. Pre-BMP data were collected between 1986 and 1989, then five manure storage facilities were built and livestock access to streams was restricted. The storage facilities made it possible for farmers to apply manure to the fields only in the spring and fall (prior to planting a cover crop), when it was most appropriate for plant growth and theoretically posed the least threat to water quality. The post-BMP sampling period extended from 1989 to 1996.

Although mean annual precipitation increased by only 1.8% between the pre- and post-BMP periods, mean annual streamflow increased by 29%. As in the Nomini Creek watershed, this was related to an increase in the number of high-intensity summer storms. Fecal coliform concentrations at the watershed outlet dropped slightly in the post-BMP period; however, concentrations increased at subwatershed sampling stations, possibly due to increased transport during summer storms. There were consistent reductions in concentrations of fecal streptococci at both the watershed and sub-watershed scale, possibly caused by fecal streptococci mortality during storage.

Prior to the implementation of BMPs, Virginia standards for fecal coliform were exceeded 86% of the time; in the post-BMP monitoring period, standards were exceeded 74% of the time. The authors concluded that "although BMP implementation can be expected to accomplish some improvement in water quality, BMP implementation alone does not ensure compliance with current water quality standards."

7.7.2.6 Conclusion

As is evident from these examples, BMP effectiveness is variable and can be strongly affected by local conditions. In the Vermont watersheds, significant gains were accomplished through a rather simple action—keeping cows out of the streams. It also should be noted that these watersheds were 60% forested and that there was very little cropland relative to pasture. Farmers relied almost exclusively on hay and pasture grass for livestock feed. In the Big Spring watershed in Iowa, cropland covered 97% of the watershed and excess fertilizer and herbicides had been applied for many years. In this situation it also may be many years before improved water quality fostered the long-term use of BMPs is measurable.

7.8   CURRENT MANAGEMENT ISSUES

Agricultural land use increases sediment and nutrient loading and may generate new sources of pathogens and chemicals, such as herbicides and pesticides, that were previously unknown in the natural environment. During the last 75 years, scientific knowledge and public awareness of the consequences of disturbances associated with agricultural land use have grown enormously. Scientists and managers have come to appreciate the challenges involved in implementing management strategies that can achieve clean water and stable aquatic ecosystems while maintaining agricultural production. New management strategies must be both economically feasible and practical. At the same time, devising management strategies that effectively and simultaneously address all the water quality issues associated with farming is a complex, perhaps intractable, task. Chemical compounds travel by different hydrologic pathways depending on their chemical characteristics (see Chapter 3) (Blanchard and Lerch, 2000). Strategies that reduce the export of one agricultural chemical may inadvertently increase the export of a different compound or the same substance in a different form. Strategies that are successful in one region may be far less so in another because of differences in topography, climate, and soils. As noted throughout this volume, the time lag between the implementation of a mitigation strategy and measurable and sustained improvement in water quality can be considerable because agricultural chemicals have accumulated in watershed soils, wetlands, and stream sediments.

7.8.1   EROSION AND SEDIMENT

Conservation tillage (no-till or low-till farming) minimizes soil disturbance and leaves crop residue on the field, reducing soil exposure to rainfall impact (Mostaghimi et al., 1992). However, many of these practices would not be possible without the increased availability and use of herbicides and commercial fertilizers containing nitrogen and phosphorus. Before herbicides were used, farmers needed to plant corn in hills that were far enough apart to permit cultivation for weed control between the rows. This sometimes meant cultivation in two directions, so farmers were compelled to till up and down slopes in addition to tilling on the contour. This exposed more soil to rainfall and inevitably increased erosion. No-till systems rely on herbicides to control weed competition and fertilizers to sustain the growth of crops at high densities (Giere et al., 1980; Knox, 2001). Unfortunately this may trade one nonpoint source pollution problem—soil erosion and sediment transport—for another—pesticide and dissolved fertilizer export.

7.8.2   NUTRIENTS

During the last 50 years, agricultural operations throughout the country have become increasingly specialized. Livestock production has changed—larger numbers of animals are maintained in CAFOs while animal feed (primarily corn and grain) is grown elsewhere. Farmers in the eastern United States import 83% of the grain they feed to livestock, while the Midwestern states export more than half of the grain produced in that region (McDowell et al., 2002). As a result, farms in the East often generate much more manure than they need or can use for fertilizer (Natural Resources Conservation Service, 1997; Sharpley et al., 2001). Because transporting large quantities of manure is both difficult and expensive, disposal and treatment has become a daunting and ubiquitous problem. At the same time, commercial fertilizers are often applied in excess of crop requirements. Applications of manure and commercial fertilizer in excess of crop requirements becomes a continuing source of nitrogen and phosphorus that fuels eutrophication in receiving waters (Sharpley et al., 1994; Shepard, 2000; Stevenson, 1986).

Applied in appropriate amounts, manure is a valuable source of nitrogen, phosphorus, and micronutrients. It also adds organic matter that can improve soil texture and is beneficial to soil structure and water holding capacity (Sharpley et al., 1994). However, it is difficult for farmers to calculate the actual quantity of the nutrients provided by manure and other organic sources because only a portion of the nutrients is immediately available to plants and that portion is variable. More nutrients become available as the organic matter decays. This may take several years. This time lag and uncertainty leads farmers who seek maximum crop yields (hay, field corn, soybeans, or alfalfa) to apply commercial fertilizers on fields that are already rich in nitrogen and phosphorus from other sources such as manure, nitrogen-fixing legumes, and crop residues.

Trachtenberg and Ogg (1994) analyzed data from the 1990 Farm Costs and Returns Survey conducted by the USDA. The survey documented nitrogen demand (crop requirements) and nitrogen supply (nitrogen from livestock, prior plantings of legumes, and fertilizer applications) for 6140 farms raising conventional crops such as corn, soybeans, and wheat. They found that approximately 25% of all fertilizer purchases on these farms was unnecessary. Better analysis of available soil nitrogen on each farm (such as the routine use of a late spring soil test) would save farmers money and help to protect the environment. In a study examining livestock operations in Wisconsin, Shepard (2000) found that, on average, farmers applied 43 kg/ha (38 lb/acre) of nitrogen and 83 kg/ha (74 lb/acre) of phosphorus more than corn crops required. Of the 1928 farmers surveyed, 50% applied excess nitrogen and 90% applied excess phosphorus. A relatively small number of farmers were likely responsible for the greatest proportion of environmental damage (14% of farmers applied ≥357 kg N/ha and 12% of farmers applied ≥242 kg P/ha; one standard deviation above the mean rate of application). This study also found that while farmers commonly spread manure on cropland, only 36% attempted to credit the amount of nitrogen and phosphorus supplied by that manure in calculating how much commercial fertilizer was needed to optimize crop yields and production costs. Of those that did, only 3% (20 out of 694) did so accurately. Farmers face many obstacles in managing manure and nutrients on their land. Fertilizer dealers recommend applications that ignore manure inputs; equipment is designed for waste disposal, not resource management; and water resource protection programs have emphasized the construction of expensive manure storage structures rather than an accurate accounting of the nutrients from all sources (Trachtenberg and Ogg, 1994).

Napier (2000) criticizes reliance on voluntary educational programs, namely the Information, Subsidy, and Technical Assistance (ISTA) approach that has guided federal farm conservation programs since the Dust Bowl era of the 1930s. In a study of 1011 farmers in three watersheds (located in Minnesota, Illinois, and Ohio), Napier found that while conservation measures were commonly used, other farming practices negated their effectiveness. He noted that "farmers could adopt no-till practices and remain significant agricultural polluters by applying manure during the winter and by over-applying inorganic fertilizers during the growing season." He also found that access to and funding of agricultural conservation programs did not necessarily correlate with the implementation of effective conservation practices. It was not a lack of information that kept farmers from adopting comprehensive conservation programs, as assumed by the ISTA approach. Frequently conservation recommendations did not adequately address local agricultural concerns. Napier recommends "whole farm planning," examining all farm management practices that might affect the generation of nonpoint source pollution, and providing a variety of management options so that farmers might implement a total management program appropriate to their individual needs. This is the guiding principle of the Watershed Agricultural Program in the New York City watersheds (National Research Council, 2000).

Since the major portion of phosphorus entering streams is associated with sediment, efforts to control phosphorus often focus on soil erosion control (Staver and Brinsfield, 2001). To that end, no-till and low-till farming decrease overland flow and the transport of sediment and sediment-associated phosphorus. However, if the amount of phosphorus applied each year continues to exceed plant requirements, phosphorus stored in the soil will increase (see Box 7.4). Under these conditions, phosphorus sorption sites on soil particles can become saturated, leading to an increase in the export of SRP in both subsurface flow and overland flow (McDowell and Sharpley, 2001; Sharpley, 2003). An increase in subsurface flow can also increase the amount of soluble nitrate that enters streams (McDowell et al., 2002). Thus controlling erosion and reducing the transport of sediment-associated agricultural chemicals such as phosphorus and herbicides such as atrazine (Blanchard and Lerch, 2000) may lead to increases in dissolved nitrate and phosphorus unless application rates are carefully controlled (Staver and Brinsfield, 2001). Sharpley (2003) found that long-term application of manure (9 to 20 years) resulted in phosphorus enrichment of the upper 5 cm of soil compared to deeper soil layers. He recommended occasional plowing as a "last resort" to mix phosphorus-enriched surface soils with lower soil layers and increase phosphorus retention in the soil profile as a whole.

Box 7.4   Mass Balance Models

Mass balance models (Cassell and Clausen, 1993; Cassell et al., 2001, 2002) or nutrient budgets (Baker and Richards, 2002) are an effective way to (1) identify and quantify input, output, and soil storage components of nutrients and (2) forecast changes that might occur under different management strategies or (3) changes in land use that might alter inputs and outputs. Mass balance models are used because nutrient loading, rather than concentration, is needed to predict downstream impacts (see Box 3.2). Figure 7.10 displays the components of a mass balance model for phosphorus.

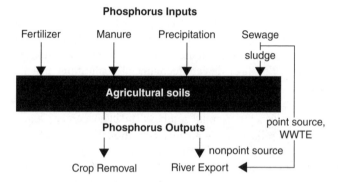

FIGURE 7.10 Mass balance model for phosphorus (modified from Baker and Richards, 2002).

In an ideal situation, nutrient inputs match nutrient removal in harvested crops and stream export is minimal. When inputs are greater than outputs, soil storage will increase:

$$\Sigma \text{ Inputs} - \Sigma \text{ Outputs} = \Delta \text{ Soil Storage}$$

Export from soil storage will vary with the volume of water flowing through the soil, which in turn varies in relation to rainfall, snowmelt, and evapotranspiration.

Two recent papers (Sharpley et al., 1994, 2001) argue that programs for the management of agricultural phosphorus must take into account the environmental costs of phosphorus enrichment and eutrophication of receiving water bodies. This would require a comprehensive accounting of phosphorus from all sources, including the amount stored in soil. The authors recommend the use of soil testing to identify areas already enriched with phosphorus. Among those sites, management efforts should focus on areas where steep slopes and highly erodible soils make erosion and the transport of sediment-associated phosphorus most likely. A subsequent article (McDowell et al., 2002) emphasizes the necessity of coordinating management strategies for both nitrogen and phosphorus within a landscape context. It is imperative to identify and properly manage those areas in a watershed that have high levels of phosphorus and are most vulnerable to erosion in order to control inputs of sediment-associated phosphorus. At the same time it is necessary to limit nitrogen inputs to areas where soil permeability is high and the transport of dissolved nitrate in subsurface flow is most likely (Gburek et al., 1991). In all cases, the control of excess (unnecessary) nutrient inputs is a critically important task.

7.8.3 PESTICIDES

Practices designed to reduce the amount of pesticides applied to farm fields have gained broader acceptance in the last 20 to 30 years. IPM, first introduced in the 1960s, aims to achieve efficient production of food or fiber by using the smallest possible amount of synthetic pesticides. The basic premise of IPM programs is that "pesticides should be applied only when the cost associated with pesticide damage exceeds the cost of applying pesticides"—not simply on a routine preventative basis (Mostaghimi et al., 2001).

In general, IPM involves the use of multiple control tactics and integration of a knowledge of pest biology into the management system (Buhler et al., 2000). IPM may involve identifying and importing natural enemies of the particular species or enhancing and expanding the resources and habitat available for native predators (Landis and Orr, 2002). The sterile insect technique involves releasing large numbers of sterile insect pests that mate with native females, resulting in the production of infertile eggs and a reduction in the population of insect pests (Bartlett and Staten, 2002). (The University of Minnesota maintains a website with an electronic textbook that is gathering chapters from experts on IPM around the world; http://ipmworld.umn.edu/chapters/maize.htm [accessed July 2006].)

It has been more difficult to develop and implement IPM strategies for controlling weed competition with crops than insects and diseases. In light of the current level of herbicide use and risk of soil and water contamination, this is an important area of study. The widespread use of herbicides also favors the evolution of herbicide-resistant weed strains, necessitating the development of alternate management methods (Buhler, 2002). In general, IPM approaches for weed control attempt to enhance crop vigor while minimizing the growing space and resources available for weed species. Some promising strategies are crop rotation, intercropping of species of varying physical forms and phenologies, narrow row spacing, and the use of cover crops (Buhler, 2002; Buhler et al., 2000).

Riparian buffers have been shown to be effective in trapping sediment and, as a result, the pesticides that are adsorbed to sediment surfaces. Riparian buffers also slow water movement, foster infiltration, and increase residence time in the soil. This provides additional time for the chemical substances to be deactivated by biochemical and microbial processes (Arora et al., 1996; Paterson and Schnoor, 1992).

7.8.4 PATHOGENS

As with nutrient pollution, appropriate management of manure is an especially important way to limit agricultural nonpoint source pollution. Pathogens like *Cryptosporidium* oocysts die within hours if they dry out; however, when contaminated fecal matter is deposited directly into streams, the oocysts can remain infectious for as long as 176 days (Lawhorn, 1996). It is important, wherever possible, to limit direct access of livestock to streams and to provide alternative water sources. Composting manure before applying it to fields greatly reduces its pollution potential, even if it is transported by overland flow. Alternatively, spreading uncomposted manure during periods of dry weather will kill many of the infectious oocysts before they reach the water (Entry et al., 2000a,b). In contrast, spreading manure on snow-covered or frozen fields will almost certainly lead to nutrient and pathogen export to receiving waters (Meals, 1996).

7.9 SUMMARY AND CONCLUSIONS

7.9.1 HISTORY

Agricultural development between 1630 and 1930 transformed the landscape of New England, the Mid-Atlantic, and the upper Midwest (Whitney, 1994). Virgin forests were cleared for settlements and farms. Wetlands and salt marshes were drained. Second- and third-growth forests have naturally regenerated on farmland in areas where soils and topography made farming difficult and unprofitable, such as in New England and the northern regions of Michigan, Wisconsin, and Minnesota. Today farming activity is concentrated in the Midwest, some areas of the Mid-Atlantic states (Maryland, Delaware, and eastern Pennsylvania), and in river valleys and lake plains of New York and southern New England. The conversion of forests to agricultural land use—even if farms were abandoned and reverted back to forest—has substantially altered streamflow regimes, rates of soil erosion and sediment delivery, and stream and river channel form.

7.9.2 Agricultural Research

Efforts to mitigate the environmental consequences of agricultural land use began in the 1930s in response to the ravages of severe erosion and soil loss in many areas of the country—especially the Midwest. Research on agricultural land use and water quality is carried on by the NRCS and the ARS in close cooperation with land grant universities.

7.9.3 Streamflow, Sediment, and Channel Form

Removing forest cover increases streamflow. Replacing forest cover with cropland and pasture reduces infiltration rates and increases overland flow. Wetland drainage and the installation of tile drains have caused major changes in hydrologic patterns in agricultural areas. These include increased stormflow and reduced baseflow and increased movement of sediment. Channels are scoured and enlarged in some reaches, while other reaches fill with sediment. Intermittent streams may become ephemeral, or cease to exist, if baseflow is disrupted.

7.9.4 Water Chemistry

Agricultural land use is a source of nutrients, pesticides, and pathogens in waterways. Nutrients from fertilizer and manure contribute to eutrophication in the Chesapeake Bay and other estuaries, the Great Lakes, and the Gulf of Mexico. Atrazine, the most commonly used herbicide, is a ubiquitous contaminant, albeit at low concentrations.

7.9.5 Best Management Practices

Agricultural BMPs focus on source reduction and controlling the transport of sediment, nutrients, and pesticides. Effective implementation requires the long-term commitment of a large number of farmers, especially those in hydrologically sensitive areas. Unfortunately, diligent and effective efforts of a majority of farmers in a watershed may be overshadowed by a small recalcitrant group that chooses not to participate in voluntary or incentive-based programs. As emphasized throughout this book, the proportional impact of inappropriate land use on a small area may dominate streamflow and water quality characteristics at the landscape scale. The effectiveness of BMP programs is highly variable. It is influenced by the proportion of the watershed in agricultural land use, the type of farming, management practices on individual farms, topography, soils, and climate.

7.9.6 Watershed Management Challenges

The productivity of American farms and farmers has increased steadily throughout the 20th century. This increase can be attributed to (1) advances in the basic sciences (e.g., biology, chemistry, genetics), (2) applied research and development, and education, (3) systematic crop and livestock breeding, (4) the widespread use of synthetic fertilizers and pesticides, (5) irrigation, (6) mechanization, (7) digital and information technologies, and (8) soil and water conservation methods (e.g., contour tillage, strip cropping, terracing, etc.). Ironically, these exponential advances in agricultural productivity have been rewarded with little or no increase in the price of farm products. So while feeding an ever-increasing world population, farmers have been subject to constant financial pressure under which profits, or even a modest livelihood, can only be earned by increasing production efficiency. This imbalance can relegate soil and water conservation practices to a status of short-term constraints rather than the cornerstone of a long-term management and stewardship philosophy. At the same time, farmers are the first to appreciate what President Franklin Delano Roosevelt wrote in 1937: "The Nation that destroys its soil destroys itself" (a letter to all state governors on uniform soil conservation laws, February 26, 1937). This chapter has updated this caution by describing the direct local effects on farms as well as the downstream effects and externalities of agricultural nonpoint source pollution that accrue at a regional, even national, scale.

The consolidation and specialization of agricultural production has in some cases increased the size and complexity of farming by orders of magnitude. CAFOs and dairy farms with 1000 or more cows generate manure, nutrient, and pathogen management problems on the scale of a small city. The overall effect of centralized animal agriculture extends to the farms (corn, grain, and alfalfa), feed mills, and transportation systems that are needed to support these operations. Livestock operations serve as one example of the need to develop a sustainable, equilibrium condition for the environmental and economic components of agriculture in order to protect water resources, ecosystem integrity, and public health. This goal cannot be reached without the continuing support of consumers and all levels of government. If we sincerely seek to improve water quality and aquatic ecosystem integrity, more attention to the first two steps of the environmental protection hierarchy (avoid and prevent) and less reliance on the last two (mitigate and restore) to contain or minimize problems is needed.

The "whole farm planning" approach cited earlier is gaining currency where agriculture and critical water resources coexist (National Research Council, 2000). In many respects it is a return to the principles and practices that were developed and implemented by the SCS (now the NRCS) beginning in the 1930s. Whole farm planning emphasizes the design and integration of agricultural BMPs to reduce or eliminate the outcomes that are at cross purposes. For example, no-till (sometimes called conservation tillage) practices substantially reduce soil erosion and sediment transport by protecting the soil surface with crop residue (e.g., corn or wheat stubble). To be successful, however, no-till agriculture typically requires increased fertilizer and herbicide application to reduce weed competition and maximize crop yields. In other words, solving one problem—sediment and phosphorus export to surface water—may inadvertently cause another—substantial increases in nitrogen and pesticide export to groundwater. At the watershed scale, these groundwater contaminants can enter streams, rivers, and estuaries as baseflow even after overland flow during storm events has been effectively controlled at the field scale.

The scope, accuracy, and sophistication of nutrient management practices must be increased to match nutrient additions (fertilizer + manure + nitrogen-fixing legumes) with crop requirements. This requires (1) detailed soil maps, (2) soil tests to quantify the spatial and temporal variations in nutrient availability, (3) accurate and timely information about soil water content and probable weather conditions to maximize nutrient retention in the root zone, (4) physiological and phenological information about crops (and weeds), and (5) practical software tools for farmers. Other related techniques, such as IPM, should complement or reinforce soil conservation and nutrient management practices in the whole farm plan and subsequent operations. Some scientists and managers suggest that mitigation and restoration methods, such as riparian buffers, soil erosion practices, and fencing to exclude livestock from streams, wetlands, and riparian areas, could be improved. Yet after decades of research and demonstration projects have demonstrated their strengths and weaknesses, the sustained implementation and maintenance of these time-tested BMPs as part of whole farm plans and watershed management projects is perhaps more important than the pursuit of perfection.

The sheer amount, ubiquitous nature, and costly consequences of agricultural nonpoint source pollution show that planning, monitoring, and adaptive management techniques should not be reserved for special sites and circumstances, but applied across the United States. For, as William Sloane Coffin (2004) writes:

> It is a great mistake to talk, as many political leaders do, of balancing the needs of the economy with those of the environment. An economy, national or world, is a subsystem of the ecosystem. Therefore we cannot speak of growth as an unquestioned good.

REFERENCES

Agricultural Research Service, 2004, http://www.ars.usda.gov; accessed July 2006.

Anderson, J.M., Invertebrate-mediated transport processes in soils, *Agric. Ecosyst. Environ.*, 24, 5–19, 1988.

Arora, K., Michelson, S.K., Baker, J.L., Tierney, D.P., and Peters, C.J., Herbicide retention by vegetative buffer strips from runoff under natural rainfall, *Trans. Am. Soc. Agric. Eng.*, 39, 2155–2162, 1996.

Ayers, M.A., Kennen, J.G., and Stackelberg, P.E., Water quality in the Long Island–New Jersey coastal drainages, New York, and New Jersey, 1996–1998, Circular 1201, U.S. Geological Survey, Reston, VA, 2000.

Baker, D.B. and Richards, R.P., Phosphorus budgets and riverine phosphorus export in northwestern Ohio watersheds, *J. Environ. Qual.*, 31, 96–108, 2002.

Bardgett, R.D., Anderson, J.M., Behan-Pelletier, V., Brussard, L., Coleman, D.C., Ettema, E., Moldenke, A., Schimel, J.P., and Wall, D.H., The influence of soil biodiversity on hydrological pathways and the transfer of materials between terrestrial and aquatic ecosystems, *Ecosystems*, 4, 421–429, 2001.

Barnes, B.V., Zak, D.R., Denton, S.R., and Spurr, S.H., *Forest Ecology*, 4th ed., John Wiley & Sons, New York, 1998.

Bartlett, A.C. and Staten, R.T., The sterile insect release method and other genetic control strategies, in *Radcliffe's IPM World Textbook*, Radcliffe, E.B. and Hutchinson, W.D. (eds.), University of Minnesota, St. Paul, 2002, http://ipmworld.umn.edu/; accessed July 2006.

Beaulac, M.N. and Reckhow, K.H., An examination of land use — nutrient export relationships, *Water Resourc. Bull.*, 18, 1013–1024, 1982.

Bennett, H.H., *Soil Conservation*, McGraw-Hill, New York, 1939.

Blanchard, P.E. and Lerch, R.N., Watershed vulnerability to losses of agricultural chemicals: interactions of chemistry, hydrology, and land-use, *Environ. Sci. Technol.*, 34, 3315–3322, 2000.

Boesch, D.F., Brinsfield, R.B., and Magnien, R.F., Chesapeake Bay eutrophication: scientific understanding, ecosystem restoration, and challenges for agriculture, *J. Environ. Qual.*, 30, 303–320, 2001.

Bormann, F.H., Likens, G.E., Siccama, T.E., Pierce, R.S., and Eaton, J.S., The export of nutrients and recovery of stable conditions following deforestation at Hubbard Brook, *Ecol. Monogr.*, 44, 255–277, 1974.

Bowman, I., *Forest Physiography*, John Wiley & Sons, New York, 1911.

Brady, N.C. and Weil, R.R., *The Nature and Property of Soils*, 13th ed., Macmillan, New York, 2002.

Brugam, R.B., Human disturbance and the historical development of Linsley Pond, *Ecology*, 59, 19–36, 1978.

Brunger, E., Dairying and urban development in New York State, 1850–1900, *Agric. Hist.*, 29, 169–174, 1955.

Buhler, D.D., Challenges and opportunities for integrated weed management, *Weed Sci.*, 50, 273–280, 2002.

Buhler, D.D., Liebman, M., and Obrycki, J.J., Theoretical and practical challenges to an IPM approach to weed management, *Weed Sci.*, 48, 274–280, 2000.

Burkhart, M.R. and James, D.E., Agricultural-nitrogen contributions to hypoxia in the Gulf of Mexico, *J. Environ. Qual.*, 28, 850–859, 1999.

Canter, L.W., *Nitrates in Groundwater*, Lewis Publishers, Boca Raton, FL, 1996.

Carlson, C. and Fowler, J., *The Salt Marsh of Southern New Jersey*, Stockton Center for Environmental Research, Stockton State College, Pomona, NJ, 1979.

Cassell, E.A. and Clausen, J.C., Dynamic simulation modeling for evaluating water quality response to agricultural BMP implementation, *Water Sci. Technol.*, 28(3–5), 635–648, 1993.

Cassell, E.A., Kort, R.L., Meals, D.W., Aschmann, S.G., Dorioz, J.M., and Anderson, D.P., Dynamic phosphorus mass balance modeling of large watersheds: long-term implications of management strategies, *Water Sci. Technol.*, 43(5), 153–162, 2001.

Cassell, E.A., Meals, D.W., Aschmann, S.G., Anderson, D.P., Rosen, B.H., Kort, R.L., and Dorioz, J.M., Use of simulation mass balance modeling to estimate phosphorus and bacteria dynamics in watersheds, *Water Sci. Technol.*, 45(9), 157–166, 2002.

Cavieres, M.F., Jaeger, J., and Porter, W., Developmental toxicity of a commercial herbicide mixture in mice: I. Effects on embryo implantation and litter size, *Environ. Health Perspect.*, 110, 1081–1085, 2002.

Centers for Disease Control and Prevention, http://www.cdc.gov/; accessed July 2006.

Clausen, J.C. and Meals, D.W., Water quality achievable with agricultural best management practices, *J. Soil Water Conserv.*, 43, 593–596, 1989.

Coffin, W.S., *Credo*, Westminster John Knox Press, Louisville, KY, 2004.

Colman, G.P., Innovation and diffusion in agriculture, *Agric. Hist.*, 42, 173–187, 1968.

Cook, B., Michigan forests, Michigan State University Extension Forester, 2005, http://uptreeid.com/History/MainPage.htm; accessed July 2006.

Correll, D.L., Jordan, T.E., and Weller, D.E., Effects of precipitation and air temperature on phosphorus fluxes from Rhode River watersheds, *J. Environ. Qual.*, 28, 144–154, 1999a.

Correll, D.L., Jordan, T.E., and Weller, D.E., Nutrient flux in a landscape: effects of coastal land use and terrestrial community mosaic on nutrient transport to coastal waters, *Estuaries*, 15, 431–442, 1992.

Correll, D.L., Jordan, T.E., and Weller, D.E., Precipitation effects on sediment and associated nutrient discharges from Rhode River watersheds, *J. Environ. Qual.*, 28, 1897–1907, 1999b.

Cronan, C.S., Piampiano, J.T., and Patterson, H.H., Influence of land use and hydrology on exports of carbon and nitrogen in a Maine River basin, *J. Environ. Qual.*, 28, 953–961, 1999.

Cronon, W., *Changes in the Land: Indians, Colonists, and the Ecology of New England*, Hill and Wang, New York, 1983.

Cronon, W., *Nature's Metropolis: Chicago and the Great West*, W.W. Norton, New York, 1991.

Dance, K.W. and Hynes, H.B.N., Some effects of agricultural land use on stream insect communities, *Environ. Pollut. Ser. A*, 22, 19–28, 1980.

Danhof, C.H., The tools and implements of agriculture, *Agric. Hist.*, 46, 81–90, 1972.

Davis, M.B., Erosion rates and land-use history in southern Michigan, *Environ. Conserv.*, 3, 139–148, 1976.

DeLorenzo, M.E., Scott, G.I., and Ross, P.E., Effects of the agricultural pesticides atrazine, deethylatrazine, endosulfan, and chlorpyrifos on an estuarine microbial food web, *Environ. Toxicol. Chem.*, 18, 2824–2835, 1999.

Dunne, T. and Leopold, L.B., *Water in Environmental Planning*, W.H. Freeman, New York, 1978.

Edwards, C.A. and Lofty, J.R., The effect of direct drilling and minimal cultivation on earthworm populations, *J. Appl. Ecol.*, 19, 723–734, 1982.

Enger, L., A history of timbering in Minnesota, Minnesota Public Radio, November 16, 1998, http://news.minnesota.publicradio.org/features/199811/16_engerl_history-m/; accessed July 2006.

Entry, J.A., Hubbard, R.K., Thies, J.E., and Fuhrmann, J.J., The influence of vegetation in riparian filterstrips on coliform bacteria: I. Movement and survival in water, *J. Environ. Qual.*, 29, 1206–1214, 2000a.

Entry, J.A., Hubbard, R.K., Thies, J.E., and Fuhrmann, J.J., The influence of vegetation in riparian filterstrips on coliform bacteria: II. Survival in soils, *J. Environ. Qual.*, 29, 1215–1224, 2000b.

Evans, J.K., Gottgens, J.F., Gill, W.M., and Mackey, S.D., Sediment yields controlled by intrabasinal storage and sediment conveyance over the interval 1842–1994: Chagrin River, northeast Ohio, U.S.A., *J. Soil Water Conserv.*, 55, 264–270, 2000.

EXTOXNET, The EXtension TOXicology NETwork: a cooperative effort of University of California–Davis, Oregon State University, Michigan State University, Cornell University, and the University of Idaho, 2002, http://ace.orst.edu/info/EXTOXNET/; accessed July 2006.

Fitzpatrick, F.A., Knox, J.C., and Whitman, H.E., Effects of historical land-cover changes on flooding and sedimentation, North Fish Creek, Wisconsin, Water Resources Investigations Report 99-4083, U.S. Geological Survey, Reston, VA, 1999.

Foster, D.R. and O'Keefe, J.F., *New England Forests Through Time: Insights from the Harvard Forest Dioramas*, Harvard University Press, Cambridge, MA, 2000.

Foy, R.H. and Withers, P.J.A., The contributions of agricultural phosphorus to eutrophication, paper presented to the Fertiliser Society, London, April 27, 1995.

Fraser, R.H., Barten, P.K., and Pinney, D.A.K., Predicting stream pathogen loading from livestock using a geographical information system-based delivery model, *J. Environ. Qual.*, 27, 935–945, 1998.

Gallagher, D.L., Dietrich, A.M., Reay, W.G., Hayes, M.C., and Simmons, G.M., Jr., Ground water discharge of agricultural pesticides and nutrients to estuarine surface water, *Ground Water Monit. Remediat.*, 16, 118–129, 1996.

Garrison, P.J. and Asplund, T.R., Long-term (15 years) results of NPS controls in an agricultural watershed upon a receiving lake's water quality, *Water Sci. Technol.*, 28(3–5), 441–449, 1993.

Garry, V.F., Harkins, M.E., Erickson, L.L., Long-Simpson, L.K., Holland, S.E., and Burroughs, B.L., Birth defects, season of conception, and sex of children born to pesticide applicators, Red River Valley of Minnesota, USA, *Environ. Health Perspect.*, 110(suppl. 3), 441–449, 2002.

Garry, V.F., Schreinemachers, D., Harkins, M.E., and Griffith, J., Pesticide appliers, biocides, and birth defects in rural Minnesota, *Environ. Health Perspect.*, 104, 394–399, 1996.

Gates, P.W., Large-scale farming in Illinois, 1850–1870, *Agric. Hist.*, 6(1), 14–25, 1932.

Gaumnitz, L., Restoring a watershed to life, *Wisc. Nat. Resourc. Mag.*, February/March 2002, http://www.wnrmag .com/stories/2002/feb02/coonval.htm; accessed July 2006.

Gburek, W.J., Folmar, G.J., and Schnabel, R.R., Ground water controls on hydrology and water quality within rural upland watersheds of the Chesapeake Bay basin, in *Towards a Sustainable Coastal Watershed: The Chesapeake Experiment, Proceedings of a Conference*, Publication 149, Chesapeake Research Consortium, Solomons, MD, 1994, pp. 665–677.

Gburek, W.J., Urban, J.B., and Schnabel, R.R., Effects of agricultural land use on contaminant transport in a layered fractured aquifer, in *Hydrological Basis of Ecologically Sound Management of Soil and Groundwater*, Publication 202, International Association of Hydrological Sciences, Wallingford, UK, 1991, pp. 33–42.

Gebert, W.A. and Krug, W.R., Streamflow trends in Wisconsin's Driftless Area, *Water Resourc. Bull.*, 32, 733–744, 1996.

Giere, J.P., Johnson, K.M., and Perkins, J.H., A closer look at no-till farming, *Environment*, 22(6), 15–41, 1980.

Goolsby, D.A. and Battaglin, W.A., Long-term changes in concentration and flux of nitrogen in the Mississippi River basin, USA, *Hydrol. Process.*, 15, 1209–1226, 2001.

Goolsby, D.A., Battaglin, W.A., Aulenbach, B.T., and Hooper, R.P., Nitrogen flux and sources in the Mississippi River basin, *Sci. Total Environ.*, 248, 75–86, 2000.

Goolsby, D.A., Battaglin, W.A., Aulenbach, B.T., and Hooper, R.P., Nitrogen input to the Gulf of Mexico, *J. Environ. Qual.*, 30, 329–336, 2001.

Gottschalk, L.C., Effects of soil erosion on navigation in upper Chesapeake Bay, *Geogr. Rev.*, 35, 219–238, 1945.

Grant, A., Bibliography of papers and reports on saltmarsh restoration, 2002, http://www.uea.ac.uk/~e130/ bibliography.htm; accessed July 2006.

Hackensack Riverkeeper, The Hackensack River: a true come-back story, 2005, http://www.hackensackriver- keeper.org; accessed July 2006.

Hayes, T.B., Collins, A., Lee, M., Mendoza, M., Noriega, N., Stuart, A.A., and Vonk, A., Hermaphroditic demasculinized frogs after exposure to the herbicide atrazine at low ecologically relevant doses, *Proc. Natl. Acad. Sci. USA*, 99, 5476–5480, 2002.

Helms, D., Coon Valley, Wisconsin: a conservation success story, reprinted from *Readings in the History of the Soil Conservation Service*, Soil Conservation Service, Washington, D.C., 1992, pp. 51–53, http://www.nrcs.usda.gov/about/history/articles/CoonValley.html; accessed July 2006.

Hibbert, A.R., Forest treatment effects on water yield, in *International Symposium on Forest Hydrology*, Sopper, W.E. and Lull, H.W. (eds.), 1967, pp. 527–543.

Holden, L.R., Graham, J.A., Whitmore, R.W., Alexander, W.J., Pratt, R.W., Liddle, S.K., and Piper, L.L., Results of the national alachlor well water survey, *Environ. Sci. Technol.*, 26, 935–943, 1992.

Hornbeck, J.W., Martin, C.W., and Eager, C., Summary of water yield experiments at Hubbard Brook Experimental Forest, New Hampshire, *Can. J. For. Res.*, 27, 2043–2052, 1997.

Hurt, R.D., *American Agriculture: A Brief History*, Iowa State University Press, Ames, IA, 1994.

Inamdar, S.P., Mostaghimi, S., Cook, M.N., Brannan, K.M., and McClellen, P.W., A long-term, watershed evaluation of the impacts of animal waste BMPs on indicator bacteria concentrations, *J. Am. Water Resourc. Assoc.*, 38, 819–833, 2002.

Inamdar, S.P., Mostaghimi, S., McClellan, P.W., and Brannan, K.M., BMP impacts on sediment and nutrient yields in the Coastal Plain region, *Trans. Am. Soc. Agric. Eng.*, 44, 1191–1200, 2001.

Johnson, L.B., Richards, C., Host, G.E., and Arthur, J.W., Landscape influences on water chemistry in Midwestern stream ecosystems, *Freshwater Biol.*, 37, 193–208, 1997.

Jordan, T.E., Correll, D.L., and Weller, D.E., Effects of agriculture on discharges of nutrients from Coastal Plain watersheds of Chesapeake Bay, *J. Environ. Qual.*, 26, 836–848, 1997a.

Jordan, T.E., Correll, D.L., and Weller, D.E., Nonpoint source discharges of nutrients from Piedmont watersheds of Chesapeake Bay, *J. Am. Water Resourc. Assoc.*, 33, 631–644, 1997b.

Jordan, T.E., Correll, D.L., and Weller, D.E., Relating nutrient discharges from watersheds to land use and streamflow variability, *Water Resourc. Res.*, 33, 2579–2590, 1997c.

Jordan, T.E., Correll, D.L., Peterjohn, W.T., and Weller, D.E., Nutrient flux in a landscape: the Rhode River watershed and receiving waters, in *Watershed Research Perspectives*, Correll, D.L. (ed.), Smithsonian Institution Press, Washington, D.C., 1986, pp. 57–75.

Judd, R.W., *Common Lands, Common People: The Origins of Conservation in Northern New England*, Harvard University Press, Cambridge, MA, 1997.

Knox, J.C., Human impacts on Wisconsin stream channels, *Ann. Assoc. Am. Geogr.*, 67, 323–342, 1977.

Knox, J.C., Historical valley floor sedimentation in the Upper Mississippi Valley, *Ann. Assoc. Am. Geogr.*, 77, 224–244, 1987.

Knox, J.C., Agricultural influence on landscape sensitivity in the upper Mississippi River valley, *Catena*, 42, 193–224, 2001.

Landis D.A. and Orr, D.B., Biological control: approaches and applications, in *Radcliffe's IPM World Textbook*, Radcliffe, E.B. and Hutchinson, W.D. (eds.), University of Minnesota, St. Paul, 2002, http://ipm-world.umn.edu/chapters/maize.htm; accessed July 2006.

Lawhorn, B., Human cryptosporidium and cryptosporidiosis, Texas Agricultural Extension Service, Texas A&M University system, 1996, http://agpublications.tamu.edu/pubs/vm/15162.pdf; accessed July 2006.

Lear, L.J., Bombshell in Beltsville: the USDA and the challenge of "Silent Spring," *Agric. Hist.*, 66, 151–170, 1992.

Lenat, D.R. and Crawford, J.K., Effects of land use on water quality and aquatic biota of three North Carolina Piedmont streams, *Hydrobiologia*, 294, 185–199, 1994.

Leopold, A., Coon Valley: an adventure in cooperative conservation, *Am. For.*, 41, 205–208, 1935.

Leopold, L.B., Land use and sediment yield, in *Man's Role in Changing the Face of the Earth*, Thomas, W.L., Jr. (ed.), University of Chicago Press, Chicago, 1956, pp. 639–647.

Leopold, L.B., *Water, Rivers and Creeks*, University Science Books, Sausalito, CA, 1997.

Lichtenberg, E. and Shapiro, L.K., Agriculture and nitrate concentrations in Maryland community water system wells, *J. Environ. Qual.*, 26, 145–153, 1997.

Likens, G.E. and Davis, M.B., Post-glacial history of Mirror Lake and its watershed in New Hampshire, U.S.A.: an initial report, *Int. Ver. Theor. Angew. Limnol. Verh.*, 19, 982–993, 1975.

Lindsey, B.D., Breen, K.J., Bilger, M.D., and Brightbill, R.A., Water quality in the lower Susquehanna River basin, Pennsylvania and Maryland, 1992–1995, Circular 1168, U.S. Geological Survey, Reston, VA, 1998, http://pubs.usgs.gov/circ/circ1168/; accessed July 2006.

Liu, H., Libra, R.D., and Rowden, R.D., Big Spring demonstration project, 2002, http://extension.agron.iastate.edu/Waterquality/projects/bigspring.html; accessed July 2006.

Lowrance, R., Hubbard, R.K., and Williams, R.G., Effects of a managed three zone riparian buffer system on shallow groundwater quality in the southeastern Coastal Plain, *J. Soil Water Conserv.*, 55, 212–220, 2000.

Lowrance, R., Vellidis, G., and Hubbard, R.K., Denitrification in a restored riparian forest wetland, *J. Environ. Qual.*, 24, 808–815, 1995.

Lowrance, R., Vellidis, G., Wauchope, R.D., Gay, P., and Bosch, D.D., Herbicide transport in a managed riparian forest buffer system, *Trans. Am. Soc. Agric. Eng.*, 40, 1047–1057, 1997.

Magilligan, F.J., Historical floodplain sedimentation in the Galena River basin, Wisconsin and Illinois, *Ann. Assoc. Am. Geogr.*, 75, 583–594, 1985.

Martin, C.W. and Hornbeck, J.W., Logging in New England need not cause sedimentation of streams, *Northern J. Appl. For.*, 11, 17–23, 1994.

McCorvie, M.R. and Lant, C.L., Drainage district formation and the loss of Midwestern wetlands, 1850–1930, *Agric. Hist.*, 67(4), 13–39, 1993.

McDowell, R.W. and Sharpley, A.N., Phosphorus losses in subsurface flow before and after manure application to intensively farmed land, *Sci. Total Environ.*, 278, 113–125, 2001.

McDowell, R.W., Sharpley, A.N., and Kleinman, P.J.A., Integrating phosphorus and nitrogen decision management at watershed scales, *J. Am. Water Resourc. Assoc.*, 38, 479–491, 2002.

McGuinness, J.L., Harrold, L.L., and Edwards, W.M., Relation of rainfall energy streamflow to sediment yield from small and large watersheds, *J. Soil Water Conserv.*, 26, 233–235, 1971.

Meals, D.W., Watershed-scale response to agricultural diffuse pollution control programs in Vermont, USA, *Water Sci. Technol.*, 33(4–5), 197–204, 1996.

Meals, D.W., Water quality response to riparian restoration in an agricultural watershed in Vermont, USA, *Water Sci. Technol.*, 43(5), 175–182, 2001.

Meals, D.W. and Hopkins, R.B., Phosphorus reductions following riparian restoration in two agricultural watersheds in Vermont, USA, *Water Sci. Technol.*, 45(9), 51–60, 2002.

Meyer, D.R., The national integration of regional economies, 1860–1920. In *North America: The Historical Geography of a Changing Continent* (pp. 321–346), Mitchell, R.D. and Groves, P.A. (eds.), Hutchinson, London, 1987.

Miller, G.A. and Brown, S.S., Big Spring: farming from the ground "water" up: evolution of a water quality project, 1998, http://extension.agron.iastate.edu/waterquality/bigsprep.html; accessed July 2006.

Minnesota Historical Society, Forest chronology, Forest History Center, St. Paul, MN, 2002.

Mitchell, P., Relationship between beef production and waterborne parasites (*Cryptosporidium* spp. and *Giardia* spp.) in the North Saskatchewan River basin, Alberta, Canada), Alberta Agricultural Research Institute, Alberta, Canada, 2002, http://www1.agric.gov.ab.ca/$department/deptdocs.nsf/all/wat6400/$FILE/ parasite_1_9.pdf; accessed July, 2006.

Mitsch, W.J. and Wilson, R.F., Improving the success of wetland creation and restoration with know-how, time, and self-design, *Ecol. Applic.*, 6, 77–83, 1996.

Mostaghimi, S., Brannan, K.M., Dillaha, T.A., and Bruggeman, A.C., Best management practices for nonpoint source pollution control: selection and assessment, in *Agricultural Nonpoint Source Pollution: Watershed Management and Hydrology*, Ritter, W.F. and Shirmohammadi, A. (eds.), CRC Press, Boca Raton, FL, 2001.

Mostaghimi, S., Younos, T.M., and Tim, U.S., Crop residue effects on nitrogen yield in water and sediment from two tillage systems, *Agric. Ecosyst. Environ.*, 39(3–4), 187–196, 1992.

Napier, T., Use of soil and water protection practices among farmers in the North Central region of the United States, *J. Am. Water Resourc. Assoc.*, 36, 723–735, 2000.

National Research Council, *Watershed Management for Potable Water Supply: Assessing the New York City Strategy*, National Academies Press, Washington, D.C., 2000.

Natural Resources Conservation Service, 2005, http://www.nrcs.usda.gov; accessed July 2006.

Natural Resources Conservation Service, Water quality and agriculture: status, conditions and trends, Working Paper 16, Natural Resources Conservation Service, Washington, D.C., 1997, http://www.nrcs.usda.gov/technical/land/pubs/WP16.pdf; accessed July 2006.

North Appalachian Experimental Watershed, http://www.oardc.ohio-state.edu/branches/NAEWS.htm; accessed July 2006.

Novak, J.M., Watts, D.W., Stone, K.C., Johnson, M.H., and Hunt, P.G., Pesticides and metabolites in the shallow groundwater of an eastern Coastal Plain watershed, *Trans. Am. Soc. Agric. Eng.*, 41, 1383–1390, 1998.

O'Keefe, J.F. and Foster, D.R., An ecological history of Massachusetts forests, in *Stepping Back to Look Forward: A History of the Massachusetts Forest*, Foster, C.H.W. (ed.), Harvard Forest, Petersham, MA, 1998, pp. 19–66.

Pasture Systems and Watershed Management Research Unit, http://www.ars.usda.gov/main/site_main.htm?modecode=19020000; accessed July 2006.

Paterson, K.G. and Schnoor, J.L., Fate of alachlor and atrazine in a riparian zone field site, *Water Environ. Res.*, 64, 274–283, 1992.

Patric, J.H., Soil erosion in the eastern forests, *J. For.*, 74, 671–677, 1976.

Patric, J.H., Effects of wood products harvest on forest soil and water relations, *J. Environ. Qual.*, 9, 73–80, 1980.

Patric, J.H., Evans, J.O., and Helvey, J.D., Summary of sediment yield data from forested land in the United States, *J. For.*, 82, 101–104, 1984.

Percy, D.O., Ax or plow?: significant colonial landscape alteration rates in the Maryland and Virginia Tidewater, *Agric. Hist.*, 66(2), 66–74, 1992.

Peterjohn, W.T. and Correll, D.L., Nutrient dynamics in an agricultural watershed: observations on the role of a riparian forest, *Ecology*, 65, 1466–1475, 1984.

Pionke, H.B. and Glotfelty, D.E., Nature and extent of groundwater contamination by pesticides in an agricultural watershed, *Water Res.*, 23, 1031–1037, 1989.

Pionke, H.B. and Urban, J.B., Effect of agricultural land use on ground-water quality in a small Pennsylvania watershed, *Ground Water*, 23, 68–80, 1985.

Pionke, H.B., Gburek, W.B., Sharpley, A.N., and Schnabel, R.R., Flow and nutrient export patterns for an agricultural hill-land watershed, *Water Resourc. Res.*, 32, 1795–1804, 1996.

Postel, S. and Richter, B., *Rivers for Life: Managing Water for People and Nature*, Island Press, Washington, D.C., 2003.

Rabalais, N.N., Turner, R.E., and Wiseman, W.J., Jr., Hypoxia in the Gulf of Mexico, *J. Environ. Qual.*, 30, 320–329, 2001.

Raup, H.M., The view from John Sanderson's farm: a perspective for the use of the land, *For. Hist.*, 10, 2–11, 1966.

Richards, R.P., Calhoun, F.G., and Matisoff, G., The Lake Erie Agricultural Systems for Environmental Quality Project: an introduction, *J. Environ. Qual.*, 31, 6–16, 2002.

Roldán, M.K., Utilizing GIS for mapping reforestation of an agricultural landscape, 1939–1993, in Coon Creek watershed, Wisconsin, Master's thesis, Saint Mary's University of Minnesota, 2002.

Rosza, R., Human impacts on tidal wetlands, in *Tidal Marshes of Long Island Sound: Ecology, History, and Restoration*, Dreyer, G.D. and Niering, W.A. (eds.), Bulletin 34, Connecticut College Arboretum, New London, CT, 1995.

Roth, N.E., Allan, J.D., and Erickson, D.L., Landscape influences on stream biotic integrity assessed at multiple spatial scales, *Landscape Ecol.*, 11, 141–156, 1996.

Rothrock, J.A., Barten, P.K., and Ingman, G.L., Land use and aquatic biointegrity in the Blackfoot River watershed, Montana, *J. Am. Water Resourc. Assoc.*, 34, 565–581, 1998.

Rowden, R.D., Liu, H., and Libra, R.D., Results from the Big Spring water quality monitoring and demonstration projects, *Hydrogeol. J.*, 9, 487–497, 2000.

Sanderson, J.T., Seinen, W., Giesy, J.P., and van den Berg, M., 2-Chloro-s-triazine herbicides induce aromatase (CYP-19) activity in H295R human adrenocortical carcinoma cells: a novel mechanism for estrogenicity?, *Toxicol. Sci.*, 54, 121–127, 2000.

Sartz, R.S., Folklore and bromides in watershed management, *J. For.*, 67, 366–371, 1969.

Sartz, R.S., Effect of land use on the hydrology of small watersheds in southwestern Wisconsin, Publication 96, International Association of Scientific Hydrology, Wallingford, UK, 1970, pp. 286–295.

Sartz, R.S., Thirty years of soil and water research by the Forest Service in Wisconsin's Driftless Area — a history and annotated bibliography, General Technical Report NC-44, USDA Forest Service, North Central Forest Experiment Station, St. Paul, MN, 1978.

Sartz, R.S. and Tolsted, D.N., Effect of grazing on runoff from two small watersheds in southwestern Wisconsin, *Water Resourc. Res.*, 10, 354–356, 1974.

Sartz. R.S., Curtis, W.R., and Tolsted, D.N., Hydrology of small watersheds in Wisconsin's Driftless Area, *Water Resourc. Res.*, 13, 524–530, 1977.

Schottler, S.P., Eisenreich, S.J., and Capel, P.D., Atrazine, alachlor, and cyanazine in a large agricultural river system, *Environ. Sci. Technol.*, 28, 1079–1089, 1994.

Sea Grant, http://www.nsgo.seagrant.org; accessed July 2006.

Sharpley, A.N., Soil mixing to decrease surface stratification of phosphorus in manured soils, *J. Environ. Qual.*, 32, 1375–1384, 2003.

Sharpley, A.N., Chapra, S.C., Wedepohl, R., Sims, J.T., Daniel, T.C., and Reddy, K.R., Managing agricultural phosphorus for protection of surface waters: issues and options, *J. Environ. Qual.*, 23, 437–451, 1994.

Sharpley, A.N., Kleinman, P., and McDowell, R., Innovative management of agricultural phosphorus to protect soil and water resources, *Commun. Soil Sci. Plant Anal.*, 32, 1071–1100, 2001.

Shepard, R., Nitrogen and phosphorus management on Wisconsin farms: lessons learned for agricultural water quality programs, *J. Soil Water Conserv.*, 55, 63–68, 2000.

Shukla, S., Mostaghimi, S., Lovern, S.B., and McClellan, P.W., Impact of agrichemical facility best management practices on runoff water quality, *Trans. Am. Soc. Agric. Eng.*, 44, 1611–1672, 2001.

Simmons, D.L. and Reynolds, R.J., Effects of urbanization on base flow of selected south-shore streams, Long Island, New York, *Water Resourc. Bull.*, 18, 797–805, 1982.

Sinicrope, T.L., Hine, P.G., Warren, R.S., and Niering, W.A., Restoration of an impounded salt marsh in New England, *Estuaries*, 13, 25–30, 1990.

Smithsonian Environmental Research Center, http://www.serc.si.edu; accessed July 2006.

Sovell, L.A., Vondracek, B., Frost, J.A., and Mumford, K.G., Impacts of rotational grazing and riparian buffers on physicochemical and biological characteristics of southeastern Minnesota, USA, streams, *Environ. Manage.*, 26, 629–641, 2000.

Staver, K.W. and Brinsfield, R.B., Agriculture and water quality on the Maryland eastern shore: where do we go from here?, *Bioscience*, 51, 859–868, 2001.

Stearns, F.W., History of the Lake States forests: natural and human impacts, in *Lakes States Regional Forest Resources Assessment: Technical Papers,* Vasievich, J.M. and Webster, H.H. (eds.), General Technical Report NC-189, USDA Forest Service, North Central Experiment Station, St. Paul, MN, 1997, pp. 8–29.

Stevenson, F.J., *Cycles of Soils: Carbon, Nitrogen, Phosphorus, Sulfur, Micronutrients*, John Wiley & Sons, New York, 1986.

Swank, W.T. and Crossley, D.A., Jr., Introduction and site description, in *Forest Hydrology and Ecology at Coweeta*, Swank, W.T. and Crossley, D.A., Jr. (eds.), Springer-Verlag, New York, 1988, pp. 3–16.

Trachtenberg, E. and Ogg, C., Potential for reducing nitrogen pollution through improved agronomic practices, *Water Resourc. Bull.*, 30, 1109–1118, 1994.

Trimble, S.W., Perspectives on the history of soil erosion control in the eastern United States, *Agric. Hist.*, 59, 162–180, 1982.

Trimble, S.W., A sediment budget for Coon Creek Basin in the Driftless Area, Wisconsin, 1853–1977, *Am. J. Sci.*, 283, 454–474, 1983.

Trimble, S.W., Decreased rates of alluvial sediment storage in the Coon Creek Basin, Wisconsin, 1975–93, *Science*, 285, 1244–1246, 1999.

U.S. Environmental Protection Agency, Condition of the Mid-Atlantic estuaries, EPA 600-R-98-147, U.S. Environmental Protection Agency, Washington, D.C., 1998.

U.S. Environmental Protection Agency, Mid-Atlantic integrated assessment (MAIA) estuaries 1997–1998. Summary report: environmental conditions in the Mid-Atlantic estuaries, U.S. Environmental Protection Agency, Washington, D.C., 2002b.

U.S. Environmental Protection Agency, Nonpoint source pollution: the nation's largest water quality problem, 2002a, http://www.epa.gov/OWOW/NPS/facts/point1.htm; accessed July 2006.

U.S. Geological Survey, The quality of our nation's waters—nutrients and pesticides, Circular 1225, U.S. Geological Survey, Reston, VA, 1999, http://water.usgs.gov/pubs/circ/circ1225/html; accessed July 2006.

U.S. Geological Survey, Selected findings and current perspectives on urban and agricultural water quality by the National Water-Quality Assessment Program, 2001, http://water.usgs.gov/pubs/FS/fs-047-01; accessed July 2006.

Vaithiyanathan, P. and Correll, D.L., The Rhode River watershed: phosphorus distribution and export in forest and agricultural soils, *J. Environ. Qual.*, 21, 280–288, 1992.

Verry, E.S., Forest harvesting and water: the Lake States experience, *Water Resourc. Bull.*, 22, 1039–1047, 1986.

Watt, M.K., A hydrological primer for New Jersey watershed management, Water Resources Investigations Report 00-4140, U.S. Geological Survey, Reston, VA, 2000.

Whitney, G.G., *From Coastal Wilderness to Fruited Plain: A History of Environmental Change in Temperate North America 1500 to the Present*, Cambridge University Press, 1994.

Wilkin, D.C. and Hebel, S.J., Erosion, redeposition, and delivery of sediment to Midwestern streams, *Water Resourc. Res.*, 18, 1278–1282, 1982.

Wilson, H.F., The rise and decline of the sheep industry in northern New England, *Agric. Hist.*, 9, 12–40, 1935.

Wischmeier, W.H. and Mannering, J.V., Effect of organic matter content of the soil on infiltration, *J. Soil Water Conserv.*, 20, 150–152, 1965.

Wischmeier, W.H. and Smith, D.D., Predicting rainfall erosion losses—a guide to conservation planning, Agricultural Handbook 537, U.S. Department of Agriculture, Washington, D.C., 1978.

Wisconsin Department of Natural Resources, 2005, http://dnr.wi.gov/org/land/forestry/Look/faqaboutforests.asp; accessed July 2006.

Wolman, M.G., A cycle of sedimentation and erosion in urban river channels, *Geogr. Ann.*, 49A(2–4), 385–395, 1967.

Younus, M., Hondzo, M., and Engel, B.A., Stream temperature dynamics in upland agricultural watersheds, *J. Environ. Eng.*, 126, 518–526, 2000.

8 Forest Conversion to Urban and Suburban Land Use

> ...historically most urban development has involved transforming streams into drains or sewers. The primary goal for urban waterway management for most of the 20th century was the safeguarding of humans from floods and disease. Although such a goal must remain the first priority, traditional approaches to waterway management...have been at the expense of other goals, such as public amenity and ecosystem health.
>
> **Walsh et al., 2005**[*]
> *The Urban Stream Syndrome: Current Knowledge and the Search for a Cure*

8.1 INTRODUCTION

Urban development amplifies the hydrologic changes associated with forest management and agriculture. (During the development process, trees are removed, excavation and grading compacts soils, and impervious surfaces [roads, roofs, driveways, etc.] are created.) Wherever the infiltration capacity of the soil is reduced or eliminated, overland flow increases and groundwater recharge is reduced. By design, the movement of stormflow is further accelerated with the construction of urban storm drain systems. The quantity, timing, and quality of streamflow can be substantially altered by urbanization.

Changes in streamflow and sediment loading also cause alterations in channel form and stream habitat conditions (Leopold, 1973, 1997; Waananen, 1969; Wolman, 1967). Stormwater carries sediment and a mix of pollutants that are unique to urban areas. Water samples from streams in urban areas typically show elevated concentrations of nutrients, pesticides, metals, and organic contaminants (Ayers et al., 2000; U.S. Geological Survey, 1999). The loss of forests, wetlands, and marshes also destroys ecosystems that might store and process pollutants from atmospheric deposition.

Walsh et al. (2005b) describe the net result as the "urban stream syndrome," a condition of ecological degradation in streams that includes "a flashier hydrograph, elevated concentrations of nutrients and contaminants, altered channel morphology [form], and reduced biotic richness, with increased dominance of tolerant species." As the world population grows and urban areas expand, research is being directed toward developing urban design principles and best management practices (BMPs) to minimize the impact of development on streams as well as techniques for rehabilitation and restoration of urban streams (Booth, 2005; Lloyd et al., 2002; Walsh et al., 2005a).

8.2 RESEARCH IN URBAN HYDROLOGY

Scientific research examining the effects of urbanization on hydrology and water quality is a relatively new field in comparison to programs in forestry and agriculture. Some of the first papers to systematically describe changes in streamflow, sediment delivery, and channel form with urbanization were published in the 1960s and 1970s (Leopold, 1968, 1973; Wolman, 1967; Wolman and Schick, 1967). Many universities, government agencies, and research institutes are currently involved in this emerging field and cooperative efforts among these groups are common. Several important resources for research and literature on urbanization effects on streamflow and water quality are described below.

---

[*] Reprinted with permission of the North American Benthological Society.

8.2.1 THE U.S. GEOLOGICAL SURVEY

The U.S. Geological Survey (USGS) has completed many studies that relate changes in streamflow to urban land use. Studies in the northeastern area include Krug and Goddard (1986) in Wisconsin, and Sloto (1988) in Pennsylvania. Other papers have investigated groundwater recharge in urban areas (Brown et al., 1997) and water quality in urbanized watersheds (Arnold et al., 1998; U.S. Geological Survey, 1999). Important USGS urban studies from other areas of the country include Konrad and Booth (2002). The USGS National Water Quality Assessment (NAWQA) program, described in detail in Chapter 9 (Box 9.1), is a major source of data and analysis involving mixed land use watersheds with large urban areas.

8.2.2 THE BALTIMORE ECOSYSTEM STUDY

The Baltimore Ecosystem Study (BES) is part of the Long-Term Ecological Research (LTER) program of the National Science Foundation, founded in the 1980s. The Stream and Watershed Studies program of the BES has two primary goals: (1) to evaluate long-term nutrient fluxes and budgets in urban watersheds and (2) to investigate the response of benthic macroinvertebrate communities to urban development. The primary study site is the 17,150 ha Gwynns Falls watershed, which "traverses a gradient from the urban core of Baltimore, through older residential (1900–1950) and suburban (1950–1980) zones, rapidly suburbanizing areas and a rural/suburban fringe" (Baltimore Ecosystem Study, 2005).

8.2.3 THE WATER CENTER

Studies by scientists affiliated with The Water Center at the University of Washington have resulted in many publications investigating the relationships between urbanizing watersheds and changes in streamflow, channel form, and aquatic habitat. The study sites are located in the Puget Sound lowlands, an area between the Cascade and Olympic Mountain ranges that includes the cities of Seattle and Tacoma, Washington. As in many areas of the northeastern United States, the climate is temperate and humid, and soils are derived from surficial deposits left after the retreat of the glaciers 12,000 to 14,000 years ago. They include low permeability tills, lacustrine clays, and glacial outwash (Booth et al., 2003). The findings of these studies have led to new recommendations for BMPs and stream restoration.

8.2.4 COOPERATIVE RESEARCH CENTRE FOR FRESHWATER ECOLOGY

Scientists based at the Cooperative Research Centre for Freshwater Ecology, Water Studies Centre, Monash University, Victoria, Australia, specialize in research evaluating the influence of effective versus total impervious area on aquatic ecosystems, stormwater management, and the restoration of urban watersheds. Scientists from this group have published comprehensive papers on urban stream ecology, one of which (Walsh et al., 2005b) was written in cooperation with several well-known urban ecologists from the northeastern and southern United States. The papers discussed later in this chapter include Walsh (2000) and Walsh et al. (2001, 2005a,b).

8.3 CHANGES IN STREAMFLOW, SEDIMENT, CHANNEL FORM, AND TEMPERATURE FOLLOWING URBAN DEVELOPMENT

8.3.1 STREAMFLOW

8.3.1.1 Watershed Surface Conditions

Removal of the forest reduces evapotranspiration and increases the amount of water available for overland and subsurface flow. At the same time, urban development reduces soil water storage

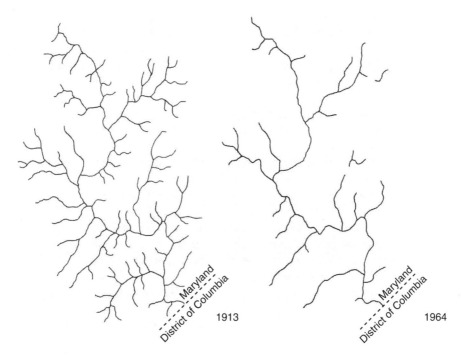

FIGURE 8.1 Drainage net of Rock Creek upstream of the District of Columbia–Maryland line in 1913, before modern urbanization, and again in 1964 (Dunne and Leopold, 1978; image courtesy of USGS).

(Brown, 1988). During construction, many dips, swales, and ephemeral first-order streams that provide temporary storage (detention) areas for stormwater may be eliminated. Natural stream networks become less complex with fewer, but larger tributaries (Figure 8.1). There is a concurrent increase in the density of artificial drainage networks—sewers, road culverts, and channelized streams (Dunne and Leopold, 1978; Paul and Meyer, 2001). The increase in impervious area is an obvious and measurable alteration in watershed surface condition; however, urban development brings about changes in open or pervious areas as well (Booth et al., 2004). The influence of impervious area on groundwater recharge, streamflow, and water quality also varies, depending on its landscape position and connections to the stream network.

8.3.1.1.1 Open Land in Urban Areas

The loss of organic matter and soil compaction substantially reduces the infiltration capacity of soils when forests are converted to suburban lawns and urban parklands (Booth et al., 2002; Booth and Jackson, 1997). Burges et al. (1998) found substantial differences in the soil profiles of forested and suburban watersheds in the Puget Sound lowlands, near Seattle, Washington. Both watersheds are underlain by dense glacial till. In the forested watershed, the till was covered with 0.8 to 1.0 m of organic and mineral soil. In the developed watershed, the soil beneath lawns was only 10 cm deep and contained little or no organic matter. The maximum flow rate per unit area from pervious areas in the developed watershed was 10 times greater than that measured in the forested watershed during a 24-hour, 50-year R.I. rain event.

8.3.1.1.2 Total Impervious Area and Effective Impervious Area

Total impervious area (TIA) is defined as the "proportion of a (watershed) covered by surfaces impermeable to water" (Walsh et al., 2005a). As noted earlier, it is the area of the watershed covered by roads, parking lots, driveways, and buildings. TIA is often used as an indication of the degree of urbanization in a watershed and TIA thresholds (usually between 10% and 20%) have repeatedly been linked with measurable, sometimes marked, declines in stream habitat conditions and biotic populations (Horner et al., 1996; Schueler, 1994; Schueler and Claytor, 1996).

Effective impervious area (EIA) is the proportion of impervious area directly connected to receiving waters by drainage pipes or sewers (Walsh et al., 2005a). Traditionally the primary goals of urban water management have been flood control and waste disposal (Chocat et al., 2001; Walsh, 2000). Nineteenth-century discoveries regarding the transmission of deadly waterborne diseases, such as typhoid and cholera, spurred the construction of sewer systems in urban areas. This led to great improvements in public sanitation and dramatic advances in public health. There was little concern, however, for the effects of urban stormwater on receiving waters (Walsh, 2000). That changed during the 20th century as urban streams and rivers became increasingly degraded and major urban rivers—the Chicago River, the Cuyahoga in Cleveland, the Charles in Boston, for example—became little more than transport systems for sewage, residential wastewater, and industrial waste.

Since the 1970s, improved wastewater treatment (Box 8.1) has dramatically reduced pollutant loading from point sources; however, urban storm drainage systems remain problematic because of the acceleration of stormflow and pollutant delivery. Several studies have identified the proportion of EIA in a watershed as the most important link between urban development and the degradation of urban streams (Booth and Jackson, 1997; Hatt et al., 2004; Leopold, 1968; Walsh et al., 2005a):

> The direct connection of impervious surfaces to streams means that even small rainfall events can produce sufficient [stormflow] to cause frequent disturbance through regular delivery of water and pollutants; where impervious surfaces are not connected to streams, small rainfall events are intercepted and infiltrated (Walsh et al., 2005a).

8.3.1.2 Stormflow

Urban development increases the magnitude of peak flows for high-frequency floods and decreases the response or lag time, also called the time of concentration or time to peak (Figure 8.2) (Dunne and Leopold, 1978; Leopold, 1968; Paul and Meyer, 2001; Walsh et al., 2005b). This has been repeatedly demonstrated by (1) studies comparing urban and rural watersheds (Cherkauer, 1975; Focazio and Cooper, 1995), (2) long-term studies of precipitation and flood frequency in watersheds undergoing urban development (Leopold, 1994; Seaburn, 1969; Yorke and Herb, 1978), and (3) hydrologic modeling studies (Bledsoe and Watson, 2001; Inman, 1988; Sloto, 1988).

In Wisconsin, Cherkauer (1975) compared the streamflow response of an urban stream and a rural stream to the same rainfall event. The urban watershed had an area of 7.5 km² that was 65% developed with residential areas, industrial sites, and parking lots. The rural watershed was slightly larger (9.7 km²) and was 95% agricultural land. The peak discharge in the urban stream caused by the 2.2 cm rainfall was 260 times higher than that in the rural stream (0.475 m³/sec versus 0.0018 m³/sec, respectively).

Focazio and Cooper (1995) examined mean annual storm hydrographs at several sites in the Chickahominy River watershed (653 km²) in Virginia. The Upham Brook subwatershed (98 km²) is an urbanized tributary of the Chickahominy. The Chickahominy River at Atlee has a 161 km² watershed that includes a less densely developed suburban residential area. The Upham Brook and Atlee subwatersheds join above a predominantly rural section of the Chickahominy River watershed before reaching the stream gauge at Providence Forge. When normalized for watershed area, mean annual streamflow (1989 to 1991) was greater at the urban subwatershed (0.021 m³/sec, Upham Brook) than both the residential (0.014 m³/sec, Atlee) and mixed land use sites (0.019 m³/sec, Providence Forge). Stormflow hydrographs varied with antecedent conditions but exhibited consistent patterns in relation to dominant land use and watershed area. Time to peak, stormflow volume, peak discharge, and slope of the rising and recession limbs of the hydrograph all vary in ways expected for a spectrum of urban, residential, and mixed (predominantly rural) land use. Figure 8.3

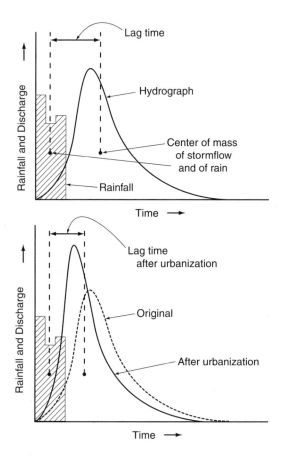

FIGURE 8.2 Hypothetical unit hydrographs before and after urbanization (Leopold, 1968; USGS Circular 554).

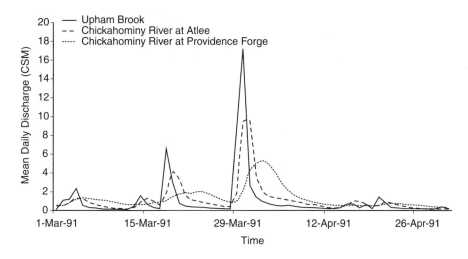

FIGURE 8.3 Normalized mean daily discharge (ft^3/sec/mi^2) for three gauging stations in the Chickahominy River watershed, Virginia, used by Focazio and Cooper (1995). Solid line, Upham Brook (urbanized, 38 mi^2); dashed line, Chickahominy River at Atlee (residential, 62 mi^2); dotted line, Chickahominy River at Providence Forge (rural, but includes Upham and Atlee subwatersheds, 252 mi^2).

shows these patterns in mean daily discharge data normalized with respect to watershed area. What might be regarded as an experimental design flaw in the original paper actually adds to the value of the Focazio and Cooper study. The watersheds are not independent in a statistical sense because the urban and residential watersheds are located within the larger rural watershed. In fact, they comprise 40% of the total area: rural = 64,752 ha, residential = 16,188 ha, urban = 9,713 ha. If the urban and residential watersheds were reforested (or measured before conversion of forests to residential and commercial use) the stormflow volume and peak discharge measured from the rural watershed would have been even lower and the time to peak even longer. The response of the rural watershed shows the ability of forests to offset or counterbalance the effect of other land uses.

Leopold (1968, 1973, 1994) studied the effects of urban development on streams in the Piedmont region of Maryland for more than 25 years. He recorded streamflow in Watts Branch, a tributary of the Potomac River, as housing units increased (from 140 in 1950 to 2060 in 1984), public and commercial buildings with large paved parking areas were constructed, and a section of interstate highway was built upstream of the stream survey sites. The size of the mean annual flood increased from 22 m^3/sec between 1958 and 1973 to 27 m^3/sec between 1973 and 1987. In a nearby stream, Seneca Creek, the mean annual flood increased from 65 m^3/sec during 1931 to 1960 to 99 m^3/sec in the period 1961 to 1991, following substantial urbanization (Leopold, 1994).

The effect of paving and sewer construction is seen most directly in the increase in stormflow volume and peak discharge caused by relatively small precipitation events. In forested areas, water from small rainstorms is often completely absorbed in the soil, producing little if any increase in streamflow, especially during the growing season. As EIA increases, water from even minor storms is directed to overland flow and channeled to streams. The magnitude of streamflow from small storms (recurrence interval 1 year or less) can increase by a factor of 10 following the conversion of a forested watershed to one with 20% impervious area (Hollis, 1975). The increase in overland flow leads to more high-frequency floods. Bankfull discharge was equaled or exceeded an average of 4.9 times per year on Watts Branch between 1958 and 1967. This increased to 7.4 times per year for the period from 1968 to 1977 (Leopold, 1994). The persistent increase in the magnitude and frequency of small floods disrupts the dynamic equilibrium between streamflow and the stream channel. This leads to changes in channel form (Leopold, 1994; Olsen et al., 1997; Wolman and Miller, 1960) (see discussions of hydraulics, open channel flow, flood frequency, and dynamic equilibrium in Chapter 2). Urban development has less effect on floods from large storms of extended duration (recurrence interval greater than 10 years). During high-intensity, long-duration storms, soils in some rural areas can become saturated and produce overland flow, much like urban watersheds (Hollis, 1975; Leopold, 1968).

8.3.1.3 Baseflow

The increase in overland flow and stormflow that accompanies urban development typically causes a reduction in groundwater recharge. The effect of urban development on baseflow appears to be more variable than the effect on overland flow. In addition, alteration in baseflow is influenced by local geology, topography, and soil characteristics that control the proportion of streamflow derived from groundwater in predevelopment conditions. The localized effects of wells and water withdrawals from streams and rivers, as well as wastewater discharge, may also influence the flow regime. Simmons and Reynolds (1982) studied the proportion of baseflow and stormflow in streams on Long Island, New York, in the 1970s and early 1980s. Approximately 95% of the streamflow in rural areas (85% vacant or agricultural land) was derived from groundwater. This pathway of flow is typical in regions with large areas of highly permeable sandy glacial outwash, such as Long Island, New York, and Cape Cod, Massachusetts. The proportion of flow from groundwater was reduced to 20% in streams from urbanized areas with sewer systems (50% residential, 20% streets and parkways, 10% commercial-industrial, and 20% open).

In other studies, a reduction in baseflow did not correlate with urban area (Yorke and Herb, 1978) or urban area was associated with an increase in low flows and the transition of ephemeral to perennial streams (Hollis, 1975). Discharges from wastewater treatment plants during low flow periods may augment baseflow in urban areas (Paul and Meyer, 2001). In a study of 10 watersheds (3 urban, 3 suburban, and 4 rural) in the Puget Sound basin, Konrad and Booth (2002) found that Q_{min}, the minimum flow for a 7-day period during the year, did not follow a consistent pattern in relation to development. The most reliable streamflow indicators of urbanization were Q_{max}, the annual maximum instantaneous discharge or the "magnitude of the largest flood in a stream during the year," and T_{Qmean}, the number of days during the year that the mean annual discharge (Q_{mean}) was exceeded. T_{Qmean} decreases with urban development because urban development increases peak discharge, and the duration of stormflow events is inversely proportional to peak discharge (Booth, 2005).

8.3.2 SEDIMENT: THE CONSTRUCTION PHASE

Studies in the 1950s and 1960s quantified the exponential increases in sediment export from watersheds that included highway and urban construction sites without effective erosion control measures. Wolman and Schick (1967) (see also Wolman, 1967) analyzed the sediment yield from urban construction sites in the Piedmont region of Maryland. Sample sites were located in regions of deep soils and generally moderate (less than 10%) slopes. Sediment concentrations measured in stream waters downstream of construction sites ranged from 3,000 to more than 150,000 mg/L. In rural forested and agricultural watersheds, the maximum stream water sediment concentration was 2,000 mg/L. Sediment yields from construction sites ranged from approximately 7,000 to 490,000 kg/ha/yr. Wolman and Schick's review of studies in Maryland and other southeastern states showed that these rates greatly exceeded the sediment export rate from forested (18 to 50 kg/ha/yr) and forested and agricultural (810 to 2,800 kg/ha/yr) watersheds. The construction sites included those for residential subdivisions, commercial development, and new highways. Sediment concentrations varied widely in relation to the proportional area of disturbed soils. When construction sites are a small part of a larger watershed area, the sediment entering a stream is diluted by cleaner water coming from rural or previously developed land. Figure 8.4 shows the increase in sediment load following early settlement and agricultural development followed by a decline as abandoned farmlands reverted to woodlands and then the rapid increase as these woodlands were cleared and developed. This increase is followed by a rapid decrease when construction is completed and roads are paved. A comparable study in the Detroit metropolitan area found that sediment loading from

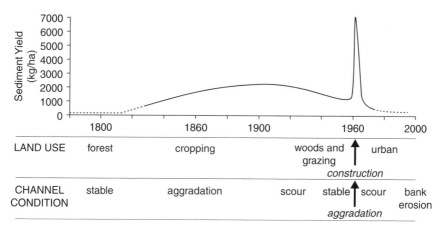

FIGURE 8.4 The cycle of land use changes, sediment yield, and channel behavior in a Piedmont region beginning prior to the advent of extensive farming and continuing through a period of construction and subsequent urban landscape (modified from Wolman, 1967; reprinted with the permission of Blackwell Publishing).

construction sites averaged 155,000 kg/ha/yr, compared to 5,800 kg/ha/yr for southeast Michigan as a whole—a 26-fold increase (Thompson, 1970).

8.3.3 STREAM CHANNEL FORM

Stream channel form is altered in response to persistent changes in sediment supply and streamflow patterns (Graf, 1975; Leopold, 1973, 1992; Wolman, 1967). In urbanizing watersheds, channel changes follow a predictable sequence (Paul and Meyer, 2001) (Figure 8.5). Initially channels fill with sediment from construction sites. This can be avoided or reduced with erosion and sediment control BMPs. When construction is completed—buildings are finished and roads and parking lots are paved—the supply of sediment diminishes and increased flooding causes channel erosion and increases in cross-sectional area (Wolman, 1967). The pattern of channel enlargement is highly variable because the rates and patterns of change are highly dependent on the geologic substrate and channel gradient (Booth and Henshaw, 2001). Bank erosion may increase the width of the streams. More frequent overbank flow may create wider systems of braided channels or the depth of the stream may increase by incision—the erosion of streambed material (Arnold et al., 1982; Booth, 1990; Doyle et al., 2000; Krug and Goddard, 1986). Sediment eroded from upstream channel reaches may be deposited downstream in areas where flow velocity decreases (in broader sections of the channel, in wetlands, or at inlets to lakes and estuaries), smothering benthic organisms (Roesner et al., 2001).

Leopold's (1973) Watts Branch study was originally designed to investigate lateral migration (meandering) of stream channels, not the effects of urban development per se. Fourteen monumented cross sections on the stream, a tributary of the Potomac River, were resurveyed every other year from 1953 to 1972. During the first 10 years of the study, channel cross-sectional area remained relatively constant. In 1962, however, sediment originating from erosion at construction sites and

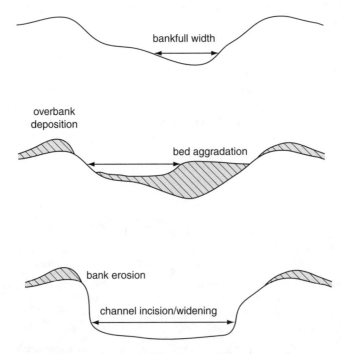

FIGURE 8.5 Channel changes associated with urbanization. Erosion from urban construction increases sediment loading. This leads to stream bed aggradation and sediment deposition on streambanks. As construction is completed and urban streets are paved, the supply of sediment is reduced and the magnitude and frequency of stormflows increase. This causes bank and channel erosion (redrawn from Paul and Meyer, 2001).

deposited on riverbanks began to cause noticeable changes. Average channel width decreased from approximately 9 m to about 6 m, while the median flow depth increased from an average of 0.9 m to 1.14 m. The stream became narrower, with steeper and higher banks. Beginning in 1966, after about 10 years of upstream residential development, the stream channel cross section increased as the streambanks and channel bed eroded away. This was caused by a sustained increase in the magnitude and frequency of stormflow (Dunne and Leopold, 1978; Leopold 1973, 1994).

In areas where riparian vegetation has not been removed in the course of development, the increase in flooding can cause the destruction of riparian trees. Streambanks are undercut by erosion, leaving tree roots exposed and susceptible to deterioration (Arnold et al., 1982). Streamside trees may fall into the channel and be washed downstream—further destabilizing the stream channel. Intact riparian forests can provide a continuing supply of large woody debris that provides habitat and slows erosion in the stream channel (Booth 1991; Martin and Hornbeck, 1994), while plant roots help to stabilize streambanks (Gordon et al., 1992).

Hammer (1972, 1973) measured streams in 78 watersheds (28 rural and 50 with varying proportions of urban development) north of Philadelphia. He used regression analysis to relate stream channel cross-sectional area to watershed area for the rural streams. He then calculated the expected size of the stream channels in the urban watersheds as a function of their area. Not surprisingly, most of the urban streams had a larger cross-sectional area than would be predicted by watershed area alone. Other regression models showed that stream channel enlargement was predicted by several variables identified with urban development: area of sewered streets and impervious areas such as highways and parking lots, residential areas, and golf courses.

Arnold et al. (1982) noted the development of enlarged and braided channels on reaches of Sawmill Brook in central Connecticut, an area undergoing rapid urbanization in the 1960s and 1970s. Aerial photographs taken in 1934, 1966, 1970, and 1975 show pronounced changes in land use. The area had been primarily farmland in the 1930s. As agriculture declined, many open fields reverted to forest cover. At the same time, areas of concentrated urban development appeared. By 1975, residential areas, industrial parks, and highways comprised 30% of the watershed. By the 1970s, streambank erosion had caused large numbers of old trees to topple into the stream, forming a debris dam. Stormflow was forced into overflow channels that had been ephemeral but now became permanent parts of the stream system. Although the sediment load from construction had decreased, sediment from collapsing and eroding streambanks continued to move through the stream system. The mass erosion of banks in the upper reaches of the stream caused sediment to accumulate in downstream gravel bars. Overall, the Sawmill Brook was subjected to major fluctuations in size and shape over a relatively short time period. It should be noted that this region is affected by hurricanes and coastal storms. The streamside trees had survived 1938 and 1955 hurricanes (the average age of streamside trees was 73 years) but could not withstand the persistent change in flow patterns caused by urban development.

Studies in the Puget Sound lowlands corroborate the eastern experience with urbanization—increased peak discharge and enlarged channel size. Booth (1990) also found that streams enlarged in different ways to accommodate increased stormflow from urban watersheds. While some expanded laterally, others, especially those in areas of steep slopes, showed rapid incision. Some channels vulnerable to incision appeared to cross a threshold where the relatively slow expansion in width and depth became a rapid and catastrophic downcutting that produced a channel many times larger than that needed to convey stormflow. The streams most susceptible to incision were "low-order, high gradient streams with loose sandy streambed material." Incision was also observed to be the most common streambed disturbance in urban areas near Indianapolis, Indiana. Here the channel slopes are low (0.1% to 1.0%) but the bed material is loam and sandy loam underlain by alluvial sand and gravel, and has little inherent stability (Doyle et al., 2000). (While sandy soils deposited in flat terrestrial areas generally evince low erodibility, sandy soils in streams are easily transported by streamflow [E.S. Verry, personal communication, 2005].)

Several rating systems have been designed to assess stream channel instability (Doyle et al., 2000; Johnson et al., 1999; Natural Resources Conservation Service, 1998; Olsen et al., 1997; Pfankuch, 1975; Rosgen, 1994; Simon and Downs, 1995). These provide a systematic approach for rating many of the characteristics described above and are used to evaluate the current stability of stream channels and the susceptibility of these channels to future degradation.

8.3.4   WATER TEMPERATURE IN URBAN STREAMS

Water temperatures in small streams will increase when the shade provided by riparian forests is removed. As noted in earlier chapters, the forest canopy shades the stream and moderates water temperatures (Beschta and Taylor, 1988; National Research Council, 2000). In urban environments there is an additional thermal burden. When soil is replaced with an impervious surface, the increase in surface temperature can be as much as 12°C (Galli, 1990). Stormwater flows over these surfaces, rather than through the soil, and is heated in the process. In addition, the proportion of baseflow from cooler groundwater sources is typically reduced. Stormwater that is temporarily stored in detention ponds to reduce flooding may be exposed to sunlight and heat. Wastewater from industrial processes can increase stream water temperatures as well. A study in Maryland that monitored temperatures in urban streams found that temperature standards issued by the Maryland Department of the Environment were violated 10% of the time under baseflow conditions when impervious surfaces covered 12% of the watershed area. This increased to 67% of the time when the impervious area reached 60% (Galli, 1990). (Recall the inverse relationship between water temperature and dissolved oxygen concentration, biological oxygen demand [BOD], and the direct relationship with the metabolic rate of aquatic organisms discussed in Chapter 4.)

8.4   WATER CHEMISTRY IN URBAN STREAMS

Urban development introduces an unprecedented assortment of chemical pollutants into streams, rivers, lakes, reservoirs, estuaries, and ultimately the oceans. Nutrients and pesticides are common pollutants in urban areas (U.S. Geological Survey, 1999). Transportation-related pollutants include polycyclic aromatic hydrocarbons (PAHs) (Stepenuck et al., 2002; Van Metre et al., 2000) and road salt (Cherkauer, 1975; National Research Council, 1991). (While roads, trucks, and automobiles also create water quality and ecological problems in suburban and rural areas, the problem is most acute in urban areas.) Additional pollutants originating in urban areas include metals (Sanudo-Wilhelmy and Gill 1999), fecal contaminants (Mallin et al., 2000; Stepenuck et al., 2002), volatile organic compounds (VOCs) (Thomas, 2000), polychlorinated biphenyls (PCBs) (Ayers et al., 2000), and antibiotics, steroids, and hormones (Kolpin et al., 2002).

8.4.1   THE CLEAN WATER ACT IN URBAN AREAS

The Clean Water Act (CWA) was passed into law in 1977 as an amendment to the 1972 Federal Water Pollution Control Act. It has two primary objectives. The first is to regulate the discharge of pollutants into the nation's waterways; the second is to raise water quality standards to levels that are safe for fishing and swimming. The CWA prohibits the discharge of pollutants from point sources into the waters of the United States without authorization through a National Pollutant Discharge Elimination System (NPDES) permit. The regulations promulgated under the CWA have effectively controlled most point sources of pollution (U.S. Environmental Protection Agency, 2002, 2006a).

In the Hudson–Raritan estuary, for example, new and upgraded wastewater treatment plants that began operation after the passage of the CWA reduced the discharge of untreated wastewater from 19,700 L/sec (450 million gal/day) to less than 220 L/sec (5 million gal/day) between 1970 and 1988. Additional measures implemented between 1989 and 1993 by New York City water pollution control programs led to further reductions in untreated wastewater and coliform bacteria concentrations. These measures included a coastal surveillance program that led to the identification and abatement of illegal connections to the storm sewer system and reduced the amount of sewage that bypassed wastewater treatment plants. In addition, increased treatment of waters from combined sewer overflows (CSOs) (Box 8.2), new sewer connections, and water conservation programs helped to reduce pollutant loads. Fecal coliform concentrations in surface waters around New York City were 19 to 199 cells/100 mL in most locations in 1993. No counts in New York Harbor were higher than 2000 cells/100 mL (New York State standards are less than 200 cells/100 mL for swimming and 200 to 2000 cells/100 mL for fishing). Fecal coliform concentrations in the Hudson and East Rivers declined by 78% and 63%, respectively (Brosnan and O'Shea, 1996).

Metal pollution in the Hudson River estuary also declined dramatically between the 1970s and 1990s as inputs from sewage and industrial operations were reduced. Cadmium fluxes were reduced from 81.5 to 1.8 kg/day, copper from 630 to 217 kg/day, nickel from 518 to 43 kg/day, and zinc from 924 to 285 kg/day. Median concentrations in the estuary declined by 55% to 89% for cadmium, 36% to 56% for copper, 53% to 85% for nickel, and 53% to 90% for zinc (Sanudo-Wilhelmy and Gill, 1999).

Box 8.1   Primary, Secondary, and Tertiary Wastewater Treatment

I. Primary
   a. Screens: remove floating debris.
   b. Grit chamber: allows sand and gravel to settle and be removed.
   c. Sedimentation tank: raw primary biosolids (sludge) settle to the bottom of the tank and are removed by pumping.
II. Secondary
   d. Biosolid decomposition: the sludge is transferred to a facility where it is processed in a trickling filter or an aeration tank. These bring the sludge in contact with decomposing bacteria that transform the organic matter into inert by-products. The partially treated wastewater is moved to a second sedimentation tank. Here, excess bacteria are removed.
   e. Chlorine disinfection: effluent from the sedimentation tank is treated with chlorine to kill pathogens and reduce odor. Following this the effluent is released into receiving waters.
III. Tertiary treatment systems employ a variety of advanced chemical and environmental engineering methods that are capable of removing nutrients and other contaminants. It is possible with tertiary treatment techniques (membrane filtration, carbon adsorption, distillation, and reverse osmosis) to produce effluent that consistently meets drinking water standards (U.S. Environmental Protection Agency, 1998).

Unfortunately, the most commonly available wastewater treatment cannot remove or transform all pollutants (Paul and Meyer, 2001). Between 1972 and 1991, $200 billion was spent constructing new wastewater treatment plants in the United States and $154 billion was spent on operation and maintenance expenses at existing plants (Box 8.3). By 1996, most wastewater treatment plants provided secondary treatment, which removes about 85% of the organic matter in wastewater and uses chlorination to kill harmful bacteria, but it does not achieve major reductions in nutrients, metals, and other toxic chemical compounds (Litke, 1999; U.S. Environmental Protection Agency, 1998). Tertiary wastewater treatment, which removes more pollutants, is less widely implemented in the United States (25% of operations in 1996). Most of the tertiary treatment facilities are located in states near the Great Lakes and on the East Coast (Litke, 1999).

Currently, urban water pollution comes primarily from wastewater effluent and from nonpoint sources, not addressed by the original CWA regulations (EPA, 2002). Nonpoint source pollutants are washed off building surfaces, industrial sites, and roadways by precipitation and carried by overland flow to streams, rivers, lakes, and estuaries. Residential septic systems release nutrients and other pollutants that contaminate groundwater. Persistent toxic chemicals such as PCBs are still found in streambed sediments, despite current bans on their use (Garabedian et al., 1998; Myers et al., 2000; Wall et al., 1998). Urban areas are also a concentrated source of air pollution. These may be deposited locally or at great distances from the source (U.S. Geological Survey, 1999).

Box 8.2   CSOs, SSOs, and Stormwater Drains

Combined sewer overflows (CSOs) carry sewage and stormwater draining from city streets in the same pipe. Under dry weather conditions, sewage is delivered to wastewater treatment plants. Following rainstorms, however, the combined volume of wastewater and storm water often exceeds the capacity of treatment facilities and CSOs are designed to overflow, releasing storm water and untreated wastewater (sewage, industrial waste, toxic materials, and floating debris) directly into streams, rivers, lakes, and coastal estuaries. This raises the levels of bacteria, pathogens, and toxic substances and can cause health problems for humans and wildlife — macroinvertebrates, fish, aquatic birds, and mammals.

   The vast majority of the 772 communities in the United States with CSO drainage systems are found in the Northeast — New England, New York, New Jersey, Pennsylvania, Ohio, Indiana, and Illinois. More than 100 communities in New England use CSO systems. Programs to eliminate CSOs have begun and have achieved notable progress in some areas (Boston Harbor). It is estimated that the cost of eliminating CSOs in New England alone could reach $4 billion (U.S. Environmental Protection Agency, 2004a,b).

   Sanitary sewer overflows (SSOs) carry wastewater only. They can release wastewater if water leaks from the pipes during heavy rains. Storm drains transport overland flow nonpoint source contaminants from vehicles and pet waste from urban streets directly to streams and rivers. The water in these drains may also be contaminated by leaking sewer pipes and illegal sewer connections (Massachusetts Water Resources Authority, 2005).

Two amendments to the CWA regulate nonpoint source pollutants as part of the NPDES program. A 1987 amendment attempted to mitigate the effects of nonpoint source pollution through regulations on (1) medium and large municipal separate storm sewer systems in cities with populations of 100,000 or more; (2) stormwater from construction sites that disturb 5 or more acres (2.02 ha) of land; and (3) stormwater from 10 different categories of construction activities. Phase II of the stormflow regulations, passed in 1999, included the operation of separate storm sewer systems in cities with populations of less than 100,000 and construction sites of less than 5 acres (2.02 ha) in the regulatory program. There are six parts of the Phase II program: (1) public education and outreach, (2) public participation, (3) illicit discharge detection and elimination, (4) construction site (runoff) control, (5) post-construction (runoff) control, and (6) pollution prevention (Pennington et al., 2003).

8.4.2   Nutrients

While agricultural loading of fertilizer and manure is the largest source of nutrient pollution overall, cities and suburban areas, which cover a relatively small proportion of land area, also contribute substantial amounts of nitrogen and phosphorus (Fisher et al., 2000; Meals and Budd, 1998; Osborne and Wiley, 1988; U.S. Geological Survey, 1999). Nutrient inputs from cities and suburbs in New England, the Mid-Atlantic, and Midwest cause water quality problems at both local and regional scales, adding to the nutrient loading in estuaries along the Atlantic Coast, the Great Lakes, Lake

Champlain, and the Gulf of Mexico (Boesch et al., 2001; Goolsby et al., 2001; Meals and Budd, 1998; Myers et al., 2000). Concentrated inputs of nutrients, such as wastewater effluent, can generate local concentrations of nitrate and phosphorus substantially higher than those found in surface waters in agricultural areas (Garabedian et al., 1998).

8.4.2.1 Nitrogen

Nitrogen sources in urban areas include wastewater effluent, fertilizer, septic systems, and atmospheric deposition. Wastewater treatment converts ammonia nitrogen to nitrate, which is less toxic to fish, but still a major pollutant (U.S. Geological Survey, 1999). In a USGS study of water quality in the Connecticut, Housatonic, and Thames River basins (Garabedian et al., 1998), the highest nitrate concentration (10 mg/L) was measured downstream of a wastewater treatment facility.

Septic systems in unsewered residential areas are a major source of nitrate groundwater contamination (Bowen and Valiela, 2001; Thomas, 2000). Septic systems are a particular problem in areas with highly permeable sandy soils. These soils are found throughout the Coastal Plain and in scattered areas of glacial outwash deposits throughout New England, New York, and the upper Midwest (Myers et al., 2000; Randall, 2001). The Cape Cod aquifer was formed from deposits of glacial outwash and unconsolidated glacial moraine. This layer of sand and gravel is hundreds of meters thick and provides the sole source of drinking water on the Cape. A study completed on the Cape (Persky, 1986) found there was a significant correlation ($r^2 = 0.80$; $p \le .05$) between housing density and median nitrate levels in wells. The Cape Cod Planning and Economic Development Commission established a drinking water standard of 5.0 mg/L for public supply wells. This level was exceeded in five of nine sample areas where housing density was greater than one unit per acre. Thomas (2000) measured nitrate concentrations in groundwater samples from drinking and monitoring wells in a newer residential area near Detroit, Michigan. She found that 26% of the groundwater samples, recharged after 1953, had nitrate concentrations above background levels (greater than 2 mg/L). Possible sources of contamination other than septic systems include pet waste and lawn fertilizers.

A substantial portion of the nitrogen entering water bodies in the Northeast comes from atmospheric deposition; urban areas are an important source. While pollution controls such as catalytic converters on automobiles have successfully reduced the nitrogen emissions per vehicle (Nixon et al., 1996), increasing human population density and total miles driven means that nitrogen emissions and deposition in urban areas (and in agricultural areas from volatilization of manure and fertilizer) will continue to be a problem. "The highest deposition rates of nitrogen (greater than 2 tons/mi^2 [7 kg/ha]) occur in a broad band from the Upper Midwest through the Northeast" (U.S. Geological Survey, 1999). In 1991 it was estimated that up to 25% of the nitrogen input to the Chesapeake Bay was the result of atmospheric deposition (Fisher and Oppenheimer, 1991; Nixon et al., 1996).

8.4.2.2 Phosphorus

Between 1940 and 1970, much of the phosphorus causing eutrophic conditions in the Great Lakes and elsewhere came from the use of phosphate detergents. States around the Great Lakes and on the East Coast were among the first in the nation to ban phosphate in detergents. These products were gradually eliminated by detergent manufacturers and as of 1994, phosphate detergents were no longer in use. The addition of secondary wastewater treatment to municipal systems typically removes 10 to 20% of the influent phosphorus and reduces effluent concentrations to 3 to 5 mg/L (Litke, 1999). Since 1975, a growing number of wastewater treatment plant discharge permits have limits on phosphorus concentration in their effluent (generally from 0.5 to 1.5 mg/L). Tertiary treatment that is capable of removing 99% of influent phosphorus has been added to some wastewater treatment plants. The EPA recommended stream water concentrations to avoid eutrophication are 0.05 mg phosphorus/L for streams flowing into lakes and 0.1 mg phosphorus/L for flowing waters. Again, states near the Great Lakes and on the East Coast have been the most active in this area. Michigan and Indiana have more than 100 plants with tertiary treatment facilities. New York, Pennsylvania, Illinois, Wisconsin, and Minnesota each have between 50 and 100 (Litke, 1999).

Even with tertiary treatment, the sheer volume of wastewater treatment plant effluent still contributes to nutrient loading in many watersheds. The USGS measured phosphorus, ammonium, and nitrate concentrations in four streams in New England at sampling sites that received more than 10% of flow from wastewater effluent. The median concentrations of these nutrients were 2 to more than 10 times higher at these sites than at those without wastewater treatment plants. The median phosphorus concentration measured in urban streams in this study was nine times greater than the median for forested streams and three times greater than that measured in streams in agricultural areas (Garabedian et al., 1998).

Box 8.3   The Cleanup and Recovery of Boston Harbor

In 1985, the U.S. Environmental Protection Agency (EPA) and the Department of Justice filed a lawsuit against the Commonwealth of Massachusetts for violating the federal CWA by failing to properly treat wastewater flowing into Boston Harbor. Judge A. David Mazzone ordered the construction of new wastewater treatment facilities and set a timetable for the cleanup (1986 to 1999) (U.S. Environmental Protection Agency, 2005c). The newly formed Massachusetts Water Resources Authority (MWRA) supervised the implementation of the Boston Harbor Project, a 13-year program that has substantially improved the condition of the harbor, harbor islands, and surrounding beaches.

At the time of the lawsuit in the 1980s, the harbor tributary rivers (Charles, Mystic, and Neponset) and 88 CSOs discharged "an estimated 12.5 billion liters of partially treated or raw combined sewage" into Boston Harbor each year. Some wastewater received primary treatment in old facilities that were in poor condition. The MWRA improved primary treatment while constructing the new Deer Island Treatment Facility, which entered service in 1997. In addition to providing secondary treatment at Deer Island, the MWRA also increased the pumping capacity from 2650 million L/day in 1989 to 3410 million L/day in 1998. This increased the amount of wastewater that could be treated and reduced discharges from CSOs. The discharge of sewage sludge was discontinued in 1991. Sludge is now transported to a facility that uses it to produce fertilizer pellets. The current plan calls for closing 36 CSOs by 2008. This will reduce annual discharge to 1.5 billion L. Ninety-five percent of the discharge from remaining CSOs will be treated.

The MWRA reduced metals in effluent from more than 450 kg/day in 1989 to approximately 90 kg/day in 2000 primarily by enforcing regulations requiring pretreatment of industrial waste. The decrease in the discharge of sludge and other solids to the harbor reduced the input of adhered metals as well. By the year 2000, effluent discharges within Boston Harbor had ceased. Discharges from the Deer Island treatment facility are now piped through a 15.3 km outfall tunnel into Massachusetts Bay. The mixing and dilution are greater and flushing rates are more rapid at this discharge site.

As a result of the harbor cleanup, bacteria counts have fallen and the eight harbor beaches are swimmable most of the time. The bacterial concentrations at the beaches vary in relation to seasons and storm events yet the specific causes of high counts are not always apparent. Metals in sediments are 50% lower than in the 1980s. The amount of organic matter in sediments has decreased as well and is reflected in decreased biochemical oxygen demand. Between 1991 and 2000, both the abundance and diversity of benthic macroinvertebrates increased. In addition, fish caught in the harbor displayed fewer abnormalities (liver disease and tumors) than in the 1980s.

Pollution problems persist in the Charles, Mystic, and Neponset Rivers. They still receive nonpoint source pollution, CSO inflows, and loading from upstream water treatment plants. In addition, large amounts of decaying organic matter have accumulated in bottom sediments. Haline stratification leads to anoxic conditions in river bottom sediments and reduces decomposition rates. There is also nonpoint source pollution from the urban area that rings Boston Harbor. Results to date have been promising but more source reduction and mitigation is needed (Massachusetts Water Resources Authority, 2002). In 2006 the EPA raised the rating of the Charles River to a B+ (from a D in 1995), meaning that boating standards were met 97% of the time and swimming standards were met 50% of the time (Associated Press, 2006).

8.4.2.3 Pesticides

In agricultural areas, pesticides frequently adsorb to sediment particles before reaching streams and ponds. This can help to slow pesticide movement and allow for chemical breakdown to occur before the pesticides reach streams or groundwater. Impervious surfaces and storm drains in urban areas can create a "continuous pathway" that transmits a large proportion of pesticides from source to sink. In contrast, pesticide residues that enter streams or storm detention basins have the opportunity to adsorb to sediments and organic matter, preventing further transport (Larson et al., 1997). Prometon, one of the herbicides found most frequently in urban streams, is a nonselective herbicide used to kill weeds and grass in parking lots, from which it may be channeled directly to streams through storm drains (Garabedian et al., 1998; U.S. Geological Survey, 1999).

Samples of stream water at 64 sites in urban and agricultural areas of New York State indicated that the insecticides diazinon (Box 8.4) and carbaryl were most closely associated with urban land use (Phillips et al., 2002). This contrasted with agricultural areas of the state, where herbicides such as atrazine, metolachlor, alachlor, and cyanazine were the most common contaminants. This pattern has been confirmed nationwide by the USGS (1998). In addition, banned organochlorine insecticides such as DDT, chlordane, and dieldrin were most frequently found in sediments of urban streams and in the tissue of fish taken from those streams. Agricultural use of chlordane and aldrin (the parent material of dieldrin) was restricted in the 1970s. However, these chemicals continued to be used in urban areas until the mid-1980s for termite control. The chemicals accumulate in animal tissue and can affect the nervous system (Agency for Toxic Substances and Disease Registry, 1995, 2002). In 1999 the USGS NAWQA program reported that 40% of urban stream sediments in 20 NAWQA study areas had concentrations of organochlorine pesticides that exceeded national guidelines for the protection of aquatic life (U.S. Geological Survey, 1999).

Although drinking water standards for individual pesticides are seldom exceeded, complex mixtures of insecticides and herbicides are commonly found in urban streams (Ator et al., 1998; Garabedian et al, 1998; Lindsey et al., 1998; Peters et al., 1998; U.S. Geological Survey, 1999). Concentrations that exceed drinking water standards are most frequently found in stormwater samples. An analysis of stream water in the predominantly residential and urban Deer Creek watershed in western Pennsylvania found 25 pesticides and pesticide metabolites. These included the insecticide diazinon and four herbicides—prometon, atrazine, simazine, and metolachlor. Maximum concentrations of diazinon (0.097 µg/L) were within drinking water standards, but exceeded the EPA water quality guideline for aquatic life (0.08 µg/L) (Anderson et al., 2000).

Box 8.4   Diazinon

Diazinon is an organophosphate insecticide. Organophosphates entered widespread use during the 1960s and 1970s as replacements for the banned organochlorides (Larson et al., 1997).

Diazinon has been used since the 1960s in urban and agricultural areas throughout the United States to kill cockroaches, ants, fire ants, grubs, fleas, mites, aphids, beetles, and ticks (Ohio State University Extension, 2001). In urban areas it has been applied to lawns and gardens and also in office buildings, schools, and warehouses (U.S. Environmental Protection Agency, 2000b). It is an ingredient in pest strips and also has been used to kill fleas and ticks in kennels (EXTOXNET, 1996). The drinking water standard for diazinon is 0.6 μg/L; the guideline for the protection of aquatic life is 0.08 μg/L (Anderson et al., 2000). No maximum contaminant load (MCL) has been established. The breakdown rate depends upon stream water pH. In acidic water, the half-life of diazinon is 12 hours, however, in a neutral solution (pH 7.0) the half-life of this compound can be as long as 6 months (EXTOXNET, 1996).

Studies conducted by the USGS NAWQA in the 1990s found that diazinon, in combination with other pesticides, was a frequent contaminant of urban streams. Diazinon was detected in stream water at 11% of the 40 agricultural indicator sites and at 70% of the urban indicator sites between 1992 and 1996. Of the detections at the urban sites, 50% of the samples had concentrations greater than 0.025 μg/L, 10% were greater than 0.25 μg/L, and 5% were greater than 0.43 μg/L (U.S. Geological Survey, 1998). Diazinon was detected in all of the air samples taken over the Mississippi River from New Orleans, Louisiana, to St. Paul, Minnesota, in 1994. The highest concentrations were measured near major metropolitan regions. Diazinon has also been detected in rain and fog (U.S. Environmental Protection Agency, 2001; U.S. Geological Survey, 1998). In general, diazinon remains in the top half inch of the soil. However, it was detected in 54 wells in California (Anderson et al. 2000; EXTOXNET, 1996; Wall et al., 1998).

Diazinon kills insects by inhibiting an enzyme, acetylcholinesterase, essential to the proper functioning of the nervous system. Between 1994 and 1998, diazinon was associated with more bird mortality than any other pesticide. Ducks, geese, hawks, songbirds, and woodpeckers were among the reported fatalities (U.S. Environmental Protection Agency, 2001). Diazinon has also been found to be highly toxic to bees and fish (although it does not accumulate in fish tissue). In 1988 the EPA restricted the use of diazinon, eliminating its application on golf courses and sod farms due to bird mortality (EXTOXNET, 1996). In 2000 the EPA completed an agreement with permitted users to phase out diazinon for indoor use and on lawns. The chemical will be phased out over a 4 year period because it "poses risks to humans, birds, and other forms of wildlife." Diazinon is still being used on more than 40 agricultural crops and on imported bananas, citrus, olives, peppers, tomatoes, cattle, and sheep (U.S. Environmental Protection Agency, 2000b). The goal of the residential use restriction for diazinon (especially on urban and suburban lawns) is to protect drinking water supplies.

8.4.3 ROAD SALT

Sodium chloride (NaCl) is the most common deicing agent applied in the Northeast. It is reliable, easy to handle and store, and far less expensive than alternative compounds, although these are more environmentally benign (National Research Council, 1991). A 1991 report published by the Transportation Research Board of the National Research Council (NRC) showed that the highest road salt application rates in the United States were found in New York and New England—areas with cold, snowy winters and high traffic volume.

Thomas (2000) found that groundwater in 90% of monitoring wells in the Detroit metropolitan area had elevated salinity levels (Box 8.5). Chemical analysis indicated that this was the result of human activity rather than natural factors. A survey of 2000 lakes and ponds in Massachusetts between 1983 and 1984 (Mattson et al., 1992) found that more than 98% exceeded chloride concentrations predicted from an EPA survey of eastern lakes (Brakke et al., 1988).

The six regions (of the state) exhibit perfect rank correlation ($\alpha < 0.01$) between median chloride and population density, and this positive relationship suggests that the high levels of sodium and chloride found in inland lakes are probably due to anthropogenic inputs such as road salt. The amount of salt applied to roads varies depending on snowfall, but can be as high as 230,000 metric tons per year on state roads alone, and has had a measurable impact on public water supplies (Mattson et al., 1992).[*]

Massachusetts has a long-term monitoring program for public water supplies. Application rates have been reduced and sodium chloride alternatives (sand and a mixture of sodium chloride and calcium chloride) are used in some areas where concentrations have been especially high (as much as 95 mg/L) (National Research Council, 1991). The state has also published BMP guidelines to help reduce pollution from road salt storage facilities (Massachusetts Department of Environmental Protection, 1997). Similar practices have been adopted in the New York City water supply system (National Research Council, 2000). A study in Nassau County (Long Island), New York, found elevated concentrations of road salt in monitoring wells following winter storms. Road salt was only applied during a few winter storms, yet the effect on groundwater persisted for several months (Brown et al., 1997).

Road salt damages trees and other vegetation in urban and rural areas. Sodium chloride reduces the osmotic potential in the soil, making it more difficult for plants to extract water. Dehydration causes the stomates in leaves to close. This limits gas exchange and photosynthesis, and slows growth. Plants growing in soils that are contaminated with sodium chloride can die from dehydration, just as they would in a drought, even though soil water may be plentiful (Kozlowski and Pallardy, 1997). In addition, salt spray and splash from passing vehicles can cause direct cellular damage to the twigs and foliage of trees and other plants.

Road salt readily dissolves in water, producing sodium (Na^+) and chloride (Cl) ions. Sodium ions attach to soil particles, frequently displacing the nutrient cations necessary for plant growth—calcium (Ca^{2+}), potassium (K^+), and magnesium (Mg^{2+})—from soil cation exchange sites (Mason et al., 1999). Na^+ ions form weak ionic bonds with soil particles. However, the massive influx of Na^+ ions from road drainage increases the rate of exchange with more tightly held ions (Barnes et al., 1998; Brady and Weil, 2002).

Because plant species differ in their tolerance to salt, this contaminant can change community composition and animal habitat. A study in western Massachusetts investigated the effects of road salt on a 70 ha wetland near the Massachusetts Turnpike and home to several state-listed rare plant and animal species (Richburg et al., 2001). *Phragmites australis*, a large, salt-tolerant, invasive reed, commonly seen near highways throughout the Northeast, had invaded the wetland area. Water samples from shallow wells (60 cm deep) had elevated salt concentrations (Na^+ > 112 mg/L, Cl > 54 mg/L) at distances up to 300 m from the turnpike. *Phragmites* became established along the turnpike, then expanded into dense colonies throughout the bog. This vegetation change occurred because other plant species were damaged by the salt and then crowded out by the salt-tolerant invasive *Phragmites*.

Field and laboratory studies (Blasius and Merritt, 2002) indicate that benthic macroinvertebrates are fairly tolerant of road salt concentrations—much higher than those commonly found in streams—even during spring snowmelts when large volumes of stormwater with high salt concentrations enter streams during a short period of time. Most invertebrate species in this study survived concentrations of up to 10,000 mg/L for 24 hours.

Road salt in solution flows through soil and can contaminate municipal and household drinking water wells. Elevated sodium concentrations are a health concern for people with or at risk for hypertension. For patients on restricted sodium diets, the American Heart Association recommends the use of distilled water if their household water contains more than 20 mg sodium/L. Sodium concentrations measured in domestic wells in the Detroit area study (Thomas, 2000) were as high as 150 mg/L.

There are, unfortunately, other water quality problems associated with road salt applications. Iron cyanide compounds (sodium ferro-cyanide [$Na_4Fe(CN)_6$] or yellow prussiate, and ferric ferrocyanide [$Fe_4(Fe_3(CN)_6)_3$] or Prussian blue) are frequently added to road salt to prevent it from

---

[*] Reprinted with permission of the American Water Resources Association.

clumping and caking. These compounds dissolve with the salt, yielding iron-cyanide complexes in water. Iron-cyanide complexes display little or no toxicity to humans and aquatic life. However, it is possible, in the presence of light, for these complexes to disassociate and release highly toxic free cyanide (HCN or CN). Researchers have not found widespread evidence of toxic levels of cyanide in water contaminated by road salt, but the number of studies has been limited (Paschka et al., 1999).

Box 8.5   Shallow Groundwater in an Urban Area

Mullaney and Grady (1997) investigated water quality in a shallow, unconfined aquifer in Manchester, Connecticut. The area includes a variety of land uses—medium and high density residential housing, commercial and industrial development, low density residential housing and forest land. The glacial sand and gravel aquifer provides baseflow for area streams and supplies drinking water wells. A municipal sewer system serves 92% of households in the area.

Five sites were selected along the groundwater gradient. One site had a single well; the other four sites had nested clusters of three wells. Each cluster had a well that was completed near the bottom of the aquifer, one in the center, and the third near the top. Water samples were collected up to seven times during 1994 and 1995 to reflect various seasonal conditions.

Sodium (Na^+) levels were higher in all 13 study wells than in 10 wells in undeveloped areas in Connecticut with similar aquifer composition (23 mg/L versus 6 mg/L, median value). The median chloride (Cl^-) concentration was 46 mg/L versus 8 mg/L in the rural wells. This was attributed to road salt applications. Chloride concentrations were highest in water closest to the surface. Nitrate and nitrite concentrations were also higher (3.8 mg/L median concentration compared to 0.4 mg/L in 10 rural wells) and in this case there was no change in concentration with depth. The most probable source of elevated nitrogen was lawn fertilizer in residential areas. Four pesticides, two herbicides, and two insecticides, were detected in the shallowest wells at two sites. These included dichlorodiphenyldichloroethylene (DDE), a breakdown product of DDT, and dieldrin (used for termite control, but banned in Connecticut since 1987). VOCs were detected in 12 of the 14 sample wells at depths ranging from 1 to 18 m below the water table. Chloroform, trichloroethane (TCE), and methyl *tert*-butyl ether (MTBE) were the most commonly found compounds. The most probable source of chloroform is chlorinated water leaking from the municipal water distribution system. TCE is a solvent used for dry cleaning and degreasing. MTBE, a gasoline additive, is carried with overland flow from city streets. The use of MTBE was mandated by amendments to the Clean Air Act in 1990 (Garabedian et al., 1998; Mullaney and Grady, 1997; Zogorski et al., 2001). The additive added oxygen to fuel and was formulated to reduce carbon monoxide emissions and smog. All detections of VOCs were below EPA MCLs except TCE, with concentrations as high as 11 μg/L (EPA MCL = 5 μg/L). This was the only violation found of EPA drinking water standards. It was clear, however, and a cause for concern, that urban land use had a pervasive effect on water chemistry throughout the aquifer.

8.4.4   METALS

There are many nonpoint sources of metal pollution in urban areas. Metal roofing releases zinc, copper, and cadmium (Davis et al., 2001). Asphalt shingles and painted siding on older buildings are a source of lead (Van Metre and Mahler, 2003). Brake pads on vehicles contain and release nickel, chromium, lead, and copper; tires are an additional source of zinc (Davis et al., 2001; Paul and Meyer, 2001). Atmospheric emissions and deposition are another source of metals, particularly mercury (Gao et al., 2002; Landis et al., 2002; Sanudo-Wilhelmy and Gill, 1999).

Pitt et al. (1993, 1995) analyzed 87 urban stormwater samples from a variety of source areas in the Birmingham, Alabama, area and found metals in almost all of them. Roofing with galvanized drainage components was the primary source of zinc in stormwater from commercial and industrial areas; a result confirmed in other studies (Bannerman et al., 1993). Stormwater from parking areas had the highest concentrations of nickel. Roads and vehicle service areas produced the highest concentrations of cadmium and lead. Analyses of water samples (including all pollutants) showed that 8 of the 12 samples of roof drainage, 8 of the 16 water samples from parking areas, and 2 of the 5 samples from vehicle service areas were moderately to highly toxic.

Metals in atmospheric deposition may have local and regional sources. Gao et al. (2002) investigated trace elements associated with airborne particulate matter over the New York–New Jersey harbor bight. They found that atmospheric concentrations of nickel, zinc, copper, lead, and cadmium in fine atmospheric particles were approximately 2000 times higher than that expected from the natural abundance of these particles in crustal soil. The primary sources of atmospheric metals in this area are metal processing plants, fossil fuel combustion, and solid waste incineration. Landis et al. (2002) found that mercury concentrations in precipitation over the Chicago–Gary, Indiana urban area were twice as high as those measured at reference sites. Sanudo-Wilhelmy and Gill (1999) measured dramatic decreases in copper, cadmium, nickel, and zinc following treatment of point sources to the Hudson River estuary, but found no reduction in concentrations of lead and mercury. The authors concluded that lead had accumulated in watershed soils over the decades before leaded gasoline was banned and that the mercury originated from air emissions in Canada and the Midwest. Both elements were leaching out of the watershed soils and entering waterways during storm events.

Metals tend to persist in streambed sediments and aquatic ecosystems. Streambed sediments in the Hudson River basin watersheds with more than 10% urban land use had the highest proportion of sites with samples that exceeded sediment limits for copper, lead, mercury, nickel, and zinc (Wall et al., 1998). Metals are commonly found in the livers and tissue of fish sampled in NAWQA studies in basins throughout the Northeast (Anderson et al., 2000; Lindsey et al., 1998; Myers et al., 2000; Wall et al., 1998).

8.4.5 PATHOGENS

Fecal coliform bacteria and associated pathogens are a ubiquitous concern in urban and agricultural areas. A study of bacteriological water quality in a North Carolina estuarine system (Mallin et al., 2000) found that fecal coliform bacteria abundance was positively correlated with measures of watershed population and the percentage of developed land within a watershed. At the time of the study, point sources and problems with domestic septic systems had largely been controlled. The greatest indicator of fecal coliform abundance was the percentage of impervious surface area within the watershed. In regression analysis, this parameter explained 95% of the variability in fecal coliform abundance among the five watersheds. Streams in the Bradley Creek watershed, the most heavily developed area, had fecal coliform counts of 98 colony forming units (CFU)/100 mL (geometric mean) compared to a mean of 13 CFU/100 mL in the least developed watershed. The Bradley Creek area had been closed to shell fishing for many years. Mallin et al. theorized that connected impervious areas allowed pollutants (e.g., pet waste) to concentrate and be carried by stormwater directly to streams.

Contaminated stormwater can be a direct threat to human health. A *Cryptosporidium* outbreak in March and April of 1993 caused more than 100 deaths and sickened more than 400,000 people in Milwaukee, Wisconsin (Fox and Lytle, 1996; National Research Council, 2000). *Cryptosporidium* oocysts were passed through a water treatment plant that processed water from Lake Michigan. Severe spring storms increased stormwater inflows to the lake and increased turbidity and bacterial counts in water entering the treatment plant.

8.4.6 Polycyclic Aromatic Hydrocarbons

Polycyclic aromatic hydrocarbon (PAH) sources include automobile exhaust, degradation of tires, oil and fuel spills, industrial solvents, air-blowing of asphalt, domestic heating (coal, oil, gas, and wood), refuse incineration, and biomass burning (Gustafson and Dickhut, 1997; Paul and Meyer, 2001; Van Metre et al., 2000). Exposure to PAHs causes varied and wide-ranging toxic effects in a variety of organisms—birds, invertebrates, fish, and mammals, including humans. "Fish exposed to PAH contamination have exhibited fin erosion, liver abnormalities, cataracts, and immune system impairments leading to increased susceptibility to disease" (U.S. Environmental Protection Agency, 2005b). PAHs are potent carcinogens.

Van Metre et al. (2000) analyzed sediment cores from 10 lakes and reservoirs: one in the state of Washington, two in Texas, two in Minnesota, and five on the East Coast (New York, New Jersey, Virginia). Cores with the earliest sediments were dated back to 1800. The shortest time record went back to 1968. All sediment cores were collected between 1996 and 1998. Modest to substantial increases in PAH concentrations over time were found in sediments in all lakes and reservoirs that were sampled. Contamination of the most recent sediments at all sites exceeded sediment quality guidelines. Areas with the greatest increase in the percentage of urban land use demonstrated the largest increases in PAH concentrations. In addition, the assemblage of PAH compounds changed during the last 40 years. PAHs in sediments deposited prior to 1960 were primarily from uncombusted sources—oil seeps and petroleum spills. Since then the proportion of PAH compounds from combusted sources—primarily vehicle exhaust—have increased. An in-depth examination of the two watersheds surrounding the sites in Texas indicated that the increase in PAHs was directly proportional to the increase in vehicle miles traveled for the entire metropolitan area during the last 30 years. Van Metre et al. (2000) suggested that urban sprawl elsewhere was affecting water quality within the watershed, since PAH contaminants enter the water supply with atmospheric deposition.

A study in the Chesapeake Bay region (Gustafson and Dickhut, 1997) showed that PAH gas-phase concentrations were as much as 50 times higher in air samples from urban sites relative to rural sites. In addition, the urban sites showed the greatest concentrations during the summer months. There was little seasonal variation in samples taken from the rural sites. This pattern suggested that PAHs were being volatilized from contaminated surfaces such as unshaded roads in urban areas. High temperatures on these surfaces will increase volatilization rates. Again, atmospheric deposition in streams and rivers and directly into the bay is an area-wide source.

8.4.7 Volatile Organic Compounds

Volatile organic compounds are carbon-based compounds that evaporate when exposed to air. Common chemical compounds in this group include carbon tetrachloride, trichloroethylene (TCE), tetrachloroethylene (also known as perchloroethylene [PCE]), trichloroethane (TCA), benzene, and toluene. These compounds are found in fuel, fuel-combustion products, dry cleaning fluid, metal degreasers, refrigerants, paint, paint thinners, adhesives, deodorants, drain cleaners, septic tank degreasers, pesticides, and magic markers. Products containing VOCs are used in homes and extensively in manufacturing. Effects of low-level exposure are unknown, however, exposure to or ingestion of high concentrations of VOCs can cause cancer (Cleanupdate, 1998).

Volatile organic compounds are detected in groundwater, like other soluble contaminants. A study on Cape Cod, Massachusetts, found a contaminant plume of TCA and PCE emanating from a military landfill (McCobb and LeBlanc, 2002). In areas of low density residential development, VOC contaminants can enter the groundwater from septic systems and leach fields. Thomas (2000) found that 29 out of 30 monitoring wells located in the recharge areas of an unconfined sand and gravel aquifer in the Detroit metropolitan area contained trace amounts of VOCs. While no drinking water standards were exceeded, these data show that contaminants are moving into and throughout the groundwater system.

The serious health hazards associated with VOCs are the reason that drinking water standards are set at very low concentrations. The EPA maximum contaminant level goal (MCLG) is zero (nondetect) for benzene (a VOC used in making plastics and synthetic fabrics and as a solvent in printing and dry cleaning). The maximum contaminant level (MCL) for benzene, defined as the "lowest level to which water systems can be reasonably required to remove this contaminant should it occur in drinking water," is 5 µg/L (U.S. Environmental Protection Agency, 2000a, 2005a).

8.4.8 POLYCHLORINATED BIPHENYLS

Polychlorinated biphenyls were manufactured in the United States prior to 1977 and used as coolants and lubricants in electrical equipment such as transformers and capacitors. The manufacture of PCBs was prohibited when it was found that the chemical (commercially known as Aroclor) accumulated in sediments in stream and river beds, and caused health problems. PCBs do not break down in the environment. Like DDT, they are taken up by vegetation and accumulate in the tissues of herbivores and higher level predators such as fish, birds, and mammals that consume contaminated fish. PCBs have been shown to cause liver cancer in rats (Agency for Toxic Substances and Disease Registry, 2000).

The Michigan Maternal/Infant Cohort Study (Fein et al., 1984; Jacobson and Jacobson, 1990, 1996; Johnson et al., 1998) found that infants born to women who had consumed at least 11.8 kg of PCB-contaminated Lake Michigan fish in the 6 years prior to giving birth had lower birth weights and displayed developmental delays in the first year of life when compared to a control group whose mothers had consumed none of the contaminated fish. This group of children continues to suffer from cognitive deficits (lower IQ scores, memory and attention deficits, and difficulty in acquiring reading skills) as shown by follow-up assessments at ages 4 and 11 years. The study was controlled for other variables such as socioeconomic status and maternal age and education.

Many sites in the Northeast are contaminated with PCBs. Legal actions have been taken against General Electric, which used PCBs and discharged them into the Hudson River near Glens Falls, New York, and into the Housatonic River near Pittsfield, Massachusetts (Garabedian et al., 1998; U.S. Environmental Protection Agency, 2006b).

8.4.9 ANTIBIOTICS, STEROIDS, AND HORMONES

Organic wastewater contaminants, including antibiotics, prescription and nonprescription drugs, steroids, and hormones, are the source of increasing concern (Kolpin et al., 2002). Wastewater treatment plants are not designed to remove these chemical compounds. They also may enter stream systems from veterinary pharmaceuticals used at animal feedlots and for pets. Little is known about their effects on human health or about the possible synergistic effects with the other compounds found in streams. A nationwide survey found that one or more of these compounds were found in 80% of 139 sampled streams. In fact, the occurrence of multiple compounds was the norm. The median number of compounds detected was 7, with a maximum of 38 in one stream. The streams included several in the Boston, New York, Philadelphia, and Chicago metropolitan areas (Kolpin et al., 2002).

8.5 URBAN DEVELOPMENT AND AQUATIC ECOSYSTEMS

Many studies have documented declines in invertebrate and fish populations associated with urban development. Declines in biodiversity and population numbers in streams and wetlands have been related to general measures of development, such as human population density, urban land cover, and percent TIA and EIA (Booth and Reinelt, 1993; Hicks and Larson, 1996; Horner et al., 1996; Jones and Clark, 1987; Klein, 1979; Morley and Karr, 2002; Walsh et al., 2005a; Weaver and Garman, 1994). Declines are initiated by persistent changes in streamflow and sediment associated with urban development. Major disruptions in habitat conditions for stream life occur even at relatively low levels of urbanization and before there are detectable changes in water chemistry due to pollution (Booth, 1991; Hart and Finelli, 1999; Horner et al., 1996; Waters, 1995). The loss of riparian vegetation that shaded streambanks and increasing areas of paved roads (Box 8.6), parking lots, and roofs lead to higher water temperatures and, as a result, chronic or acute stress on aquatic species (Galli, 1990). Episodic and chronic pollutant loads coupled with erratic stream-flow patterns also reduce the time available for stream ecosystems to recover (Booth, 1991; Klein, 1979; Townsend and Scarsbrook, 1997).

Box 8.6 Roads and the Degradation of Water Resources and Aquatic Habitat

Forman and Deblinger (2000) examined the ecological impact of a 25 km stretch of Route 2 (a four-lane artery) between Route 95 (Cambridge, Massachusetts) and Route 495 (Littleton, Massachusetts). Traffic volume ranged from 34,000 to 50,000 vehicles per day at various points along the road. This stretch of road crosses over 2 rivers and 13 streams. Alteration of streams and wetlands and road salt contamination were the most obvious effects of this road. Almost all the streams were channelized in the vicinity of the road crossing. Dunne and Leopold (1978) list three primary disadvantages of stream channelization:

(1) channel instability or effects of channel readjustment to the imposed conditions; (2) downstream effects especially increased bank erosion, bed degradation or aggradation; and (3) esthetic degradation, especially the change in stream biota and the visual alteration of riparian vegetation, of stream banks and channel pattern or form.

Engineered channels extended from 30 m to 400 m upstream and from 30 m to 500 m downstream of the road. Nine wetlands that are crossed by or within 500 m of Route 95 have been drained or otherwise altered. Road salt damage to vegetation was confined to wetlands, streams, and ponds within 10 m of the road. The salt concentration in a municipal reservoir at the intersection of Route 95 and Route 2 has been elevated for many years. The adverse ecological effects on aquatic systems extended from 200 to 1500 m from the various road crossings.

Wang et al. (2000) examined the effect of increasing urban land cover in predominantly agricultural watersheds in southeastern Wisconsin. They compared fish sampling data from 1997 with data from 1974 to 1977 in 43 test watersheds. The mean agricultural land cover had decreased from 54% to 43% and the mean urban land cover had increased from 24% to 31% over about two decades. They also sampled fish communities in four reference watersheds where the proportion of urban (4% to 5%) and agricultural (59% to 65%) land cover had remained relatively constant since the 1970s. The mean number of fish species decreased 15%, fish density by 41%, and mean index of biotic integrity (IBI) scores from 24 to 16 in the test watersheds. Pollution intolerant species were absent in streams located in the most urbanized watersheds (more than 10% impervious surface in the 1970s). In more rural streams of the test watersheds (less than 10% impervious surface in the 1970s) there was little change in the number of tolerant versus intolerant species. However, there were significant declines in both the number of species and the number of individuals. The biotic communities in the reference streams were essentially unchanged between the 1970s and 1990s. A second study of macroinvertebrate communities in the same area (Wang and Kanehl, 2003) found that high macroinvertebrate index scores (measuring Ephemeroptera, Plecoptera, and Trichoptera [EPT] abundance, EPT taxa, filterers, and scrapers) occurred only in watersheds with an EIA of less than 7%.

The Ohio Environmental Protection Agency (Ohio EPA) conducts annual chemical, physical, and biological surveys of streams at 250 to 350 sampling sites in 10 to 15 parts of the state. An analysis of 110 of these sites found correlations between IBI scores and measures of urbanization. Eighty-five percent of sites classified as "impacted by urban land use and pollution sources" received poor or very poor IBI scores. IBI scores were poor or very poor for 40% of suburban sites and for 5% to 10% of rural sites. Biological impairment was most severe when urban development included industrial and commercial sites and streams received stormwater inputs from CSOs (Yoder and Rankin, 1996). Miltner et al. (2004) also used Ohio EPA data to track changes in IBI scores during 10 years of rapid development, measured by changes in impervious area, in the Columbus, Ohio, metropolitan area. During this time, population and urban land use in the study watersheds more than doubled. Fish sampling data for 2000 were compared to data from 1996, 1994, 1993, 1992, and 1991. Stream sites affected by urban stressors such as CSOs and wastewater treatment plants were removed from the analysis, leaving 123 sampling sites. There were significant declines in IBI scores when impervious area exceeded 13.8% of the contributing area upstream of the sampling sites. Pollution intolerant species such as silver shiners (*Notropis photogenis*) and hornyhead chubs (*Nocomis biguttatis*) were not present. Some streams showed declines when only 4% of the upstream contributing area was developed. This was attributed to the effects of poorly regulated construction sites. Significantly, there were a few sites where IBI scores did not decline despite urban land use greater than 30% in 2000. These were streams with protected riparian areas and undeveloped floodplains, or streams that received significant groundwater inputs.

Morley and Karr (2002) examined the relation between the benthic-index of biotic integrity (B-IBI) and various watershed land cover attributes at different spatial scales. The spatial scales included (1) the entire subwatershed above each sampling point, (2) the riparian area 200 m on either side of the stream and extending along the entire drainage network above the sampling point, and (3) the local riparian area 200 m on either side of the stream for 1 km above each sampling point. Land cover categories included (1) percent forested land, (2) percent urban land (urban forest + urban grassy + intense urban, defined as 100% paved or bare soil), (3) percent intense urban, and (4) percent impervious area. Sample sites were selected within a specific range of elevation (5 to 140 m), slope (0.4 to 3.2%), and watershed area (5 to 69 km^2) and excluded locations near bridges or culverts, or those influenced by impoundments or construction sites. B-IBI scores declined as both urban land, intense urban land, and impervious area increased. These relationships were all statistically significant ($p < .001$, $n \geq 31$). The strongest relationship was between B-IBI score and the percent urban land at all three spatial scales: subwatershed ($r = 0.73$, $p < .001$, $n = 34$), riparian ($r = 0.75$, $p < .001$, $n = 34$), and local ($r = 0.71$, $p < .001$, $n = 31$). There was considerable variability in B-IBI scores across the 45 monitoring sites (range 10 to 48). Good or excellent conditions (B-IBI ≥ 38) were observed at only 10% of the sites. The best site, indicated by the presence of stoneflies and other intolerant taxa and predators, was in one of the least developed areas. Streams in the poorest condition—low biological diversity and no intolerant taxa—were located in the most urbanized areas; however, a few stream sites in good condition were found in watersheds with urban land cover as high as 50%.

In a study that included many of the same Puget Sound basin sites, Booth et al. (2004) concluded that while B-IBI scores generally declined as TIA increased, TIA could not "be used to predict biological condition at a given site." They found that biologic conditions were "highly variable" among watersheds where impervious area ranged from approximately 10% to 40%. In watersheds with TIAs of greater than 40%, variability declined and B-IBI scores were uniformly low, regardless of riparian condition (Figure 8.6).

An EPA sponsored study (Hession et al., 2003) assessed the effects of the presence or absence of riparian forest in urban watersheds on fish and benthic macroinvertebrates. Hession et al. gathered data at forested and nonforested sections of 12 streams in 18 small watersheds (1 to 16 km^2) north and west of Philadelphia, Pennsylvania. Watershed impervious cover ranged from 1% to 66%.

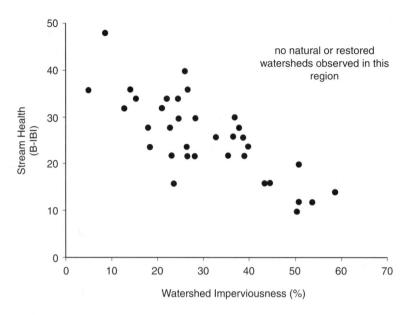

FIGURE 8.6 Stream health as measured by a benthic macroinvertebrate index (B-IBI) versus watershed imperviousness. Scores are variable up to approximately 30% impervious area. B-IBI scores decline consistently above 30% watershed imperviousness (Booth, 2005; reprinted with the permission of the North American Benthological Society).

Impervious cover in 15 of the 18 watersheds was greater than 30%. Hession et al. reached the following conclusions:

1. All macroinvertebrate metrics correlated significantly with the amount of watershed impervious cover. Macroinvertebrate metrics declined with increasing impervious area.
2. The presence of forested or nonforested riparian buffers did not significantly affect the relationship between most macroinvertebrate metrics and impervious surface.
3. Two macroinvertebrate metrics were significantly different for forested and nonforested reaches, but only at the lowest levels of watershed imperviousness.
4. Fish community characteristics were similar in forested and nonforested reaches.
5. Fish community metrics (i.e., the number of intolerant species) declined significantly at relatively low levels of urbanization. Riparian conditions had little effect on fish community metrics.

In this study it also appears that stream habitat declines and the effect of riparian buffers diminishes above 30% to 40% TIA.

Walsh et al. (2005a) assessed the relationship between 14 ecological indicators, including dissolved organic carbon (DOC), chlorophyll *a*, reactive phosphorus, stream algae and diatoms, biological indices of benthic macroinvertebrates, the number of families of EPT, and impervious area at 15 sites on small streams outside Melbourne, Australia. Results of linear regression models showed that EIA was a better predictor of stream condition than TIA for all of the indicators except DOC. The number of EPT families declined from 1% to 14% EIA. At about 14% EIA, scores reached a threshold and remained at a relatively low level of community composition. Current proposals to rehabilitate urban streams and minimize the adverse effects of new development focus on the design of drainage systems that reduce the proportion of EIA and maintain or increase infiltration throughout the watershed.

8.5.1 Urban Riparian Areas

Most riparian buffer studies have tested the capacity of buffers to store and process sediment and pollutants from agricultural fields and timber harvesting (Groffman et al., 2003). In developed areas, storm drainage systems (piped networks) may completely bypass riparian zones and discharge water and nonpoint source pollutants directly into streams. Other changes in soils and pathways of flow in urban areas also limit the effectiveness of riparian buffers to mitigate pollution.

In one of the few urban riparian studies, Groffman et al. (2002) examined soil nitrogen processes in riparian buffers along streams as part of the Baltimore (Maryland) Ecosystem Study. They found that urban riparian areas differed from those in forested watersheds. The depth to the saturated zone (water table) was lower due to reduced infiltration. Stream channels were incised as a result of increased peak discharge and reduced sediment supply following expansion of impervious area (Figure 8.7). This "urban drought" caused changes in the soils in the riparian buffer. Typically riparian soils are hydric soils that develop in intermittently anaerobic (anoxic) conditions because the water table is at or near the surface. Hydric soils usually contain a large proportion of organic matter in various states of decomposition. Hence the potential for denitrification, an anaerobic

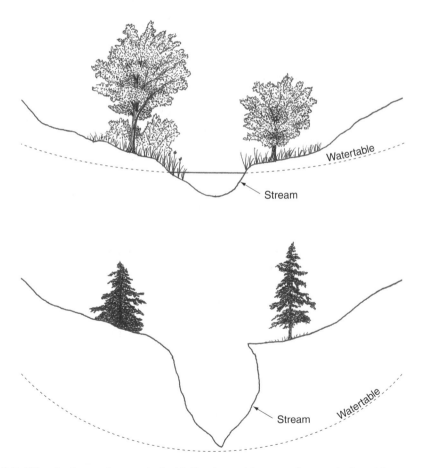

FIGURE 8.7 Urban land use reduces watershed infiltration and increases the magnitude and frequency of storm-flow events, causing stream channel erosion and in some cases incision. This can reduce the level of the water table in urban riparian areas, causing riparian wetlands to dry up and eliminating areas that remove nitrogen through denitrification (Groffman et al., 2003; reprinted with the permission of the Ecological Society of America).

process that requires organic matter, is high. Denitrification potential decreases with soil depth because soil organic matter typically decreases. The lowered water table in the urban setting increases soil oxygen and leads to aerobic conditions. Increased oxygen availability accelerates the decomposition rate of organic matter. Denitrification potential decreases markedly under these conditions, while nitrates dissolved in subsurface flow travel below the root zone of riparian buffers. The urban soils were nitrate enriched in comparison to forest soils in reference watersheds—the result of increased nitrogen loading from urban land use, increased rates of nitrification, and reduced denitrification (Groffman et al., 2002, 2003).

8.6   URBAN BMPs

Urban BMPs address three major issues: stormwater management, reduction of nonpoint source pollutants, and sediment control at construction sites. Stormwater management (decreasing overland flow and increasing watershed infiltration) is critically important because it can help to maintain the stability of stream channels and reduce the transport of pollutants from urban and suburban buildings and streets. Practices designed to regulate flow and reduce nonpoint source pollution include detention ponds, wetland creation and restoration, constructed swales, and filtering and infiltration practices (Pennington et al., 2003; Strecker et al., 2001).

8.6.1   STORMWATER MANAGEMENT

Most states and municipalities require the construction of detention ponds at new development sites. As we have seen, the expansion of watershed impervious area and drainage connection that occurs with urban development increases peak discharge and decreases the duration of individual stormflow events. Detention ponds provide temporary storage and hydraulic regulation of the outflow rate; this attenuates peak discharge. The goal of detention pond design and construction is to maintain post-development discharges at their pre-development levels.

Detention ponds can be designed for two different levels of performance. Stormwater management regulations attempt to balance the protection of urban streams against the cost of pond construction. The least expensive level of performance is the "peak standard" (Booth et al., 2002). This requires detention ponds to attenuate flow from a storm of a specified recurrence interval (25 year, 10 year, or in some instances as low as 2 year storms) (Roesner et al., 2001). The goal of the alternative "duration standard" is the construction of detention ponds that "maintain(s) the post development duration of a wide range of peak discharges at pre-development levels" (Booth et al., 2002).

While there is no doubt that detention ponds reduce peak discharges for design storms and distribute that flow over a longer time period, it is not clear that these actions reduce channel erosion and protect aquatic habitat (Booth et al., 2002; Roesner et al., 2001). There are many reasons for this outcome. First, there are challenges in detention pond planning and design. It has proven difficult to develop and implement stormwater models that can accurately predict the size of detention ponds needed to maintain post-development flows at pre-development levels. Second, once standards are developed, they are often applied to different sites without due consideration of the geologic conditions of individual streams. The rate and extent of channel erosion will differ based on substrate composition and individual channel characteristics (Booth, 1990; Roesner et al., 2001). Planning that allows for site variation is available but expensive. Third, changes in infiltration and overland flow at individual sites may be too small to require mitigation; however, the cumulative impact of many small sites over an entire watershed can cause substantial stream-flow changes (Booth and Jackson, 1997; Booth and Leavitt, 1999). Finally, and most importantly, detention ponds do not substantially increase watershed-scale infiltration. Thus the total volume of flow remains greater than that under predevelopment conditions despite changes in the shape of the individual storm hydrograph:

No constructed detention ponds, even the largest designed under [the flow duration] standard, can delay wintertime rainfall sufficiently for it to become summertime streamflow. Yet exactly this magnitude of delay does occur under predevelopment conditions, because far more of the precipitation is stored as groundwater (Booth et al., 2002).

8.6.2 Water Quality BMPs

The same structural BMPs designed for stormwater management—detention and retention ponds, constructed wetlands, grassed swales, filtration practices (e.g., surface sand filters), and infiltration practices (e.g., porous pavement)—also retain nonpoint source pollutants (Nightingale, 1987). The EPA is working cooperatively with the American Society of Civil Engineers (ASCE) to develop a monitoring protocol for data collection for urban water quality BMPs. This will make it possible to evaluate BMP efficiency—"how well a BMP or BMP system removes pollutants." It has been difficult to assess the relative performance of particular BMPs because of the variety of techniques used in earlier monitoring studies. Sample collection techniques, analyses of water quality constituents, and quantification of watershed characteristics all vary among individual studies (Strecker et al., 2001).

Pennington et al. (2003) used data from the Rouge River system and Winer's (2000) BMP performance summary to estimate whether or not it was possible to comply with water quality standards using structural BMPs. The Rouge River is a tributary of the Detroit River in southeastern Michigan. It has four separate branches and a watershed area of 466 mi^2. Forty-eight municipalities, with a combined population of 1.5 million people, are located within the watershed; 75% of the watershed area is developed. The study compiled the median concentrations of eight water quality constituents (bacteria, BOD, organic nitrogen, nitrate and nitrite, total phosphorus, total suspended solids, total copper, and total zinc) from 22 different sampling sites between 1994 and 1999, then calculated the percent reduction required to meet established water quality standards. The success of BMPs in controlling pollution was dependent on both pollutant loading and watershed surface characteristics. The effectiveness of some BMPs in removing specific pollutants declined as impervious area increased. For example, wetlands were able to achieve the efficiency required to meet water quality standards for all the pollutants listed above when watershed impervious area was less than 20%, but failed to remove sufficient percentages of bacteria and nitrate when impervious area exceeded 40%. Pennington et al. concluded that "communities are ill-advised to rely exclusively on structural BMPs to address water quality concerns ... especially in urbanized areas."

8.6.3 Sediment at Construction Sites

There are a variety of BMPs intended to control sediment loading from construction sites. For example, Ohio construction site BMPs prescribe that areas of bare soil that will be unprotected for more than 45 days be stabilized using fast-growing grasses within 7 days (Miltner et al., 2004). This is reduced to 2 days if the site is within 15 m of a stream. Sediment ponds, silt fences, protection of inlets, and permanent stabilization when construction is completed are also part of the BMP regulations. As with all BMP programs, however, the effectiveness of the BMPs depends on maintenance and enforcement. During their study of a rapidly developing watershed near Columbus, Ohio, Miltner et al. found that these types of construction site BMPs were seldom installed and that "enforcement of the general stormwater permit for a 10 county area in Central Ohio [fell] to one person."

Miltner et al. compared this situation to construction site BMPs and enforcement in Baltimore County, Maryland. The stream protection ordinance in Baltimore County requires that a forested buffer extend on both sides of a stream and that the buffer include the adjacent floodplain, slopes, and wetlands. Where deemed necessary, the forested buffer may be "extended to protect steep slopes, erodible soils, and contiguous sensitive areas." In addition, there are 14 statewide stormwater management standards enforced by the Maryland Department of the Environment. The goal of these standards is to preserve pre-development hydrologic characteristics, including stormflow volume and groundwater recharge, and also to maintain stream channel stability. Performance bonds to ensure compliance are required with the issuance of construction permits. The Maryland BMP regulations have not been in place long enough to be evaluated, but initial indications, such as the reappearance of some fish species at some sites, are encouraging.

8.6.4 NEW APPROACHES TO URBAN BMPS

8.6.4.1 Innovative Technologies

Traditionally mitigation efforts have focused on increasing stormwater and nonpoint source pollution storage within the drainage system or developing end-of-pipe storage facilities. The communities of Skokie and Wilmette, Illinois, two suburbs of Chicago (Carr et al., 2001), experimented with an "on-street" storage system. Here the sewered areas are completely developed and the land is very flat. Sewer backups and overflows and basement flooding were persistent problems during and after rainstorms in residential areas. The street storage approach was designed to reduce the amount of water flowing into sewer systems, "first in off-street areas, second on residential streets and lastly in underground storage facilities," until it could be safely released. Allowable flood elevations—the depth of water that can be stored safely on the street surface without flooding basements—were determined block by block. Berms were constructed to increase infiltration in upland areas. The Skokie experiment began by disconnecting downspouts from the drainage network in 1981. Construction and installation of 2900 flow regulators, 871 berms, 10 off-street storage facilities, 83 subsurface facilities, and some new storm and combined sewers were completed in 1999. Major reductions in sewer overflows and basement flooding, along with public acceptance and support, were indicators of the project's success.

Papiri et al. (2003) tested the efficiency of the EcoDrain (GreenTechTexas International, Houston, TX). EcoDrain is a device that is installed in the inlet of a storm drain. It directs overland flow entering storm drains to a filtration unit containing a filtration pillow filled with plant fibers and oleophagic (oil eating) bacteria that decompose organic compounds in oil and other hydrocarbons. The filtering unit can treat up to 35 L/min, releasing clean, treated water into the storm sewer system. The opening of the filtration chamber is too small to admit larger debris, such as gravel, leaves, and rubbish. This research and development work is continuing, but preliminary results are promising. At a site near a marina in Australia, the EcoDrain reduced oil and grease concentrations by 95% and metals (copper, lead, and zinc) by nearly 98%. EcoDrain filtration pillows must be removed and replaced when their treatment capacity is exhausted.

8.6.4.2 Low Impact Design

Recent innovations in urban BMPs focus on maintaining predevelopment streamflow and water quality by increasing infiltration and soil storage at new development sites. These strategies are often referred to as "low impact design" (LID). LID involves the use of new materials such as permeable pavement, disconnecting impervious areas from the drainage system, and clustering buildings to maintain forested open space (Clausen, 2004; Holman-Dodds et al., 2003).

8.6.4.3 Permeable Pavement

Permeable pavements are commercially available products commonly made up of concrete blocks with intervening spaces or a grid or lattice structure of a "plastic web-type" material. Spaces within the pavement may be filled with sand and then planted with grass or filled with gravel. This design allows infiltration of stormwater while maintaining a hard surface that is suitable for traffic. The performance of several permeable pavements was tested in parking stalls used by municipal employees in the King County, Washington, public works facility (Booth and Leavitt, 1999; Brattebo and Booth, 2003). The products tested in this pilot study included Grasspave² (Invisible Structures, Inc., Golden, CO), Gravelpave² (Invisible Structures, Inc., Golden, CO), Turfstone (Interlock Paving Systems, Inc., (Hampton, VA), and UNI Eco-Stone (UNI-Group U.S.A., Palm Beach Gardens, FL). Except for a brief period of overland flow during the most intense storm of the sampling period, "virtually all water infiltrated for every observed storm." In addition, none of the pavements showed signs of excessive wear during the 6-year test. In asphalt control areas, overland flow closely matched precipitation. Water quality in the infiltrated water in the permeable systems improved as well. Concentrations of motor oil, copper, and zinc were significantly less ($p < .01$) than in stormwater from asphalt areas.

8.6.4.4 Low Impact Design: Parking Lots

Parking lots can be major contributing areas for nonpoint source pollutants. A 2-year study in Florida (Rushton, 2001) examined the effectiveness of LID and BMPs in reducing stormwater and pollution from a 4.65 ha parking lot at the Florida Aquarium, serving 700,000 visitors annually. The parking area was divided into eight basins containing two replicates each of four experimental treatments: (1) asphalt pavement with no swale; (2) asphalt pavement with a swale; (3) cement pavement with a swale; (4) permeable pavement with a swale. Parking spaces were made 0.62 m shorter in order to allow space for the swales. Because the front ends of the cars extended over the swales, the number of parking places was not reduced. The vegetated swales were planted with native grass that did not require mowing. On average, the swales alone increased infiltration by 30% and permeable pavements with swales increased infiltration by about 50%. Metal concentrations were highest in stormwater from the asphalt basins. Permeable pavement reduced metal concentrations in stormwater by 75%. PAHs were detected in the stormwater from all basins. The permeable pavement effluent contained the fewest PAH compounds at the lowest concentrations. Water draining from the asphalt basins approached toxic levels for benzo(b)fluoranthene, a PAH compound.

8.6.4.5 A Test of Urban BMPs and LID: The Jordan Cove Study, Connecticut

Jordan Cove (Clausen, 2004) is a 10 year study that has tested the water quantity and quality effects of BMPs in urban development. The Jordan Cove site includes three study watersheds: control, traditional development, and BMP development. The control site contains 43 lots on 5.5 ha. The control site was developed in 1988. The traditional site was calibrated with the control site in 1996 and 1997; development began in the fall of 1997. The "traditional" site is an 18 lot subdivision, using traditional zoning, an 8.5-m-wide asphalt road and curb and gutter stormwater collection system. Impervious surface coverage in the traditional watershed is 32%. The BMP site was calibrated with the control site between 1996 and 1999. Development began in the spring of 1999 using BMPs "before, during, and after construction." The BMP watershed subdivision consists of 12 units on 1.7 ha. The houses are clustered on small lots (0.10 ha versus 0.15 ha at the traditional site and 0.16 ha at the control site) with "reduced lawns, low-mow, and no-mow areas." The road into the BMP subdivision is reduced in width (6.1 m versus 8.5 m) and constructed of concrete pavers rather than asphalt. Total impervious cover in the BMP watershed is 22%, compared with 32% in the traditional watershed and 29% in the control watershed.

Additional innovations in the BMP watershed development include grassed bioretention swales instead of curbs and gutters. Stormwater on each lot is directed into individual bioretention gardens. Various driveway surface materials are being tested, including concrete pavers, gravel, and asphalt. BMPs designed to prevent sediment loss and reduce overland flow during construction include silt fences, earthen berms, and post-storm maintenance. Deed restrictions help to ensure that new owners maintain the BMPs as originally designed.

Results from this study show that mean weekly flow volume on the BMP watershed decreased from 15 m³ during the calibration period to 0.36 m³ during the construction period. This was due to BMPs, including construction of a berm that retained water onsite during construction, rapid excavation of basements that subsequently retained water, and new fill that increased soil infiltration rates. Mean weekly flow volume during the 2 year period following construction was 9.71 m³, a decrease of 78% from the calibration period. In contrast, weekly streamflow volume in the traditional watershed increased from 0.1 m³ during the calibration period to 22.5 m³ during the construction period and 66.5 m³ during the post-construction period. Tests of driveway materials in the BMP watershed showed that infiltration increased significantly when pavers or crushed stone were used in place of asphalt.

Mean weekly sediment export increased significantly in the BMP watershed during the construction (55.7 g/ha to 1038.6 g/ha) and post-construction periods (55.7 g/ha to 164.9 g/ha). Far greater increases were observed in the traditional watershed during construction (47 g/ha/week to 4,241 g/ha/week). When construction was complete, sediment export decreased to 979 g/ha/week at the traditional site. The increase in sediment export was primarily the result of the major increase in streamflow—sediment concentrations did not change during construction and decreased after construction was completed.

Nitrate and organic nitrogen concentrations increased in the BMP watershed during and after construction, primarily due to the use of fertilizer on lawns and decreased streamflow (0.2 mg/L versus 0.4 mg/L post-construction). The change in loading was not statistically significant. Nitrate concentrations in the traditional watershed did not change, however, nitrate loading increased due to the increase in streamflow (0.2 g/ha/week to 14.4 g/ha/week post-construction). Phosphorus concentration and loading increased significantly in the BMP watershed (0.7 g/ha/week to 3.7 g/ha/week post-construction). In the traditional watershed, phosphorus concentration decreased, but loading increased in proportion to changes in stormflow volume (0.4 g/ha/week to 7.5 g/ha/week post-construction). The export of metals—copper, lead, and zinc—increased significantly at the traditional site, but not at the BMP site. The increased loading of sediment, nutrients, and metals at the traditional site is associated with the increase in streamflow volume, primarily stormflow.

8.7 STREAM RESTORATION

There are many examples of stream restoration projects in mixed use and urban watersheds. Hassett et al. (2005) collected data on 4700 river and stream restoration projects in the Chesapeake Bay watershed (Maryland, Virginia, and Pennsylvania). The projects fell into four major categories: (1) riparian management, including revegetation of riparian areas, fencing to exclude livestock from streams, and removal of invasive, exotic weeds; (2) streambank stabilization to reduce or eliminate erosion or slumping of bank material into the river channel (but not including stormwater management); (3) water quality management, including CSO separation; and (4) in-stream habitat improvement designed to increase the availability of suitable, diverse habitats for all life stages of target organisms.

Despite their good intentions, there are two major obstacles to the success of these projects. First (and this is true in other areas of the United States as well; see Booth [2005]) is the lack of monitoring. Of all the stream restoration projects in the Chesapeake Bay watershed, the analysis performed by Hassett et al. indicated that only 5.4% completed any monitoring of the results. This varied widely in relation to the type of project. Monitoring was recorded in only 1% of the 2991 riparian management projects, while 73% of 111 fish passage and floodplain reconnection projects were monitored. The question of whether and under what circumstances stream restoration projects (revegetating riparian areas, adding woody debris, and stabilizing streambanks) lead to the successful restoration of aquatic communities has not been satisfactorily answered.

Second, most restoration projects are conducted at the riparian buffer or stream reach scale. As discussed earlier, disturbances that affect urban streams—increased overland flow, loss of watershed storage capacity, and transport of nonpoint source pollutants—are watershed-scale phenomenon caused by reduced infiltration and evapotranspiration in urban open land and increases in impervious area and drainage connections (Booth, 2005; Walsh et al., 2005b). Results from the studies that have examined issues of scale (Booth et al., 2004; Brooks et al., 2002; Hession et al., 2003) indicate that riparian buffers are most effective where urban land area constitutes a relatively small proportion of total watershed land cover (less than 30% TIA; less than 40% urban land).

Booth (2005) describes the majority of current restoration projects as "short-term, local-scale enhancement." This is not to say that riparian restoration and channel enhancement are not important and necessary, but watershed-scale efforts are also needed. Channel improvements brought about by reach and riparian-scale actions will, most likely, require continued support and maintenance as long as the channel disturbance is a watershed-scale change in flow patterns. The same issues apply to BMPs implemented on a site-by-site rather than a regional basis (Roesner et al., 2001).

The accumulated body of research on urban watersheds and urban streams suggests that a combination of restoration strategies addressing both small-scale (e.g., protecting and revegetating riparian buffers and enhancing stream habitat) and large-scale (e.g., increasing watershed infiltration) issues will be necessary to improve the abundance, diversity, and health of biotic communities in streams in mixed use and urban watersheds. Research is needed to identify the most effective restoration strategies as measured by biological response. In particular, research is needed to assess:

> (1) response of stream structure and function to forestland conversion to urban land use under different drainage design and spatial arrangements, and (2) if structure and function can be restored by drainage retrofits in existing urban areas (Walsh et al., 2005b).

In this instance, "drainage retrofits" refer to the benefits that might accrue from disconnecting impervious areas from the drainage network, thereby reducing EIA for an entire watershed. What would the effect on urban streams be if, for example, LID principles were applied to all new development and all suburban homeowners directed roof drains from their driveway and on to the lawn?

Streams in good condition can be found in watersheds with 10% to 30% TIA (Booth et al., 2004; Morley and Karr, 2002). It is possible to have healthy streams, as indicated by the presence of a diverse biotic community, including intolerant species, in many urban communities. In heavily urbanized areas (TIA greater than 30%; urban land cover greater than 40%), restoring streams to anything approaching their pre-development ecological state is unlikely. However, urban streams and rivers in these areas are important resources for urban communities—as scenic and recreational sites. Improved wastewater treatment and stormwater management can greatly improve the condition of urban rivers and streams, while the development of riverside and waterfront parks, including bicycle and walking trails (e.g. Boston, Minneapolis, Chicago), provide opportunities for public education to promote awareness of clean water and aquatic ecosystems and to generate public support for riparian restoration and protection.

8.7.1 Urban Planning for Stream Rehabilitation

It is becoming increasingly clear that protecting and restoring streams and rivers in developed and developing areas will require regional planning. This includes (1) a substantial commitment to the conservation of open space (especially forested areas where soil infiltration is highest), (2) the protection and rehabilitation of riparian areas, (3) the consistent use of LID principles in new development, and (4) policies and regulations that substantially improve stormwater management. Scientists working in the field of stream ecology recognize that increasing communication with other fields is critical:

> The challenge for stream ecologists in furthering our understanding of streams in urban areas is to not only better understand interactions between catchments and stream processes, but to integrate this work with social, economic, and political drivers of the urban environment (Walsh et al., 2005b).

Still, while research indicates a need for preservation of large blocks of forest and concentrating development to minimize forest conversion, recent development patterns are just the opposite (Stein et al., 2005). A report on development patterns along the East Coast, published by the Pew Oceans Commission (Beach, 2002), concluded that

> Runaway land consumption, dysfunctional suburban development patterns, and exponential growth in automobile use are the real engines of pollution and habitat degradation on the coast. Some large coastal metropolitan areas are consuming land ten times as fast as they are adding new residents. Across the country driving has increased at three to four times the increase in population. If today's land consumption trends continue, more than one-quarter of the coast's acreage will be developed by 2025—up 14 percent from 1997.

8.8   SUMMARY AND CONCLUSIONS

The conversion of forest land to an urban area comprised of residential, commercial, and industrial uses causes substantial changes in watershed structure and function. The replacement of forest vegetation with buildings and transportation infrastructure simultaneously decreases the favorable influence of plants on soils, nutrient cycling, and the water balance and increases stormflow and nonpoint source pollutant loading. What was a pollutant sink becomes a pollutant source. The hydrological and ecological effect of urbanization is directly related to the proportion of impervious surfaces (roads, driveways, parking lots, roofs, etc.) and the nature of the new land use (e.g., single-family home versus a truck stop). Natural soils are graded, compacted, and subject to erosion during the land development process. This alteration reduces the infiltration capacity, permeability, and porosity (storage capacity) of soils. It also has an unfavorable influence on soil structure, organic matter content, and the amount of water available to plants.

8.8.1   Impervious Surfaces, Overland Flow, and Urban Stormwater

Overland flow occurs whenever rainfall or snowmelt rates exceed the infiltration capacity of the soil or surface. Hence the proportion of total streamflow originating as overland flow increases in relation to the area and density of urbanization. Overland flow short-circuits the temporary storage of soil water as well as the shallow subsurface and groundwater flow paths that sustain deep-rooted woody vegetation and dry weather streamflow (baseflow) in forested watersheds. It also limits a crucial opportunity for pollutant assimilation and transformation *in* the soil. Increases in overland flow (both magnitude and frequency) are a primary reason for the "flashy" flow regime so often identified with urbanization.

Any sustained increase in stormflow volume, velocity, and hydraulic force causes compensatory changes in urban streams. Channel instability is evinced by downcutting (incision), bank erosion, and changes in the size and character of the substrate (cobbles and boulders left behind in high velocity reaches, while unstable silt, sand, and gravel deposits accumulate in quiet water). These fluvial processes, along with the removal of large woody debris, reduce or eliminate another important opportunity for the amelioration of water quality.

Overland flow and high velocity streamflow have the hydraulic energy to erode soil and lift and carry a heterogeneous array of other pollutants. These include, but are not limited to, sediment, nutrients from wastewater and fertilizers, insecticides, herbicides, metals from building materials and vehicles (brake pads and tires), road salt, PAHs, VOCs, and pharmaceutical products or their residuals. In addition, as noted earlier in this chapter, many older cities still have CSOs that bypass wastewater treatment plants when the volume of stormwater and wastewater exceeds their throughput capacity and chemical process limits. The receiving waters, and ultimately estuaries and oceans, bear the ecological brunt of this effluent. CSOs are the last vestige of the unfinished environmental engineering work associated with the CWA and other 1970s-vintage efforts to control point source pollution.

In sum, the conversion of forests to urban land use generates larger quantities of lower quality water that flows more rapidly through the watershed. As a result, the aquatic biota in streams and wetlands are subjected to more frequent and acute stress. This eliminates less tolerant organisms and often favors the establishment of tolerant species. The simplification and chronic disturbance of terrestrial and aquatic ecosystems, coupled with the limited residence time of water and pollutants, can overwhelm and drastically reduce the effectiveness of biological uptake and assimilation of nonpoint source pollutants.

This familiar litany of hydrological and ecological effects occurs gradually and cumulatively over years, decades, and even centuries. Watershed managers are typically faced with a daunting task when unhealthy and economically unsustainable conditions generate the political will and financial resources needed for ecological restoration.

8.8.2 Urban Stream (Watershed) Restoration

Successful urban stream restoration projects begin with a thorough characterization of the watershed. It is necessary to monitor watershed conditions and their effects on water flow and quality during the development, implementation, and adaptation (*in situ* refinement) of watershed management plans. Many small-scale (stream reach) efforts to improve water quality with riparian forest buffers have met with limited success when the primary upstream causes of erratic streamflow and water quality were not addressed.

Changes in aquatic macroinvertebrates and fish communities are detectable with as little as 5% or 10% impervious area in a watershed—even before changes in water chemistry become apparent. This does not imply that development inexorably leads to unmanageable adverse impacts. Streams in watersheds with as much as 30% impervious area can still exist in good condition (Booth et al., 2002). The difference between EIA—the area that is directly connected to the receiving water—versus TIA has a fundamental influence on the streamflow regime, water quality, and aquatic biointegrity. The difference between EIA and TIA largely determines the volume and timing of precipitation that leaves the watershed via overland flow versus the precipitation that enters the soil and aquifers to sustain baseflow. Successful stream and watershed restoration may depend on reducing the proportion of EIA relative to TIA.

These research needs are, in effect, the working hypotheses for LID and BMPs for urban stormwater (retrofits to existing development), respectively. The LID paradigm deliberately combines forest and open space conservation with compact or cluster development, spatially distributed stormwater management (minimizing EIA and maximizing infiltration and groundwater recharge), swales and created wetlands, other bioengineering and stream stabilization methods, and improved domestic water supply and wastewater treatment systems (e.g., community water systems and package plants instead of household wells and septic systems). In other words, all feasible efforts to avoid or prevent undesirable changes at the household, neighborhood, and watershed scales are explored during the LID process and implementation (Figure 8.8 and Table 8.1). Urban stormwater BMPs and structural retrofits seek to mitigate impacts and restore ecological structure and function *whenever* and *wherever* possible (Cappiella et al., 2005). This is, in effect, a systematic dismantling or mitigation of the undesirable changes in soils, vegetation, and pathways of flow (consistent with public health and safety) that have occurred over the course of decades.

The rewards for the successful implementation of LID for new development and BMPs to enhance the ecological structure and function of existing development are more livable, attractive, and healthy communities (Platt, 2004). Vigorous urban forests, clean water, scenic vistas, and associated recreational opportunities (e.g., Central Park in New York, the Emerald Necklace in

FIGURE 8.8 A comparison of the structural, ecological, and hydrological characteristics of forests, conventional development, and low impact development (described in Table 8.1).

Boston, the Chicago lakeshore, etc.) are cultural, economic, and ecological focal points of communities, both large and small. This topic and philosophy are revisited in Chapter 10—the summary and management implications for this book. Clearly urban watershed restoration and forest conservation are paired as the defining challenges of 21st-century watershed management (Stein et al., 2005).

TABLE 8.1
A Summary of the Structural, Ecological, and Hydrological Characteristics of Forests, Conventional Development, and Low Impact Development (illustrated in Figure 8.8)

Forest (reference condition)

Deep-rooted woody vegetation maximizes interception and evapotranspiration (available soil water storage) and soil permeability.

The litter layer and organic horizon protect and enhance mineral soil permeability.

Rainfall and snowmelt rates rarely exceed infiltration capacity and generate overland flow.

Stormwater flows laterally through the root zone and shallow groundwater to streams.

Detention storage of stormwater in the soil helps to maintain dry weather (base) flow.

Riparian vegetation and large woody debris help to maintain stream channel stability.

Water quality is favorably affected by the forest "biofilter."

Conventional development

Little deep-rooted woody vegetation to intercept rainfall and promote soil permeability.

Extensive excavation, grading, and soil compaction reduces infiltration capacity.

Rainfall and snowmelt rates often exceed infiltration capacity and generate overland flow.

Stormwater from rooftops, driveways, and lawns is hydraulically connected to roads then streams (effective impervious area [EIA] ≈ total impervious area [TIA]).

Stormwater is collected and transmitted directly to streams, short-circuiting the soil.

Stream channels are destabilized by the erratic flow regime and the lack of riparian vegetation.

Water quality is adversely affected by limited infiltration, residence time, and natural filtration through the soil and shallow groundwater.

Low impact development

More deep-rooted woody vegetation to intercept rainfall and promote soil permeability.

Soil disturbance is minimized in order to maintain infiltration capacity.

Rainfall and snowmelt rates occasionally exceed infiltration capacity and generate overland flow.

Stormwater from rooftops and driveways is dispersed to permeable areas — grassed swales and shallow depressions (EIA << TIA).

Stormwater from roads is dispersed laterally — curbs and storm drains are omitted.

Riparian vegetation is protected or restored to help maintain stream channel stability.

Maximizing infiltration capacity, residence time, and natural filtration through the soil and shallow groundwater minimizes water quality degradation.

REFERENCES

Agency for Toxic Substances and Disease Registry, ToxFaqs for aldrin and dieldrin, 2002, http://www.atsdr.cdc.gov/tfacts1.html; accessed July 2006.

Agency for Toxic Substances and Disease Registry, ToxFaqs for chlordane, 1995, http://www.atsdr.cdc.gov/tfacts31.html; accessed July 2006.

Agency for Toxic Substances and Disease Registry, ToxFaqs for polychlorinated biphyenyls (PCBs), 2000, http://www.atsdr.cdc.gov/tfacts17.html; accessed July 2006.

Anderson, R.M., Beer, K.M., Buckwater, T.F., Clark, M.E., McAuley, S.D., Sams, J.I., III, and Williams, D.R., Water quality in the Allegheny and Monongahela River basins: Pennsylvania, West Virginia, New York, and Maryland, 1996–98, Circular 1202, U.S. Geological Survey, Reston, VA, 2000.

Arnold, C.L., Boison, P.J., and Patton, P.C., Sawmill Brook: an example of rapid geomorphic change related to urbanization, *J. Geol.*, 90, 155–166, 1982.

Arnold, T.L., Sullivan, D.J., Harris, M.A., Fitzpatrick, F.A., Scudder, B.C., Ruhl, P.M., Hanchar, D.W., and Stewart, J.S., Environmental setting of the upper Illinois River basin and implications for water quality, Water Resources Investigations Report 98-4268, U.S. Geological Survey, Reston, VA, 1998.

Associated Press, EPA gives lower Charles River a high B for cleanliness, April 24, 2006.

Ator, S.W., Blomquist, J.D., Brakebill, J.W., Denis, J.M., Ferrari, M.J., Miller, C.V., and Zappia, H., Water quality in the Potomac River basin, Maryland, Pennsylvania, Virginia, West Virginia, and the District of Columbia, 1992–1996, Circular 1166, U.S. Geological Survey, Reston, VA, 1998.

Ayers, M.A., Kennen, J.G., and Stackenberg, P.E., Water quality in the Long Island–New Jersey coastal drainages, New York, and New Jersey, 1996–1998, Circular 1201, U.S. Geological Survey, Reston, VA, 2000, http://nj.usgs.gov/nawqa/; accessed July 2006.

Baltimore Ecosystem Study, 2005, http://www.lternet.edu/sites/bes/; accessed July 2006.

Bannerman, R.T., Owens, D.W., Dodds, R.B., and Hornewer, N.J., Sources of pollutants in Wisconsin stormwater, *Water Sci. Technol.*, 28(3–5), 241–259, 1993.

Barnes, V., Zak, D.R., Denton, S.R., and Spurr, S.H., *Forest Ecology*, 4th ed., John Wiley & Sons, New York, 1998.

Beach, D., *Coastal Sprawl: The Effects of Urban Design on Aquatic Ecosystems in the United States*, Pew Oceans Commission, Arlington, VA, 2002, http://www.pewtrusts.com/pdf/env_pew_oceans_sprawl .pdf; accessed July 2006.

Beschta, R.L. and Taylor, R.L., Stream temperature increases and land use in a forested Oregon watershed, *Water Resourc. Bull.*, 24, 19–25, 1988.

Blasius, B.J. and Merritt, R.W., Field and laboratory investigations on the effects of road salt (NaCl) on stream macroinvertebrate communities, *Environ. Pollut.*, 120, 219–231, 2002.

Bledsoe, B.P. and Watson, C.C., Effects of urbanization on channel instability, *J. Am. Water Resourc. Assoc.*, 37, 255–270, 2001.

Boesch, D.F., Brinsfield, R.B., and Magnien, R.E., Chesapeake Bay eutrophication: scientific understanding, ecosystem restoration, and challenges for agriculture, *J. Environ. Qual.*, 30, 303–320, 2001.

Booth, D.B., Stream-channel incision following drainage-basin urbanization, *Water Resourc. Bull.*, 26, 407–417, 1990.

Booth, D.B., Urbanization and the natural drainage system—impacts, solutions, and prognoses, *Northwest Environ. J.*, 7, 93–118, 1991.

Booth, D.B., Challenges and prospects for restoring urban streams: a perspective from the Pacific Northwest of North America, *J. North Am. Benthol. Soc.*, 24, 724–737, 2005.

Booth, D.B. and Henshaw, P.C., Rates of channel erosion in small urban streams, in *Stream Channels in Disturbed Environments*, Wigmosta, M.S. and Burges, S.J., (eds.), *Land Use and Watersheds: Human Influence on Hydrology and Geomorphology in Urban and Forest Areas*, American Geophysical Union, Washington, D.C., 2001.

Booth, D.B. and Jackson, C.R., Urbanization of aquatic systems: degradation thresholds, stormwater detection, and the limits of mitigation, *J. Am. Water Resourc. Assoc.*, 33, 1077–1090, 1997.

Booth, D.B. and Leavitt, J., Field evaluation of permeable pavement systems for improved stormwater management, *J. Am. Plan. Assoc.*, 65, 314–325, 1999.

Booth, D.B. and Reinelt, L.E., Consequences of urbanization on aquatic systems—measured effects, degradation thresholds, and corrective strategies, in *Proceedings of Watershed '93: A National Conference on Watershed Management*, EPA 840-R-94-002, U.S. Environmental Protection Agency, Washington, D.C., 1993, pp. 545–550.

Booth, D.B., Hartley, D., and Jackson, R., Forest cover, impervious-surface area, and the mitigation of stormwater impacts, *J. Am. Water Resourc. Assoc.*, 38, 835–845, 2002.

Booth, D.B., Haugeraud, R.A., and Goetz Troost, K., The geology of Puget lowland rivers, in *The Restoration of Puget Sound Rivers*, Montgomery, D.R., Bolton, S., Booth, D.B., and Wall, L., (eds.), University of Washington Press, Seattle, 2003.

Booth, D.B., Karr, J.R., Schauman, S., Konrad, C.P., Morley, S.A., Larson, M.G., and Burges, S.J., Reviving urban streams, land use, hydrology, biology, and human behavior, *J. Am. Water Resourc. Assoc.*, 40, 1351–1364, 2004.

Bowen, J.L. and Valiela, I., The ecological effects of urbanization of coastal watersheds: historical increases in nitrogen loads and eutrophication of Waquoit Bay estuaries, *Can. J. Fish. Aquat. Sci.*, 58, 1489–1500, 2001.

Brady, N.C. and Weil, R.R., *The Nature and Property of Soils*, 13th ed., Macmillan, New York, 2002.

Brakke, D.F., Landers, D.H., and Eilers, J.M., Chemical and physical characteristics of lakes in the northeastern United States, *Environ. Sci. Technol.*, 22, 155–163, 1988.

Brattebo, B.O. and Booth, D.B., Long-term stormwater quantity and quality performance of permeable pavement systems, *Water Res.*, 37, 4369–4376, 2003.

Brooks, S.S., Palmer, M.A., Cardinale, B.J., Swan, C.M., and Ribblett, S., Assessing stream ecosystem rehabilitation: limitations of community structure data, *Restor. Ecol.*, 10, 156–168, 2002.

Brosnan, T.M. and O'Shea, M.L., Sewage abatement and coliform bacteria trends in the lower Hudson-Raritan estuary since passage of the Clean Water Act, *Water Environ. Res.*, 68, 25–35, 1996.

Brown, C.J., Scorca, M.P., Stockar, G.G., Stumm, F., and Ku, H.F.H., Urbanization and recharge in the vicinity of East Meadow Brook, Nassau County, New York: Part 4—Water quality in the headwaters area, 1988–1993, Water-Resources Investigations Report 96-4289, U.S. Geological Survey, Reston, VA, 1997.

Brown, R.G., Effects of precipitation and land use on storm runoff, *Water Resourc. Bull.*, 24, 421–426, 1988.

Burges, S.J., Wigmosta, M.S., and Meena, J.M., Hydrologic effects of land-use change in a zero-order catchment, *J. Hydrol. Eng.*, 3, 86–97, 1998.

Cappiella, K., Schueler, T., and Wright, T., *Urban Watershed Forestry Manual, Part 1: Methods for Increasing Forest Cover in a Watershed*, NA-TP-04-05, USDA Forest Service, Northeastern Area State and Private Forestry, Newton Square, PA, 2005.

Carr, R.W., Esposito, C., and Walesh, S.G., Street surface storage for control of combined sewer surcharge, *J. Water Resourc. Plan. Manage.*, 127(3), 162–167, 2001.

Cherkauer, D.S., Urbanization impact on water quality during a flood in small watersheds, *Water Resourc. Bull.*, 11, 987–998, 1975.

Chocat, B., Krebs, P., Marsalek, J., Rauch, W., and Schilling, W., Urban drainage redefined: from stormwater removal to integrated management, *Water Sci. Technol.*, 43(5), 61–68, 2001.

Clausen, J., *Jordan Cove Urban Watershed Section 319 National Monitoring Program Project—Annual Report*, Department of Natural Resources Management and Engineering, College of Agriculture and Natural Resources, University of Connecticut, Storrs, CT, 2004, http://www.canr.uconn.edu/jordancove/; accessed July 2006.

Cleanupdate, Volatile organic compounds (VOCs): historical use leads to water concern, 1998, http://www.bnl.gov/erd/cleanupdate/vol3no2.html; accessed July 2006.

Davis, A.P., Shokouhian, M., and Ni, S., Loading estimates of lead, copper, cadmium, and zinc in urban areas, *Chemosphere*, 44, 997–1009, 2001.

Doyle, M.W., Harbor, J.M., Rich, C.F., and Spacie, A., Examining the effects of urbanization on streams using indicators of geomorphic stability, *Phys. Geogr.*, 21, 155–181, 2000.

Dunne, T. and Leopold, L.B., *Water in Environmental Planning*, W.H. Freeman, New York, 1978.

EXTOXNET, Extension Toxicology Network, Pesticide Information Profiles, Diazinon, 1996, http://ace.orst.edu/info/extoxnet/pips/diazinon.htm; accessed July 2006.

Fein, G.G., Jacobson, J.L., Jacobson, S.W., Schwartz, P.M., and Dowler, J.K., Prenatal exposure to polychlorinated biphenyls: effects on birth size and gestational age, *J. Pediatr.*, 105, 315–320, 1984.

Fisher, D.C. and Oppenheimer, M., Atmospheric nitrogen deposition and the Chesapeake Bay estuary, *AMBIO*, 20(3–4), 102–108, 1991.

Fisher, D.S., Steiner, J.L., Endale, D.M., Stuedemann, J.A., Schomberg, H.H., Franzluebbers, A.J., and Wilkinson, S.R., The relationship of land use practices to surface water quality in the upper Oconee watershed of Georgia, *For. Ecol. Manage.*, 128, 39–48, 2000.

Focazio, M.J. and Cooper, R.E., Selected characteristics of stormflow and base flow affected by land use and cover in the Chickahominy River basin, Virginia, 1989–1991, Water Resources Investigations Report 94-4225, U.S. Geological Survey, Reston, VA, 1995.

Forman, R.T.T. and Deblinger, R.D., The ecological road-effect zone of a Massachusetts (U.S.A.) suburban highway, *Conserv. Biol.*, 14, 36–46, 2000.

Fox, R.R. and Lytle, D.A., Milwaukee's crypto outbreak: investigation and recommendations, *J. Am. Water Works Assoc.*, 88(9), 87–94, 1996.

Galli, J., Thermal impacts associated with urbanization and stormwater management best management practices, Metropolitan Washington Council of Governments, Washington, D.C., 1990.

Gao, Y., Nelson, E.D., Field, M.P., Ding, Q., Li, H., Sherrell, R.M., Gigliotti, C.L., Van Ry, D.A., Glenn, T.R., and Eisenreich, S.J., Characterization of atmospheric trace elements on $PM_{2.5}$ particulate matter over the New York–New Jersey harbor estuary, *Atmos. Environ.*, 36, 1077–1086, 2002.

Garabedian, S.P., Coles, J.F., Grady, S.J., Trench, E.C.T., and Zimmerman, M.J., Water quality in the Connecticut, Housatonic, and Thames River basins: Connecticut, Massachusetts, New Hampshire, New York, and Vermont, 1992–95, Circular 1155, U.S. Geological Survey, Reston, VA, 1998.

Goolsby, D.A., Battaglin, W., Aulenbach, B.T., and Hooper, R.P., Nitrogen input to the Gulf of Mexico, *J. Environ. Qual.*, 30, 329–336, 2001.

Gordon, N.D., McMahon, T.A., and Finlayson, B.L., *Stream Hydrology: An Introduction for Ecologists*, John Wiley & Sons, Chichester, England, 1992.

Graf, W.L., The impact of suburbanization on fluvial geomorphology, *Water Resourc. Res.*, 11, 690–692, 1975.

Groffman, P.M., Bain, D.J., Band, L.E., Belt, K.T., Brush, G.S., Grove, J.M., Pouyat, R.V., Yesilonis, I.C., and Zipperer, W.C., Down by the riverside: urban riparian ecology, *Frontiers Ecol. Environ.*, 1, 315–321, 2003.

Groffman, P.M., Boulware, N.J., Zipperer, W.C., Pouyat, R.V., Band, L.E., and Colosimo, M.F., Soil nitrogen cycle processes in urban riparian zones, *Environ. Sci. Technol.*, 36, 4547–4552, 2002.

Gustafson, K.E. and Dickhut, R.M., Particle/gas concentrations and distributions of PAHs in the atmosphere of southern Chesapeake Bay, *Environ. Sci. Technol.*, 31, 140–147, 1997.

Hammer, T.R., Stream channel enlargement due to urbanization, *Water Resourc. Res.*, 8, 1530–1540, 1972.

Hammer, T.R., *Effects of Urbanization on Stream Channels and Streamflow*, Office of Water Resources Research, U.S. Department of the Interior. Regional Science Research Institute, Philadelphia PA, 1973.

Hart, D.D. and Finelli, C.M., Physical-biological coupling in streams: the pervasive effects of flow on benthic organisms, *Annu. Rev. Ecol. Syst.*, 30, 363–395, 1999.

Hassett, B., Palmer, M., Bernhardt, E., Smith, S., Carr, J., and Hart, D., Restoring watersheds project by project: trends in Chesapeake Bay tributary restoration, *Frontiers Ecol. Environ.*, 3, 259–267, 2005.

Hatt, B.E., Fletcher, T.D., Walsh, C.J., and Taylor, S.L., The influence of urban density and drainage infrastructure on the concentrations and loads of pollutants in small streams, *Environ. Manage.*, 34, 112–124, 2004.

Hession, W.C., Charles, D.F., Hart, D.D., Horwitz, R.J., Kreeger, D.A., Newbold, J.D., Pizzuto, J.E., and Velinsky, D.J., Final report: Riparian reforestation in an urbanizing watershed: effects of upland conditions on instream ecological benefits, EPA Grant R825798, U.S. Environmental Protection Agency, Washington, D.C., 2003, http://cfpub.epa.gov/ncer_abstracts/index.cfm/fuseaction/display. abstractDetail/abstract/182/report/F; accessed July 2006.

Hicks, A.L. and Larson, J.S., The impacts of urban stormwater runoff on freshwater wetlands and the role of aquatic invertebrate bioassessment, in *Effects of Watershed Development and Management on Aquatic Ecosystems*, Roesner, L.A. (ed.), American Society of Civil Engineers, Reston, VA, 1996, pp. 386–401.

Hollis, G.E., The effect of urbanization on floods of different recurrence interval, *Water Resourc. Res.*, 11, 431–435, 1975.

Holman-Dodds, J.K., Bradley, A.A., and Potter, K.W., Evaluation of hydrologic benefits of infiltration based urban storm water management, *J. Am. Water Resourc. Assoc.*, 39, 205–215, 2003.

Horner, R.R., Booth, D.B., Azous, A., and May, C.W., Watershed determinants of ecosystem functioning, in *Effects of Watershed Development and Management on Aquatic Ecosystems*, Roesner, L.A. (ed.), American Society of Civil Engineers, Reston, VA, 1996, pp. 251–274.

Inman, E.J., Flood-frequency relations for urban streams in Georgia, Water Resources Investigations Report 88-4085, U.S. Geological Survey, Reston, VA, 1988.

Jacobson, J.L. and Jacobson, S.W., Effects of exposure to PCBs and related compounds on growth and activity in children, *Neurotoxicol. Teratol.*, 12, 319–326, 1990.

Jacobson, J.L. and Jacobson, S.W., Intellectual impairment in children exposed to polychlorinated biphenyls in utero, *N Engl. J. Med.*, 335, 783–788, 1996.

Johnson, B.L., Hicks, H.E., Cibulas, W., Faroon, O., Ashizawa, A.E., De Rosa, C.T., Cogliano, V.J., and Clark, M., Public health implications of exposure to polychlorinated biphenyls, 1998, http://www.epa.gov/ waterscience/fish/pcb99.pdf; accessed July 2006.

Johnson, P.A., Gleason, G.L., and Hey, R.D., Rapid assessment of channel stability in vicinity of road crossing, *J. Hydraul. Eng.*, 125, 645–651, 1999.

Jones, R.C. and Clark, C.C., Impact of watershed urbanization on stream insect communities, *Water Resourc. Bull.*, 23, 1047–1055, 1987.

Klein, R.D., Urbanization and stream quality impairment, *Water Resourc. Bull.*, 15, 948–963, 1979.

Kolpin, D.W., Furlong, E.T., Meyer, M.T., Thurman, E.M., Zaugg, S.D., Barber, L.B., and Buxton, H.T., Pharmaceuticals, hormones, and other organic wastewater contaminants in U.S. streams, 1999–2000: a national reconnaissance, *Environ. Sci. Technol.*, 36, 1202–1211, 2002.

Konrad, C.P. and Booth, D.B., Hydrologic trends associated with urban development for selected streams in the Puget Sound basin, western Washington, Water Resources Investigations Report 02-4040, U.S. Geological Survey, Reston, VA, 2002.

Kozlowski, T.T. and Pallardy, S.G., *Physiology of Woody Plants*, Academic Press, San Diego, CA, 1997.

Krug, W.R. and Goddard, G.L., Effects of urbanization on streamflow, sediment loads, and channel morphology in Pheasant Branch basin near Middleton, Wisconsin, Water Resources Investigations Report 85-4068, U.S. Geological Survey, Reston, VA, 1986.

Landis, M.S., Vette, A.F., and Keeler, G.J., Atmospheric mercury in the Lake Michigan basin: influence of the Chicago/Gary urban area, *Environ. Sci. Technol.*, 36, 4508–4517, 2002.

Larson, S.J., Capel, P.D., and Majewski, M.S., *Pesticides in Surface Waters: Distribution, Trends, and Governing Factors*, Ann Arbor Press, Chelsea, MI, 1997.

Leopold, L.B., Hydrology for urban land planning—a guidebook on the hydrologic effects of urban land use, Circular 554, U.S. Geological Survey, Reston, VA, 1968.

Leopold, L.B., River channel change with time: an example. Address as retiring president of the Geologic Society of America, Minneapolis, Minnesota, November 1972, *Geol. Soc. Am. Bull.*, 84, 1845–1860, 1973.

Leopold, L.B., Sediment size that determines channel morphology, in *Dynamics of Gravel Bed Rivers*, Billi, P., Hey, R.D., Thorne, C.R., and Tacconi, P. (eds.), John Wiley & Sons, Chichester, UK, 1992.

Leopold, L.B., *A View of the River*, Harvard University Press, Cambridge, MA, 1994.

Leopold, L.B., *Water, Rivers and Creeks*, University Science Books, Sausalito, CA, 1997.

Lindsey, B.D., Breen, K.J., Bilger, M.D., and Brightbill, R.A., Water quality in the lower Susquehanna River basin, Pennsylvania and Maryland, 1992–1995, Circular 1168, U.S. Geological Survey, Reston, VA, 1998.

Litke, D.W., Review of phosphorus control measures in the United States and their effects on water quality, Water Resources Investigations Report 99-4007, U.S. Geological Survey, 1999.

Lloyd, S.D., Wong, T.H.F., and Porter, B., The planning and construction of an urban stormwater management scheme, *Water Sci. Technol.*, 45(7), 1–10, 2002.

Mallin, M.A., Williams, K.E., Esham, E.C., and Lowe, R.P., Effect of human development on bacteriological water quality in coastal watersheds: managing the land-water interface, *Ecol. Applic.*, 10, 1047–1056, 2000.

Martin, C.W. and Hornbeck, J.W., Logging in New England need not cause sedimentation of streams, *Northern J. Appl. For.*, 11, 17–23, 1994.

Mason, C.F., Norton, S.A., Fernandez, I.J., and Katz, L., Deconstruction of the chemical effects of road salt on stream water chemistry, *J. Environ. Qual.*, 28, 82–91, 1999.

Massachusetts Department of Environmental Protection, Bureau of Resource Protection, Drinking Water Program, Guidelines on deicing chemical (road salt) storage, 1997.

Massachusetts Water Resources Authority, Combined sewer overflows, 2005, http://www.mwra.state.ma .us/03sewer/htmL/sewcso.htm; accessed July 2006.

Massachusetts Water Resources Authority, The state of Boston Harbor: mapping the Harbor's recovery, 2002, http://www.mwra.state.ma.us/harbor/enquad/pdf/2002-09.pdf; accessed July 2006.

Mattson, M.D., Godfrey, P.J., Walk, M.F., Kerr, P.A., and Zajicek, O.T., Regional chemistry of lakes in Massachusetts, *Water Resourc. Bull.*, 28, 1045–1056, 1992.

McCobb, T.D. and LeBlanc, D.R., Detection of fresh ground water and a contaminant plume beneath Red Brook Harbor, Cape Cod, Massachusetts, 2000, Water Resources Investigations Report 02-4166, U.S. Geological Survey, Reston, VA, 2002.

Meals, D.W. and Budd, L.F., Lake Champlain Basin nonpoint source phosphorus assessment, *J. Am. Water Resourc. Assoc.*, 34, 251–265, 1998.

Miltner, R.J., White, D., and Yoder, C., The biotic integrity of streams in urban and suburbanizing landscapes, *Landscape Urban Plan.*, 69, 87–100, 2004.

Morley, S.A. and Karr, J.R., Assessing and restoring the health of urban streams in the Puget Sound basin, *Conserv. Biol.*, 16, 1498–1509, 2002.

Mullaney, J.R. and Grady, S.J., Hydrogeology and water quality of a surficial aquifer underlying an urban area, Manchester, Connecticut, Water Resources Investigations Report 97-4195, U.S. Geological Survey, Reston, VA, 1997.

Myers, D.N., Thomas, M.A., Frey, J.W., Rheaume, S.J., and Button, D.T., Water quality in the Lake Erie–Lake Saint Claire drainages: Michigan, Ohio, Indiana, New York, and Pennsylvania, 1996–1998, Circular 1203, U.S. Geological Survey, Reston, VA, 2000, http://oh.water.usgs.gov/nawqa/; accessed July 2006.

National Research Council, Highway deicing: comparing salt and calcium magnesium acetate, Special Report 235, Transportation Research Board, National Research Council, Washington, D.C., 1991, http://www4.nas.edu/trb/onlinepubs.nsf/web/trb_special_reports?OpenDocument; accessed July 2006.

National Research Council, *Watershed Management for Potable Water Supply: Assessing the New York City Strategy*, National Academies Press, Washington, D.C., 2000.

Natural Resources Conservation Service, Stream visual assessment protocol, NWCC-TN-99-1, National Water and Climate Center, Portland, OR, 1998, http://www.nrcs.usda.gov/technical/ECS/aquatic/svapfnl.pdf; accessed July 2006.

Nightingale, H.L., Accumulation of As, Ni, Cu, and Pb in retention and recharge basins soils from urban runoff, *Water Resourc. Bull.*, 23, 663–672, 1987.

Nixon, S.W., Ammerman, J.W., Atkinson, L.P., Berounsky, V.M., Billen, G., Boicourt, W.C., Boynton, W.R., Church, T.M., Ditoro, D.M., Elmgren, R., Garber, J.H., Giblin, A.E., Jahnke, R.A., Owens, N.J.P., Pilson, M.E.Q., and Seitzinger, S.P., The fate on nitrogen and phosphorus at the land-sea margin of the North Atlantic Ocean, *Biogeochemistry*, 35, 141–180, 1996.

Ohio State University Extension, Diazinon to be phased out, 2001, http://union.osu.edu/mgarden/articles/diazinon.htm; accessed July 2006.

Olsen, D.S., Whitaker, A.C., and Potts, D.F., Assessing stream channel stability thresholds using flow competence estimates at bankfull stage, *J. Am. Water Resourc. Assoc.*, 33, 1197–1207, 1997.

Osborne, L.L. and Wiley, M.J., Empirical relationships between land use/cover and stream water quality in an agricultural watershed, *J. Environ. Manage.*, 26, 9–27, 1988.

Papiri, S., Ciaponi, C., Capodaglio, A., Collivignarelli, C., Bertanza, G., Swartling, F., Crow, M., Fantozzi, M., and Valcher, P., Field monitoring and evaluation of innovative solutions for cleaning storm water runoff, *Water Sci. Technol.*, 47(7–8), 327–334, 2003.

Paschka, M.G., Ghosh, R.S., and Dzombak, D.A., Potential water-quality effects from iron cyanide anticaking agents in road salt, *Water Environ. Res.*, 71, 1235–1239, 1999.

Paul, M.J. and Meyer, J.L., Streams in the urban landscape, *Annu. Rev. Ecol. Syst.*, 32, 333–365, 2001.

Pennington, S.R., Kaplowitz, M.D., and Witter, S.G., Reexamining best management practices for improving water quality in urban watersheds, *J. Am. Water Resourc. Assoc.*, 39, 1027–1041, 2003.

Persky, J.H., The relation of ground-water quality to housing density, Cape Cod, Massachusetts, Water Resources Investigations Report 86-4093, U.S. Geological Survey, Reston, VA, 1986.

Peters, C.A., Robertson, D.M., Saad, D.A., Sullivan, D.J., Scudder, B.C., Fitzpatrick, F.A., Richards, K.D., Stewart, J.S., Fitzgerald, S.A., and Lenz, B.N., Water quality in the western Lake Michigan drainages, Wisconsin and Michigan, 1992–95, Circular 1156, U.S. Geological Survey, Reston, VA, 1998, http://wi.water.usgs.gov/nawqa/; accessed July 2006.

Pfankuch, D.J., Stream reach inventory and channel stability evaluation: a watershed management procedure, R1-75-002, USDA Forest Service, Northern Region, Missoula, MT, 1975.

Phillips, P.J., Eckhardt, D.A., Freehafer, D.A., Wall, G.R., and Ingleston, H.H., Regional patterns of pesticide concentrations in surface waters of New York in 1997, *J. Am. Water Resourc. Assoc.*, 38, 731–745, 2002.

Pitt, R., Field, R., Lalor, M., and Brown, M., Urban stormwater toxic pollutants: assessment, sources, and treatability, *Water Environ. Res.*, 67, 260–275, 1995.

Pitt, R., Lalor, M., Field, R., and Brown, M., The investigation of source area controls for the treatment of urban stormwater toxicants, *Water Sci. Technol.*, 28(3–5), 271–282, 1993.

Platt, R.H., Toward ecological cities: adapting to the 21st century metropolis, *Environment*, 46(5), 11–27, 2004.

Randall, A.D., Hydrogeologic framework of stratified-drift aquifers in the glaciated northeastern United States (regional aquifer system analysis), Professional Paper 1415-B, U.S. Geological Survey, Reston, VA, 2001.

Richburg, J.A., Patterson, W.A., III, and Lowenstein, F., Effects of road salt and *Phragmites australis* invasion on the vegetation of a western Massachusetts calcerous lake-basin fen, *Wetlands*, 21, 247–255, 2001.

Roesner, L.A., Bledsoe, B.P., and Brashear, R.W., Are best-management-practice criteria really environmentally friendly?, *J. Water Resourc. Plan. Manage.*, 127, 150–154, 2001.

Rosgen, D.L., A classification of natural rivers, *Catena*, 22, 169–199, 1994.

Rushton, B.T., Low-impact parking lot design reduces runoff and pollutant loads, *J. Water Resourc. Plan. Manage.*, 127, 172–179, 2001.

Sanudo-Wilhelmy, S.A. and Gill, G.A., Impact of the Clean Water Act on the levels of toxic metals in urban estuaries: the Hudson River estuary revisited, *Environ. Sci. Technol.*, 33, 3477–3481, 1999.

Schueler, T., The importance of imperviousness, *Watershed Protect. Techniques*, 1, 100–111, 1994.

Schueler, T. and Claytor, R., Impervious cover as a urban stream indicator and a watershed management tool, in *Effects of Watershed Development and Management on Aquatic Ecosystems*, Roesner, L.A. (ed.), American Society of Civil Engineers, Reston, VA, 1996, pp. 513–531.

Seaburn, G.E., Effects of urban development on direct runoff to East Meadow Brook, Nassau County, Long Island, New York, Professional Paper 627-B, U.S. Geological Survey, Reston, VA, 1969.

Simmons, D.L. and Reynolds, R.J., Effects of urbanization on base flow of selected south-shore streams, Long Island, New York, *Water Resourc. Bull.*, 18, 797-805, 1982.

Simon, A. and Downs, P.W., An interdisciplinary approach to evaluation of potential instability in alluvial channels, *Geomorphology*, 12, 215–232, 1995.

Sloto, R.A., Effects of urbanization on storm-runoff volume and peak discharge of Valley Creek, eastern Chester County, Pennsylvania, Water Resources Investigations Report 87-4196, U.S. Geological Survey, Reston, VA, 1988.

Stein, S.M., McRoberts, R.E., Alig, R.J., Nelson, M.D., Theobald, D.M., Eley, M., Dechter, M., and Carr, M., Forests on the edge: housing development in America's private forests, PNW-GTR-636, USDA Forest Service, Washington, D.C., 2005, http://www.fs.fed.us/projects/fote/reports/fote-6-9-05.pdf; accessed July 2006.

Stepenuck, K.F., Crunkilton, R.L., and Wang, L., Impacts of land use on macroinvertebrate communities in southeastern Wisconsin streams, *J. Am. Water Resourc. Assoc.*, 38, 1041–1051, 2002.

Strecker, E.W., Quigley, M.M., Urbonas, B.R., Jones, J.E., and Clary, J.K., Determining urban storm water BMP effectiveness, *J. Water Resourc. Plan. Manage.*, 127, 144–149, 2001.

The Water Center, University of Washington, Seattle, WA, http://depts.washington.edu/cwws/; accessed July 2006.

Thomas, M.A., The effect of residential development on ground-water quality near Detroit, Michigan, *J. Am. Water Resourc. Assoc.*, 36, 1023–1037, 2000.

Thompson, J.R., Soil erosion in the Detroit metropolitan area, *J. Soil Water Conserv.*, 25, 8–10, 1970.

Townsend, C.R. and Scarsbrook, M.R., The intermediate disturbance hypothesis, refugia, and biodiversity in streams, *Limnol. Oceanogr.*, 42, 938–949, 1997.

U.S. Environmental Protection Agency, How wastewater treatment works ... the Basics, EPA 833-F-98-002, Office of Water (4204), 1998, http://env1.kangwon.ac.kr/project/sdwr2004/litsurv/intwebsites/epa-ost/www.epa.gov/ebtpages/watewastewmunicipalwastewatertreatment.html; accessed July 2006.

U.S. Environmental Protection Agency, National Water Quality Inventory, 2000, Office of Water (4503F), Washington, D.C., 2002, http://www.epa.gov/305b/2000report/; accessed July 2006.

U.S. Environmental Protection Agency, Benzene, 2000a, http://www.epa.gov/ttn/atw/hlthef/benzene.html; accessed July 2006.

U.S. Environmental Protection Agency, Diazinon revised risk assessment and risk mitigation measures, 2000b, http://www.epa.gov/pesticides/op/diazinon/questions.pdf; accessed July 2006.

U.S. Environmental Protection Agency, Diazinon revised risk assessment and agreement with registrants, 2001, http://www.epa.gov/pesticides/op/diazinon/agreement.pdf; accessed July 2006.

U.S. Environmental Protection Agency, Mid-Atlantic integrated assessment (MAIA) estuaries 1997–1998. Summary report: environmental conditions in the Mid-Atlantic estuaries, U.S. Environmental Protection Agency, Washington, D.C., 2002.

U.S. Environmental Protection Agency, Combined sewer overflows, 2004a, http://cfpub.epa.gov/npdes/home.cfm?program_id=5; accessed July 2006.

U.S. Environmental Protection Agency, Combined sewer overflows (CSOs) in New England, 2004b, http://www.epa.gov/region1/eco/cso; accessed July 2006.

U.S. Environmental Protection Agency, Consumer fact sheet on benzene, 2005a, http://www.epa.gov/safewater/contaminants/dw_contamfs/benzene.html; accessed July 2006.

U.S. Environmental Protection Agency, Information on the toxic effects of various chemicals and groups of chemicals, 2005b, http://www.epa.gov/region5/superfund/ecology/html/toxprofiles.htm; accessed July 2006.

U.S. Environmental Protection Agency, State of the harbor: Boston Harbor cleanup history, 2005c, http://www.epa.gov/ne/ra/bharbor/state.html; accessed July 2006.

U.S. Environmental Protection Agency, Clean Water Act, 2006a, http://www.epa.gov/; accessed July 2006.

U.S. Environmental Protection Agency, Hudson River PCBs, 2006b, http://www.epa.gov/hudson; accessed July 2006.

U.S. Geological Survey, Pesticides in surface and ground water of the United States: preliminary results of the National Water Quality Assessment Program (NAWQA), Pesticides National Synthesis Project, 1998, http://ca.water.usgs.gov/pnsp/allsum/; accessed July 2006.

U.S. Geological Survey, The quality of our nation's waters—nutrients and pesticides, Circular 1225, U.S. Geological Survey, Reston, VA, 1999, http://water.usgs.gov/pubs/circ/circ1225/html/ack.html; accessed July 2006.

Van Metre, P.C. and Mahler, B.J., The contribution of particles washed from rooftops to contaminant loading to urban streams, *Chemosphere*, 52, 1727–1741, 2003.

Van Metre, P.C., Mahler, B.J., and Furlong, E.T., Urban sprawl leaves its PAH signature, *Environ. Sci. Technol.*, 34, 4064–4070, 2000.

Waananen, A.O., Urban effects on water yield, in *Effects of Watershed Changes on Streamflow*, Moore, W.L. and Morgan, C.W. (eds.), University of Texas Press, Austin, 1969, pp. 169–182.

Wall, G.R., Riva-Murray, K., and Phillips, P.J., Water quality in the Hudson River basin, New York and adjacent states, 1992–95, Circular 1165, U.S. Geological Survey, Reston, VA, 1998, http://ny.water.usgs.gov/projects/hdsn/; accessed July 2006.

Walsh, C.J., Urban impacts on the ecology of receiving waters: a framework for assessment, conservation, and restoration, *Hydrobiologia*, 431, 107–114, 2000.

Walsh, C.J., Fletcher, T.D., and Ladson, A.R., Stream restoration in urban catchments through redesigning stormwater systems: looking to the catchment to save the stream, *J. North Am. Benthol. Soc.*, 24, 690–705, 2005a.

Walsh, C.J., Fletcher, T.D., Wong, T.H.F., and Breen, P.F., Developing predictive ecological capacity for a stormwater management decision-making framework, in *Third Australian Stream Management Conference Proceedings: The Value of Healthy Streams*, Rutherford, I., Sheldon, F., Brierley, G., and Kenyon, C. (eds.), Cooperative Research Centre for Catchment Hydrology, Monash University, Victoria, Australia, 2001.

Walsh, C.J., Roy, A.H., Feminella, J.W., Cottingham, P.D., Groffman, P.M., and Morgan, R.P., II, The urban stream syndrome: current knowledge and the search for a cure, *J. North Am. Benthol. Soc.*, 24, 706–723, 2005b.

Wang, L. and Kanehl, P., Influences of watershed urbanization and instream habitat on macroinvertebrates in cold water streams, *J. Am. Water Resourc. Assoc.*, 39, 1181–1196, 2003.

Wang, L., Lyons, J., Kanehl, P., Bannerman, R., and Emmons, E., Watershed urbanization and changes in fish communities in southeastern Wisconsin streams, *J. Am. Water Resourc. Assoc.*, 36, 1173–1189, 2000.

Waters, T.F., Sediment in streams: sources, biological effects, and control, Monograph 7, American Fisheries Society, Bethesda, MD, 1995.

Weaver, L.A. and Garman, G.C., Urbanization of a watershed and historical changes is a stream fish assemblage, *Trans. Am. Fish. Soc.*, 123, 162–172, 1994.

Winer, R., *National Pollutant Removal Performance Database for Stormwater Treatment Practices*, 2nd ed., Center for Watershed Protection, Ellicott City, MD, 2000.

Wolman, M.G., A cycle of sedimentation and erosion in urban river channels, *Geogr. Ann.*, 49A(2–4), 385–395, 1967.

Wolman, M.G. and Miller, J.P., Magnitude and frequency of forces in geomorphic processes, *J. Geol.*, 68, 54–74, 1960.

Wolman, M.G. and Schick, A.P., Effects of construction on fluvial sediment, urban and suburban areas of Maryland, *Water Resourc. Res.*, 3, 451–464, 1967.

Yoder, C.O. and Rankin, E.T., Assessing the condition and status of aquatic life designated uses in urban and suburban watersheds, in *Effects of Watershed Development and Management on Aquatic Ecosystems*, Roesner, L.A. (ed.), American Society of Civil Engineers, Reston, VA, 1996, pp. 351–385.

Yorke, T.H. and Herb, W.J., Effects of urbanization on streamflow and sediment transport in the Rock Creek and Anacostia River basins, Montgomery County, Maryland, 1962–74, Professional Paper 1003, U.S. Geological Survey, Reston, VA, 1978.

Zogorski, J.S., Moran, M.J., and Hamilton, P.A., MTBE and other volatile organic compounds—new findings and implications on the quality of source waters used for drinking-water supplies, U.S. Geological Survey, Reston, VA, 2001, http://water.usgs.gov/pubs/FS/fs10501/pdf/fs10501.pdf; accessed July 2006.

9 Mixed Land Use and Cumulative Effects

Anyone wielding a hoe or an ax knows what he is doing, but before [George Perkins] Marsh no one had assessed the cumulative effect of all axes and hoes. For him the conclusion was inescapable. Man depends upon soil, water, plants, and animals. But in obtaining them he unwittingly destroys the supporting fabric of nature. Therefore man must learn to understand his environment and how he affects it. For his own sake, not for nature's alone, man must restore and maintain it as long as he tenants the earth.

David Lowenthal, 1965,
Introduction to the 1965 reprint edition of *Man and Nature*
by George Perkins Marsh (1864)

Watershed research and analysis will increasingly center on large-scale, multi-year studies of mixed land use. This shift from small-scale mechanistic studies is motivated by the urgent need to prevent or reduce water quality degradation by identifying sources of streamflow alteration and pollutants (primarily nonpoint source). Scientists and managers need to develop and validate field and modeling methods that will accurately identify sites that have a disproportionate influence—either positive or negative—on the aquatic environment. It is especially important to quantify the relative effects of specific areas and land uses on streamflow and water quality as the foundation for watershed management efforts. Furthermore, it is necessary to understand the interactions and cumulative effects of land and resource uses and changes with respect to space (location and area/extent) and time. One example of this large-scale emphasis is the U.S. Geological Survey (USGS) National Water Quality Assessment (NAWQA) program, described in Box 9.1.

The first section of this chapter describes the wide variations in watershed conditions across the northeastern area. It is important to note that a mix of land uses is the typical condition faced by watershed managers and policy makers. In contrast, most of the studies that were summarized and discussed in Chapters 6, 7, and 8 were conducted on relatively small, homogeneous watersheds. This research was instrumental in the development of hydrologic science, aquatic ecology, and watershed management principles and practices. It is, however, difficult to extrapolate the findings of small, homogeneous watersheds to larger, heterogeneous systems; they must be studied as a system. The second section summarizes several large-scale studies that endeavor to locate and quantify the sources of streamflow disturbance and water quality impairment. They attempt to estimate the proportional contribution of a wide range of land uses, areas, and locations to overall water quality. The third section focuses on cumulative effects. It includes a discussion of interaction effects as well as the legacy of earlier land use and natural disturbance.

9.1 REGIONAL VARIATION IN WATERSHED LAND USE

Many, if not most, watersheds in the Northeast contain a mix of urban, suburban, agricultural, and forested areas. Within watersheds (larger watersheds are also referred to as catchments or basins) the proportion of the land occupied by specific land uses varies. There is a regional pattern to these variations, the result of the physical environment (topography, soils) and socioeconomic history. Urban centers have grown near coastal seaports (Boston, New York, Philadelphia) and along major transportation routes near rivers and the Great Lakes (Cleveland and Columbus, Ohio; Detroit, Chicago, and Minneapolis–St. Paul). Agriculture is concentrated in regions of low relief with rich soils. Areas too mountainous for farming or too isolated for urban development tend to have the greatest proportion of forest.

In interior New England, agricultural activity (crop production, dairy farming, and sheep grazing) was widespread in the 18th and 19th centuries. Much of this farmland was abandoned when land became available in the Midwest, along with improved transportation—canals and railroads—to eastern markets. This trend was accelerated by industrialization and led to the natural reforestation of the region (see Chapters 6 and 7 for more detail). A recent USGS study of the Connecticut, Housatonic, and Thames River watersheds (a 41,400 km^2 area extending from the Canadian border through New Hampshire and Vermont, Massachusetts, and Connecticut to Long Island Sound) (Garabedian et al., 1998; Zimmerman et al., 1996) found that 74% of the land could be classified as undeveloped and forested. Ten percent of the land was classified as urban and 12% as agriculture (the remaining 4% is open water). The agricultural areas that remain are primarily located in the floodplains and adjacent terraces of river valleys in Massachusetts and Connecticut. Villages, towns, and cities occur in the river valleys and along the seacoast.

The rich soils and low relief typical of the Midwestern states have made this region one of the most productive agricultural areas in the world. An analysis of the Lake Erie–Lake St. Clair National Water Quality Assessment Program (NAWQA) study unit, located in Michigan, Ohio, Indiana, and northwestern Pennsylvania and New York, found that agriculture occupies 75% of the land area. Eleven percent of the watershed is urbanized. Metropolitan areas in the basin include Cleveland–Akron, Ohio, Detroit, Michigan, and Buffalo, New York. Only 11% of the basin is forested, with the remaining 3% classified as open water and wetlands. The presence of large areas of prime farmland has other implications for hydrology and water quality. Most of the wetlands in this region have been drained for farmland and agricultural chemicals (commercial fertilizers and pesticides [primarily herbicides]) are routinely applied to the soils (Myers et al., 2000).

Major East Coast cities—Boston, New York, Philadelphia, Washington, and Baltimore—and their surrounding metropolitan areas grew from early settlements near natural harbors in and near the Coastal Plain region. A study of the Long Island and eastern New Jersey coastal drainages in the mid-1990s found that urban/suburban land use occupied 33% of the area, an increase of 11% since the 1970s. Agricultural land use, which decreased steadily throughout the 20th century, comprised 14% of the region. The remaining lands were undeveloped forests (31%), wetlands (16%), and open water (6%) (Ayers et al., 2000). Concentrated agricultural production is still an important activity in some rural areas of the Mid-Atlantic states, New Jersey, Pennsylvania, Delaware, and Maryland (Boyer et al., 2002).

A large-scale study (Boyer et al., 2002) investigated the relationship between land use and riverine nitrogen export in 16 major river basins from Maine to Maryland (Figure 9.1). Although forests covered 72% of the combined landscape, there was great variation in population density and the proportion of forest, agricultural, and urban land use among the individual catchments (Table 9.1). The Penobscot and Kennebec River basins in Maine have large, sparsely populated forest areas that are used for pulpwood and timber production. Areas of intensive crop and livestock production are found in the Rappahannock, Potomac, Susquehanna, Schuylkill, and Mohawk River watersheds (primarily in the Mid-Atlantic states). The Schuylkill River watershed (the Schuylkill River flows through the center of Philadelphia) is a densely populated region (293 people/km^2), as are the Charles (Boston, Massachusetts; 556 people/km^2), Blackstone (Massachusetts and Rhode Island; 276 people/km^2), and Merrimack River watersheds (southern New Hampshire, Massachusetts; 143 people/km^2). Boyer et al. (2002) found that there was a strong relation between nitrogen inputs to each watershed and land use. Nitrogen loading (in kg/km^2/yr) was directly related to the

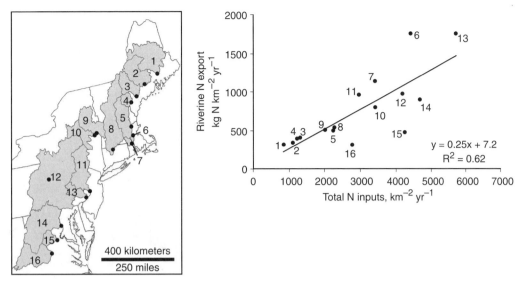

FIGURE 9.1 (Left) Nitrogen budgets were constructed for 16 major East Coast watersheds: 1. Penobscot (ME); 2. Kennebec (ME); 3. Androscoggin (ME); 4. Saco (ME); 5. Merrimack (NH, MA); 6. Charles (MA); 7. Blackstone (MA and RI); 8. Connecticut (VT, NH, MA, and CT); 9. Hudson (NY); 10. Mohawk (NY); 11. Delaware (NY, NJ, PA); 12. Schuylkill; 13. Susquehanna (PA); 14. Potomac; 15. Rappahannock; 16. James. (Slightly modified from Boyer et al., 2002; reprinted from *Biogeochemistry* 57/58:139 with kind permission of Springer Science and Business Media.) (Right) Riverine exports plotted against watershed nitrogen inputs for 16 East Coast watersheds. Exports were highest in the two watersheds with the greatest population density, the Charles River watershed in Massachusetts and the Schuylkill River watershed in Pennsylvania. (Slightly modified from Boyer et al., 2002; reprinted from *Biogeochemistry* 57/58:161 with kind permission of Springer Science and Business Media.)

percentage of agricultural land ($R^2 = 0.70$). The correlation between nitrogen loading and the combined percentage of agricultural and urban land was very high ($R^2 = 0.96$). In contrast, nitrogen loading was inversely proportional to the percentage of forested land ($R^2 = 0.77$). Sixty percent to 89% of nitrogen inputs were stored in the watershed or lost through denitrification. The nitrogen export in rivers was strongly related to total nitrogen loading in the individual basins. The greatest exports were found in the two watersheds with the greatest population density, the Charles and the Schuylkill (Table 9.2 and Figure 9.1). The Charles River watershed had the highest portion of urban land use; the Schuylkill had the highest proportion of agricultural land use, along with a substantial proportion of urban land area.

The connection between human land use and the degradation of streams, water quality, and aquatic habitat is well established. The task for scientists and watershed managers now becomes the identification and remediation of sources of pollution within individual watersheds. The Total Maximum Daily Load (TMDL) Program of the U.S. Environmental Protection Agency (EPA) requires state agencies to

> identify waters not meeting ambient water quality standards, define the pollutants and sources responsible for the degradation of each listed water, establish TMDLs necessary to secure those standards, and allocate responsibility to sources for reducing their pollutant releases (National Research Council, 2001).

This marks a paradigm shift from basing pollution control solely on effluent standards in wastewaters from point sources to a focus on the "biologic, hydrologic, and physical" condition of

TABLE 9.1
Characteristics of 16 Northeastern Coastal Watersheds

River basin[a]	Area (km²)	Persons[b] (/km²)	Mean temperature (°C)[c]	Precipitation (mm/yr)[c]	Flow (mm/yr)[d]	Land use (percent watershed area)[e]					
						Forest	Agriculture	Urban	Wetland	Water	Other
1 Penobscot	20,109	8	4.3	1075	588	83.8	1.5	0.4	5.2	6.2	3.0
2 Kennebec	13,994	9	4.3	1085	566	79.6	5.9	0.9	3.6	6.4	3.6
3 Androscoggin	8,451	17	4.6	1151	640	84.6	4.8	1.1	3.4	4.6	1.5
4 Saco	3,349	16	5.8	1218	672	87.4	3.6	0.8	3.9	3.1	1.1
5 Merrimack	12,005	143	7.4	1148	589	74.7	7.8	8.7	3.1	3.1	0.8
6 Charles	475	556	9.7	1207	583	59.3	8.4	22.2	7.2	2.5	0.5
7 Blackstone	1,115	276	9.0	1260	651	63.3	8.1	17.6	6.8	3.4	0.8
8 Connecticut	25,019	65	6.3	1160	642	79.0	9.0	4.0	4.7	2.2	1.1
9 Hudson	11,942	32	6.6	1126	622	80.8	10.4	2.7	2.5	3.4	0.2
10 Mohawk	8,935	54	6.8	1142	548	63.1	28.0	4.7	2.6	1.5	0.1
11 Delaware	17,560	85	8.7	1131	547	74.7	16.7	3.3	2.5	2.4	0.4
12 Susquehanna	70,189	54	8.9	1022	487	66.7	28.5	2.4	0.5	1.1	0.8
13 Schuylkill	4,903	293	10.6	1134	488	48.1	38.4	10.2	0.7	1.2	1.5
14 Potomac	29,940	63	11.3	985	328	60.8	34.6	2.6	0.5	0.7	0.8
15 Rappahannock	4,134	24	12.6	1045	360	61.3	35.9	1.4	0.2	0.4	0.7
16 James	16,206	24	10.1	934	407	80.6	15.6	1.4	0.6	0.7	1.1

[a] Watersheds listed in order from north to south.
[b] U.S. Census 1990.
[c] Average temperature and precipitation for 1988 to 1993.
[d] Average streamflow for water years 1988 to 1993 from USGS daily values.
[e] National land cover database for the early 1990s from Multi-Resolution Land Characteristics (MRLC) 1995.

Slightly modified from Boyer et al. (2002); reprinted from *Biogeochemistry* 57/58:141 (Table 1) with kind permission of Springer Science and Business Media.

TABLE 9.2
Overall Nitrogen Budget (kg N/km²/yr)

		Net atmospheric nitrogen deposition	Nitrogenous fertilizer use	Nitrogen fixation in forest lands	Nitrogen fixation in agricultural lands	Net nitrogen import in food and feed	Total nitrogen inputs	Streamflow nitrogen export[a]	Percent of nitrogen inputs exported in streamflow	Percent of nitrogen inputs stored or lost in basin
1	Penobscot	575	91	58	74	36	835	317	38	62
2	Kennebec	677	54	50	164	154	1099	333	30	70
3	Androscoggin	779	80	69	146	237	1310	404	31	69
4	Saco	885	42	107	96	104	1233	389	32	68
5	Merrimack	921	147	151	213	797	2228	499	22	78
6	Charles	996	197	218	187	2087	4406	1756	40	60
7	Blackstone	1040	307	260	305	1495	3407	1140	33	67
8	Connecticut	962	274	102	360	565	2262	538	24	76
9	Hudson	1033	204	103	374	271	1985	502	25	75
10	Mohawk	1075	411	70	1239	624	3420	795	23	77
11	Delaware	1212	527	201	675	352	2967	961	32	68
12	Susquehanna	1138	615	179	1147	1095	4173	977	23	77
13	Schuylkill	1143	1207	190	1225	1952	5717	1755	31	69
14	Potomac	769	1024	271	1173	1452	4689	897	19	81
15	Rappahannock	893	1030	277	1439	607	4246	470	11	89
16	James	953	361	361	703	395	2773	314	11	89

[a] Export of nitrogen in streamflow for the Charles River was 644 kg N/km²/yr. The value shown includes 1112 kg N/km²/yr of wastewater that originated within the Charles River watershed but is diverted out of the basin boundary.

Slightly modified from Boyer et al. (2002); reprinted from *Biogeochemistry* 57/58:159 (Table 6) with kind permission of Springer Science and Business Media.

receiving water bodies (National Research Council, 2001). Implementation of the TMDL program has been difficult and tentative because, as noted earlier, definitively answering questions about the sources, transport, transformation, and fate of nonpoint source pollution is a complex task that requires sophisticated monitoring and analysis.

Box 9.1   The USGS NAWQA Program

Since 1991, the USGS has been monitoring water quality in more than 50 major river basins throughout the United States. The NAWQA provides a wealth of detailed information concerning surface water and ground water quality and land use in different regions of the country. The NAWQA study units are large areas often covering sections of several states. Much of this literature is available online at http://water.usgs.gov/nawqa/ (accessed July, 2006).

Quoting from the website, the NAWQA program is designed to answer three basic questions.

1. What is the condition of our nation's streams and ground water?
2. How are these conditions changing over time?
3. How do natural features and human activities affect these conditions?

In order to answer these questions the USGS has designed a program that provides for long-term, repeated water quality measurements using study designs and methods that are "nationally consistent." Summary reports are available for each study unit (completed studies in the Northeast include Anderson et al. [2000], Ator et al. [1998], Ayers et al. [2000], Fenelon [1998], Garabedian et al. [1998], Groschen et al. [2000], Lindsey et al. [1998], Peters et al. [1998], Wall et al. [1998], Zimmerman et al. [1996]). Detailed reports of the original research studies on specific topics in each area are also available. Examples include nutrient and pesticide contamination of surface water and groundwater (Frey, 2001; Mullaney and Zimmerman, 1997) contamination of sediment samples by polychlorinated biphenyls (PCBs) and organochlorine pesticides (Long et al., 2000; Rheaume et al., 2001; Wong et al., 2000), and contamination of surface water and groundwater by pesticides and urban pollutants—volatile organic compounds (VOCs) (including methyl tertiary butyl ether [MTBE]), and polycyclic aromatic hydrocarbons (PAHs) (Sullivan et al., 1998). Many of the programs have done extensive biological sampling as well (Ayers et al., 2000). Summary statistics allow for comparisons of single study units to national patterns and trends.

9.2 WATERSHED ANALYSIS IN MIXED LAND USE WATERSHEDS

In this section we summarize several examples of large-scale watershed analysis. These studies attempt to link water quality parameters to a particular source, sometimes to a general type of land use and sometimes to a more specific location within the watershed (Box 9.2). In addition to studies in New England, New York, and Illinois, we include studies from the southeastern United States that exemplify relevant concerns and methods of analysis.

9.2.1 Lake Champlain River Basin (Vermont, New York, Quebec)

Meals and Budd (1998) conducted a study designed to estimate the proportional contributions of different land uses to phosphorus loading in Lake Champlain. New construction of advanced wastewater treatment plants, bans on phosphates in detergents, and some improved agricultural management practices helped to control increases in phosphorus concentrations during the 1980s.

However, phosphorus concentrations in several parts of the lake remains high and eutrophication poses a constant threat to water quality (Meals, 2001). The overall goal of the Lake Champlain Basin Program was to reduce total phosphorus loading. There was, however, considerable disagreement among representatives of industrial, commercial, agricultural, governmental, environmental, and academic sectors about which land uses were primarily responsible for the problem. This study evolved in response to these management needs and questions. The goal of the study was "to develop a simple, rapid, and low-cost, but credible analysis of nonpoint sources of P in the basin using existing data and accepted techniques." The model that was developed divided the Lake Champlain basin into 85 hydrologic units ranging in size from 1,200 to 60,000 ha. Phosphorus loads from these hydrologic units were estimated by first apportioning total annual discharge to various land uses within the hydrologic unit based on the percentage of total area and then multiplying these values by loading coefficients for total phosphorus (for forest, agriculture, and urban/developed land) derived from an extensive literature review. These estimates were validated by comparison with reliable total phosphorus loading data for major tributaries from an earlier study (Vermont Department of Environmental Conservation and New York State Department of Environmental Conservation, 1997).

Meals and Budd combined older (1973 to 1976) land use data with build-out scenarios to estimate land use in the entire basin at the time of the study. Total land use was estimated at 3% urban, 28% agricultural, 62% forest and wetland, and 7% open water. The total phosphorus load to Lake Champlain was estimated at 457 tonnes/yr; the measured value was 458 tonnes for an average hydrologic year. According to the model estimates, agriculture contributed 66% of the annual nonpoint source phosphorus load to Lake Champlain. Urban land contributed 18% and forest land 16%. The proportion of phosphorus from urban land is six times greater than the proportion of urban land area in the watershed. The estimated proportion of phosphorus from agriculture was about twice its proportionate area. The proportion of phosphorus from forest land was one-fourth its proportionate area. Meals and Budd concluded that better management of agricultural phosphorus should be a priority because agriculture contributes the greatest amount of the total phosphorus load to the lake. However, urban sources should be reduced as well. Towns and cities contribute far more phosphorus per unit area than other land uses, therefore nutrient removal at wastewater treatment plants would be a highly efficient way to reduce phosphorus loading.

9.2.2 CENTRAL ILLINOIS

Osborne and Wiley (1988) examined the relationships between land use/cover patterns and nitrate nitrogen and soluble reactive phosphorus (SRP) in a predominantly agricultural watershed in central Illinois. SRP represents that part of the phosphorus load that is immediately biologically available. Ninety percent of the 500 mi^2 Salt Fork River watershed is occupied by row crop agriculture, mostly corn and soybeans. The cities of Champaign–Urbana and Rantoul (populations at the time of the study 100,000 and 35,000, respectively) are also located within the watershed. Twenty-two sampling stations were sampled biweekly from December 1983 to December 1984 during both baseflow and stormflow periods. Land use/cover was interpreted from aerial photographs and digitized using the ARC/INFO geographic information system (GIS). The GIS was used to analyze land use/cover in four zones: within 100 ft (31 m), 200 ft (62 m), 400 ft (123 m), and 1000 ft (308 m) of stream channels. Aside from the agricultural land, 5% of the watershed was urban, 2.5% forest, and 2.5% wetlands and open water. All the urban areas were located near stream channels. Streams received nonpoint source urban pollution and effluent from wastewater treatment plants.

Sampling results showed that SRP concentrations increased moving downstream and that the maximum SRP concentrations (up to 6.13 mg/L) were always recorded downstream of Urbana. High concentrations were recorded downstream of other urban areas as well. SRP concentrations were greatest during baseflow periods, generally late summer to early winter, and lower during periods of increased precipitation and stormflow in the spring and early summer due to dilution. The analysis showed that the urban areas within the watershed had a much greater influence on the concentration of SRP in the streams than did agricultural land use and practices. Nitrate concentrations in agricultural regions were highest in the late winter, during snowmelt and late dormant season rains, and throughout the spring, the time when fertilizers containing nitrogen in the form of ammonia are added to fields and vegetative uptake is low. By the time overland or subsurface flow reached stream channels, most of the ammonia was converted to nitrate. As crops developed, nitrate concentrations decreased in agricultural areas, although concentrations remained higher downstream of urban areas. Maximum nitrate concentrations of ≥ 7 mg/L were recorded.

9.2.3 SOUTHEASTERN NEW YORK

The USGS did a detailed sampling of baseflow in the summer and fall of 1996 and the winter and spring of 1997 in the Croton River watershed (Heisig, 2000). The Croton River watershed has an area of 969 km². It contains 12 reservoirs that supply about 10% of the water for New York City. The sampling compared the water chemistry of 33 first- and second-order streams in various subbasins throughout the watershed. The USGS researchers then compared water quality characteristics based on the land use patterns within the various subbasins. A comparison of four of these subbasins highlights the association between land use and water quality.

1. Basin 36A is entirely forested (the only one in the study). Water chemistry in this stream was used as a reference, a basis of comparison for streams in watersheds with agricultural and various patterns of residential land use.
2. Basin 20 is characterized by high-density residential housing (394 houses/km²) and a high-density road network (10 km of road/km²). The homes have individual private septic systems; there is no municipal sewerage disposal or water treatment plant.
3. Basin 31 is an area of residential development served by municipal sewers and a wastewater treatment plant. The treated effluent is discharged at a point downstream of the basin (sampling point). Housing density and road density are less than in the unsewered basin (365 houses/km²) (5.4 km of road/km²).
4. Basin 1 is predominantly rural, with horse and dairy farms (approximately 12% of the area), fallow fields, and forest. There are houses with septic systems in this basin, however, the housing density (19.3 houses/km²) and the road density (1 km/km²) are much lower than in the two developed basins.

Water in the stream draining the unsewered residential Basin 20 had the highest levels of sodium and chloride, reflecting higher road density and road salt application. Boron and sulfate, indicators of domestic wastewater in stream water, were highest in the unsewered residential Basin 20. Nitrate levels, primarily from domestic wastewater, but also from lawn fertilizers, were also highest in the stream water from Basin 20. Nitrogen in the form of ammonia was highest in stream water from Basin 1. This implies that the animals were grazing near or were allowed direct access to streams. Because ammonia is readily converted to nitrate in biological systems, the presence of high levels of ammonia in stream water indicates a nearby source (from manure). Dissolved orthophosphate was highest in Basin 31. This was attributed to leaking sewer pipes that ran parallel to the course of the stream. Total phosphorus concentrations were highest in Basin 1 (Figure 9.2).

A comparison of water chemistry in all 33 streams showed that chloride in baseflow was closely related to road density and sewers. Sewers or the presence of riparian wetlands (areas of denitrification and vegetative uptake) downstream of residential areas reduced the nitrate concentration in streams. The mitigating effect of riparian wetlands was apparent during the summer when vegetation was actively growing.

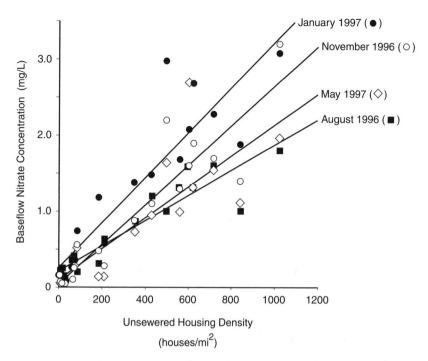

FIGURE 9.2 Relation of the concentration of nitrate in baseflow in small streams to the density of unsewered housing in the Croton watershed, New York (Heisig, 2000; figure courtesy of USGS).

Box 9.2   Mixed Land Use Studies: Sampling Design

Water quality studies in large mixed land use studies, including those described in this chapter, generally use one of three sampling strategies (Figure 9.3).

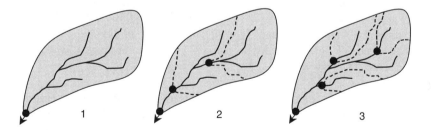

FIGURE 9.3 Three different sampling designs.

1. Whole Watersheds

This sampling strategy uses one point or station located at the outlet of the study watershed. The study of nitrogen loading and export by Boyer et al. (2002) is an example. This also may be termed a "black box" design because while the output of the entire watershed is measured, the proportional contributions of different land uses and subwatersheds must be estimated or inferred by other means.

2. Longitudinal Transects

This sampling strategy uses points along the main stem of the stream or river (e.g., Bolstad and Swank, 1997) (Figure 9.4). This "upstream to downstream" or longitudinal transect is sometimes used by water utilities and public agencies for regulatory compliance monitoring.

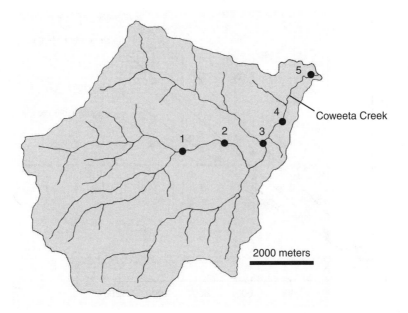

FIGURE 9.4 Water boundary and stream sampling locations in the Coweeta Creek watershed in western North Carolina; an example of a longitudinal transect sampling design. (Bolstad and Swank, 1997; reprinted with the permission of the American Water Resources Association.)

The subwatershed delineated at each sampling point includes all the upstream sampling points. The nested design means that the data gathered at each sampling station are cumulative. This approach provides more information than whole watershed sampling, especially if, as in the Bolstad and Swank study (see below), land use changes from forest to agriculture to urbanized conditions along the longitudinal transect. An abrupt change in water quality or aquatic biota at a particular sampling station may be attributed to changes in land use between it and the next upstream point. Since every sample represents a mixture of all upstream inputs, data analysis and interpretation can be complex. In addition, local variations in streamflow (lateral inflow from groundwater, shallow subsurface flow, or overland flow) between two stations can change the concentration of water quality constituents through dilution. Finally, other localized changes such as bedrock geology, soil type, and the presence or absence of wetlands may lead to substantial changes in streamflow and water quality.

3. Tributaries

This sampling strategy attempts to isolate and measure the proportional contributions of specific areas or subwatersheds to streamflow and water quality in larger heterogeneous systems. This design is most effective when the land use within each subwatershed is relatively homogeneous and differs, at least by degree, from other parts of the study area. If streamflow measurements are made when water samples are collected, the export of sediment, nutrients, or other potential pollutants can be normalized (mass/unit time/unit area). This allows for the direct comparison of land use effects as well as the estimation of influences on streamflow and water quality at the landscape scale. It also can be used to calibrate and verify export coefficients for spatially distributed water quality modeling. The Lake Oconee watershed study (Fisher et al., 2000b) most closely approximates this sampling design. While some of the points were located along the main stem of the Oconee River (Figure 9.5), others were deliberately placed on tributaries to monitor the nitrogen and phosphorus loadings associated with specific land uses (e.g., poultry operations, urban areas, etc.). The study by Rothrock et al. (1998) cited in Chapter 7 combined this sampling strategy with biological monitoring and GIS analyses to compare the effects of different land uses in the Blackfoot River watershed in Montana.

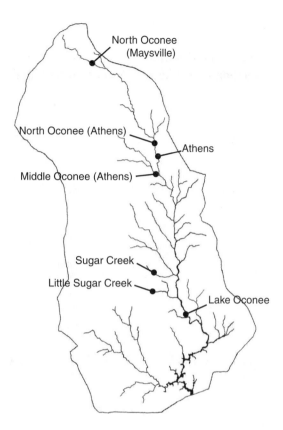

FIGURE 9.5 Selected sampling stations used to assess nitrogen and phosphorus contributions from various tributaries to Lake Oconee, Georgia. (Fisher et al., 2000; reprinted from *Forest Ecology and Management* with the permission of Elsevier.)

Combined sampling strategies (2 + 3, for example) often have the greatest utility for watershed monitoring and analysis. Whole-watershed sampling (however frequent) that represents complex mixes of land use may have limited utility for watershed planning and management. As noted throughout this book, the ability to accurately identify areas or activities that have a disproportionate, adverse impact on streamflow and water quality is critically important for prevention, mitigation, and restoration.

9.2.4 WESTERN NORTH CAROLINA

Bolstad and Swank (1997) (also see Swank and Bolstad, 1994) traced the cumulative changes in water quality in Coweeta Creek, a fifth-order stream in North Carolina that drains forested headwaters and then passes through increasingly developed downstream lowlands (Figure 9.4 and Table 9.3). The entire watershed was still more than 94% forested and generally water quality was high. However, water quality declined from upstream to downstream areas and this decline was most notable during storm events.

Table 9.4 shows the change in the mean values of selected water quality variables from upstream (Station 1) to downstream (Station 5) during baseflow. Table 9.5 shows mean values for the same variables during stormflow. Total coliform, fecal coliform, and fecal streptococcus levels typically increased from upstream to downstream and from baseflow to stormflow conditions, although results were highly variable among storms and seasons. The upstream to downstream increase for nitrate

TABLE 9.3
Characteristics of the Subwatershed Contributing Areas Above Each of the Five Sampling Stations Used in a Study Relating Land Use and Water Quality Variables, Coweeta Creek, North Carolina

Characteristics upstream of sampling station	Sampling station number				
	Upstream				Downstream
	1	2	3	4	5
Total area (ha)	1605	1798	3099	4163	4456
Forest area (ha)	1600	1782	2986	3904	4113
Agricultural area (ha)	4	13	89	155	192
Urban/suburban area (ha)	1	3	24	104	151
Total road length (km)	39.8	45.2	80.9	106.8	122.6
Unpaved road length (km)	38.6	43.9	73.4	96.4	106.5
Total road density (km/km^2)	2.49	2.51	2.61	2.60	2.75
Unpaved road density (km/km^2)	2.41	2.44	2.37	2.33	2.39
Structures/area (no./100 ha)	0.37	3.06	5.36	6.01	9.23

Note that this is a nested design; the contributing area for each sampling site includes the contributing areas for all the sampling sites upstream as well.

Bolstad and Swank (1997); reprinted with the permission of the American Water Resources Association.

TABLE 9.4
Mean Values of Selected Water Quality Variables Upstream to Downstream, Coweeta Creek, North Carolina, Sampled During Baseflow Periods

Variable	Station number				
	1	2	3	4	5
Turbidity (nephelometric turbidity units [NTU])	2.86	3.13	3.91	5.13	5.52
Total coliform (count/100 mL)	9,470	13,660	40,040	30,740	52,140
Fecal coliform (count/100 mL)	200	340	460	1130	840
Fecal streptococcus (count/100 mL)	710	1,310	2,180	1,590	1,840
Nitrate (mg/L)	0.042	0.041	0.042	0.041	0.045

Bolstad and Swank (1997); reprinted with the permission of the American Water Resources Association.

concentration measured during baseflow periods was 7%. This increased to 34% during storms. Agricultural and urban and residential land uses were sources of bacteria and nitrate. At the same time, increased amounts of overland flow transported these contaminants to streams more rapidly and effectively than subsurface flow in forested regions. Statistically significant ($p < .05$) correlations were found between water quality parameters (turbidity, fecal streptococcus, and fecal coliform) and land use characteristics (percent nonforest, structure density). Fecal streptococcus and fecal coliform also correlated with paved road density.

A study by Scott et al. (2002), in the upper Tennessee River watershed in North Carolina, examined the influence of factors, operating at different spatial and temporal scales, on water quality. The history of land use in this mountain region is somewhat similar to that in New England, however, on a shorter time scale. Forests were cleared in the late 1800s, farming was prevalent in the region until the mid-1900s. From that time to the present, farming has declined, fields have been abandoned, and second-growth forest now covers much of the region. At the same time, the

TABLE 9.5

Mean Values of Selected Water Quality Variables Upstream to Downstream, Coweeta Creek, North Carolina, Sampled During Stormflow Periods

	Station number				
Variable	1	2	3	4	5
Turbidity (NTU)	12.58	20.10	—	41.27	37.00
Total coliform (count/100 mL)	18,790	34,640	—	77,160	98,390
Fecal coliform (count/100 mL)	880	130	—	970	1260
Fecal streptococcus (count/100 mL)	450	8710	—	3260	4190
Nitrate (mg/L)	0.050	0.060	0.054	0.082	0.067

Bolstad and Swank (1997); reprinted with the permission of the American Water Resources Association.

population and residential and commercial development have increased. Data were collected at 36 stream sites representing a gradient of watershed land use from forests to farms to residential and urban. At each site, land use information was gathered and summarized at four different spatial scales: (1) the entire area draining to the sampling point, (2) within a 100 m riparian buffer on either side of the stream extending along the entire main stem to the stream source, (3) within a 100 m riparian buffer from the site to a point 2 km upstream, and (4) within the 100 m riparian buffer to a point 1 km upstream from the site. Fifty-meter reaches at each stream site were sampled between 1995 and 1999. Chemical and physical variables measured included nitrate nitrogen (NO^3), ammonium nitrogen (NH_4^+-N), SRP (dissolved phosphorus represents most of the immediately biologically available phosphorus in stream water), total dissolved solids (TDS) (Ca^{2+}, Mg^{2+}, Na^+, K^+, Cl, SO_4^2, HCO_3, and SiO_2), turbidity, mean summer temperature, proportion of coarse substrate particles, and large woody debris. Landscape variables measured included contributing area, elevation, channel slope, building and road density (combined to form an index of land use intensity), and forest cover (proportion of nonforest land) in both 1970 and in 1990.

Regression analysis showed that physical factors (channel slope, elevation, and contributing area) were significantly related ($p \leq .05$) to ammonium nitrogen, turbidity, and substrate particle size, but explained less than 50% of the variation in these parameters. Physical factors explained slightly more than 50% of the variation in stream temperature, which was related to elevation and to contributing area. More of the variation in water quality measurements (including SRP) was attributable to land use. The proportion of nonforest land along the entire main stem (spatial scale 2; see preceding paragraph) was generally a better predictor of stream characteristics than more local measurements, except for measures of large woody debris. The area of nonforest land within 100 m of the entire main stem (spatial scale 2) was closely correlated with the percent of nonforest land in the entire contributing area (spatial scale 1). Turbidity was related to contributing watershed area nonforest land in the 1970s (spatial scale 1) and to main stem nonforest land in the 1990s (spatial scale 2) and coarse woody debris was related to nonforest land along the main stem in the 1970s (spatial scale 2), suggesting legacy effects of past land use.

9.2.5 GEORGIA

A study in the upper Oconee watershed in Georgia (Fisher et al., 2000a,b) is an example of the conflicts and opportunities that arise as citizens become involved in water quality protection issues. This area has long been occupied by farms and forests. Recent population growth and urban and industrial development have increased demands on public water supplies. Citizens in the watershed are concerned about the effect of water yield and water quality on "recreation, tourism, human health, fishing, real estate values and wildlife habitat" and participate as volunteers in conservation efforts such as the Adopt-a-Stream program and river cleanups organized by Rivers Alive, a program of the Georgia Department of Natural Resources and Department of Community Affairs. Much of the concern is focused on water quality in Lake Oconee, a Georgia Power reservoir and popular local site for recreational boating, fishing, and swimming. Residents in the lake area have formed a volunteer group to monitor potential threats to water quality. The citizens group was concerned that dairy operations located west of the lake were a significant source of water pollution. This study, undertaken by the USDA Agricultural Research Service, sought to accurately quantify the primary sources of water pollution to Lake Oconee in an attempt to implement the most effective conservation efforts and minimize urban-agricultural conflict.

Sampling and analysis efforts were focused on two areas of the watershed, one in the headwaters region and a second area west of the lake, where different types of agricultural operations were concentrated (Figure 9.5). About 20,000 ha of the 100,000 ha headwaters region were cleared for agriculture. Agricultural activity in the headwaters region included 550 poultry operations producing 64 million broilers each year and maintaining 2 million egg-laying chickens. More than 33,000 beef cattle grazed here as well. Manure from the poultry farms was used as fertilizer in cattle pastures. The remaining land was used for timber production. Eight sampling sites were located at various distances downstream of the poultry farms and cattle grazing areas. The second area of concern, west of Lake Oconee, had 30 dairy farms. In addition to the dairy farms, there were 30 poultry farms and 21,000 beef cattle. Agricultural land uses took place on 26,000 of the 83,000 ha subwatershed. Timber harvesting also occurred in this area. Three sampling sites measured water quality downstream of the area west of Lake Oconee. An additional six sampling sites were located around the lake. One sampling site in a relatively undeveloped area of the watershed, the Apalachee River site, served as a reference. Samples were collected once a month in 1996 from the headwaters area and biweekly in 1995 in the areas west of Lake Oconee and at other sites.

Water sampling revealed that poultry operations in the headwaters region had relatively little effect on the water quality. The poultry farms were not near the intake for the municipal water supply, and although phosphorus, nitrogen, and fecal coliform levels were high at the sampling site nearest the poultry farms, concentrations were lower downstream. This may have been caused by dilution as the volume of streamflow increased. The nitrogen concentration in the Oconee River, downstream from the city of Athens, was nearly twice that immediately upstream (0.96 mg/L versus 0.53 mg/L) (Table 9.6). The phosphorus concentration showed a similar increase (0.092 mg/L downstream of Athens versus 0.048 mg/L upstream). These concentrations were higher than those from sampling sites near the poultry operations. Although phosphorus and nitrogen levels were elevated in streams from the dairy farm area west of Lake Oconee, the total streamflow input of these two streams was only 2.5% of the volume of water coming into the lake from the Oconee River. The Oconee River contributes 70% of the water flowing into Lake Oconee, underscoring the influence of the urban area (Athens) on water quality. While inputs of nutrients, and especially fecal coliform, from agricultural areas are a concern, urban sources of nutrients also must be addressed to protect water quality in the lake.

9.3 CUMULATIVE EFFECTS

The study of cumulative effects focuses on the consequences of the interaction of human activities and natural disturbances over space and time (MacDonald, 2000; Sidle and Hornbeck, 1991). While the immediate impact of a particular activity—draining one wetland, developing one residential area, cutting one stand of trees—on streamflow and water quality may be small, the combined effects of many individual actions across a watershed landscape can produce substantial changes in the hydrologic regime and biogeochemical cycling (Childers and Gosselink, 1990; Gosselink et al., 1990; MacDonald, 2000; Preston and Bedford, 1988). Human activity may be exacerbated or mitigated by climatic trends and weather events. The interaction of different disturbances increases the difficulty

TABLE 9.6
Annual Means of Monthly Water Quality Sample Results from Selected Sampling Stations

North of Lake Oconee (1996)	North Oconee (Maysville) (poultry)	North Oconee (above Athens)	Oconee River (below Athens, just above Lake Oconee)	Apalachee River (1996) (reference site)
Turbidity (Hach)	27	23	24	14
Phosphorus (mg/L)	0.08	0.05	0.09	0.03
Nitrogen (mg/L)	0.86	0.53	0.96	0.67
Fecal coliform (mpn/100 mL)	1270	613	639	364

West of Lake Oconee (1995)	Sugar Creek (dairy)	Little Sugar Creek (dairy)	Apalachee River (1995) (reference site)
Turbidity (Hach)	17	18	19
Phosphorus (mg/L)	0.10	0.09	0.03
Nitrogen (mg/L)	0.56	0.66	0.49
Fecal coliform (mpn/100 mL)	1041	915	163

Note that phosphorus and nitrogen concentrations are lower at the site directly above Athens than at the headwater site (North Oconee River, Maysville) near the poultry farming area. Nitrogen and phosphorus concentrations increase downstream of the urban area (Oconee River, below Athens).

mpn, most probable number.

Fisher et al. (2000).

of locating and ranking sources of water quality degradation. Cumulative effects are for the most part additive, although there is the potential for synergistic effects (the combined effect being greater than the sum of individual effects) in some instances of chemical interactions or when increased temperatures lead to changes in the form or toxicity of some pollutants (MacDonald, 2000).

The Clean Water Act requires that the potential cumulative effects of proposed development be considered during environmental impact statements and permit reviews (Abbruzzese and Leibowitz, 1997; Bedford and Preston, 1988; Council on Environmental Quality, 1997). Implementing these requirements has been difficult because of problems in defining and measuring cumulative effects processes, as well as the wide range of phenomena that may be included under the cumulative effects rubric (Coburn, 1989; Grant and Swanson, 1991). Writing in 1991, Grant and Swanson state that

> Cumulative watershed effects have been referred to as the UFOs of hydrology—phenomenon that are not well documented or explained but which a sizeable fraction of scientists, managers, and publics believe to exist.

Recent papers present specific methods designed to facilitate uniform systematic cumulative effects analyses that will meet regulatory requirements (Council on Environmental Quality, 1997; MacDonald, 2000). In this section we focus on research studies rather than regulatory efforts. These studies cover a wide range of topics and approaches. Some examine the multiple effects of different facets of a particular land use, such as forest management or agriculture, and some examine the combined effects of different land uses or activities on a watershed, a particular site, or on a particular aquatic or wetland species. While there is a temporal element to all research on cumulative effects, some studies are more specifically concerned with the legacy of past land use and disturbance. In general, "assessment of cumulative impacts requires a landscape approach and large-scale analysis" (Childers and Gosselink, 1990).

9.3.1 Forest Management: Timber Harvesting and Road Building

Early studies of cumulative effects focused on forest management (National Council of the Paper Industry for Air and Stream Improvement, 1984, 1986a,b). Larson (1984) described the problem of potential cumulative effects resulting from multiple timber harvesting operations and associated road construction. Timber harvesting increases streamflow by reducing evapotranspiration. Road building can increase streamflow by compacting soil surfaces, reducing soil infiltration, and channeling overland flow to streams. In addition, the streamflow increase from current timber harvesting operations and road building operations may add to residual streamflow increases caused by earlier harvests and road construction in the same watershed (see Chapter 6).

In the 1990s, the Minnesota Environmental Quality Board (MEQB), in response to a citizen petition, developed a Generic Environmental Impact Statement (GEIS) specifically to address potential statewide cumulative effects of increased timber harvesting. The purpose of the GEIS was to "assess a number of separate but related activities whose cumulative impacts cannot be adequately addressed through site-specific environmental impact statements" (Grigal and Bates, 1997). To accomplish this statewide analysis, the MEQB used computer modeling to assess the impacts of three different levels of harvesting (base scenario, 9.0 million m^3/yr; medium scenario, 11.0 million m^3/yr; high scenario, 15.5 million m^3/yr) on 10 major issues. The model estimated potential impacts on (1) forest productivity; (2) the forest resource base; (3) forest soils; (4) forest health (risks of insect infestation and disease); (5) biodiversity at genetic, species, and ecosystem levels; (6) forest wildlife and fish; (7) water quality; (8) recreation; (9) economics and management; and (10) aesthetics and historic and cultural resources. The water quality predictions focused on changes in sedimentation and nutrient loading to lakes, rivers, streams, and wetlands. This included an examination of the potential effects of fertilization to offset nutrient depletion of forest soils caused by forest management. The potential impacts on forest soils were examined in relation to nutrient cycling, erosion, compaction, and overall site productivity. The model used data from 13,356 forest inventory plots—each representing approximately 500 ha. Overall, "the projected impacts at the medium and high levels differed in degree rather than in type compared to those occurring at the base level." The models predicted that harvesting more than 12.5 million m^3/yr would be sustainable only if "substantial investments were made in forest management."

9.3.2 Progressive Loss of Forested Wetlands

The size, shape, type, and location of wetlands in a watershed can have a substantial influence on streamflow regimes and flood frequency. Wetlands can be sinks or sources for pollutants; this has direct consequences for water quality in associated rivers and streams (Hemond and Benoit, 1988; Preston and Bedford, 1988).

Childers and Gosselink (1990) (see also Gosselink et al. [1990]) documented the effects of cumulative wetland loss in the Tensas River Basin in Louisiana. This is a region of bottomland (wetland) hardwood forest. In 1937 the forest area in individual parishes ranged from 51 to 73%. In 1987 forest area had been reduced to 15% throughout the region. This change occurred incrementally, many small plots over a long period of time. While the adverse impact of clearing one small area may not be measurable, the cumulative effects of the progressive clearing appear to be large. Progressive loss of riparian forest (less than 15% of streambanks forested by 1987) increased erosion and sediment delivery to streams, while conversion to agriculture increased nutrient inputs. Channel "improvements" in the late 1950s and early 1960s led to large increases in the magnitude of peak discharge in some streams. Research in other areas has shown that changes in hydrology that increase flooding also limit the ability of wetlands to act as nutrient storage sites (Peverly, 1982). Water moves through the wetlands too quickly for nutrients to be affected by wetland processes, namely infiltration into wetland soils and vegetative uptake. Higher energy floods carry sediment and sediment associated particles through riparian wetlands into streams. In the Tensas basin, the three study sites showed highly significant and positive relations for both concentration and loading between stream water level and total phosphorus, total suspended solids, and turbidity. This indicated increased soil erosion typical of highly disturbed landscapes. As noted earlier, in undisturbed or minimally disturbed landscapes, concentration is reduced during stormflow events due to dilution. In this case, the EPA standard (0.1 mg/L) for phosphorus "was exceeded 96 percent of the time based on monthly sampling-between 1978 and 1986" (Childers and Gosselink, 1990). In fact, the phosphorus concentration was high enough that nitrogen:phosphorus ratios were reduced and it was suspected that nitrogen was now the limiting nutrient, and hence a greater concern in these ecosystems.

O'Brien (1988) (see also Winter, 1988, 2001) examined the regional variations in wetland hydrology based on surficial geology and concludes that the influence of wetlands on streamflow and floods varies in relation to the geologic setting in which the wetlands were formed. In many regions, wetlands develop in topographic depressions on soils with relatively low permeability—the prairie pothole wetlands of Iowa, Minnesota, and Wisconsin are an example. In New England, while some wetlands are located on relatively low permeability till or clay soils, many are located on patches of highly permeable sandy outwash deposits laid down when glacial melt water accumulated behind mountain ridges oriented northeast-southwest. Sixty percent to 70% of the wetlands surveyed in Massachusetts in 1981 were situated on highly permeable deposits or were directly connected to aquifers. Wetlands on till soils tended to be groundwater recharge areas regardless of landscape position. Wetlands on permeable outwash soils tended to be groundwater recharge sites in uplands, but acted as groundwater discharge sites if they were located on floodplains or near the outlet of a watershed. The status of a wetland as a groundwater discharge or recharge area also may change with water table elevation and the time of year. During storm events, wetlands in the headwaters can generate streamflow, while wetlands in lower reaches serve as flood storage areas. A wetland's response to storm events is strongly influenced by its hydrogeologic setting and the permeability of the organic mat that develops in wetland sites. In addition, the hydrologic role of the various types of wetlands depends in part on the size of the flood and antecedent soil moisture conditions. Thus the cumulative effects of wetlands loss will vary depending on their location, physical characteristics, and hydrologic function. O'Brien (1988) concludes that "wetlands may have far greater effects than their areal percentage in the drainage basin would indicate. Consequently, alteration of wetlands could produce effects disproportionate to their size."

9.3.3 AGRICULTURAL CHEMICALS AND TILE DRAINS

As we have seen, agricultural land use may introduce large quantities of potential pollutants—excess nutrients and pesticides—into the environment. The widespread installation of tile drains in agricultural areas, especially in poorly drained soils in the Great Lakes region, can exacerbate the impact of these pollutants.

Tile drains in croplands accelerate the movement of stormflow from fields to streams. Drains are usually installed 0.6 to 1.2 m below the soil surface and at 10 to 20 m intervals and are used extensively throughout the Midwest (Fenelon, 1998; Myers et al., 2000). These structures ensure that streams in agricultural regions receive repeated inputs of agricultural drainage at regular spatial intervals, often over a wide area, creating large-scale changes in streamflow patterns and water quality:

> Drainage may contribute continuous or repeated inputs (e.g. water, nutrients, pesticide residue) so that the time span or spatial proximity between each input is less than that required for an environmental system to assimilate or recover from a series of drainage events (Spaling and Smit, 1995).

Tile drains improve soil aeration, accelerating rates of nitrification. They also eliminate environments such as wetlands that act as sinks for nutrients and provide denitrification sites. By design, the installation of tile drains lowers the water table. So by reducing the amount of water stored in

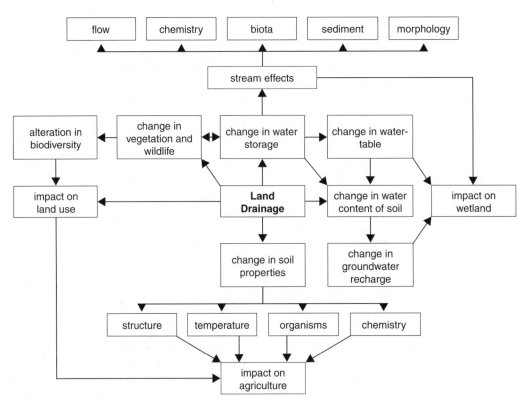

FIGURE 9.6 Cumulative effects of tile drains in an agricultural landscape. (Spaling and Smit, 1995; reprinted from *Agriculture, Ecosystems & Environment* with the permission of Elsevier.)

the soil, they decrease baseflow between precipitation events. Thus a common method for increasing agricultural productivity has the potential to repeatedly disrupt streamflow, water quality, and aquatic habitats in many ways (Figure 9.6) (Spaling, 1995).

Schwab et al. (1980), working in Ohio, measured mean annual nitrate outputs of 12.1 kg/ha/yr from undrained plots versus 18.7 kg/ha/yr from drained plots. A study in Indiana reports similar results (Fenelon, 1998). This is a region of poorly drained clay soils. The low permeability clays limit the infiltration of nitrogen to aquifers that are below the clay layer. During stormflow events, tile drains carry nitrates dissolved in shallow groundwater, which would otherwise be detained or retained in the undisturbed soils, directly to streams. When water stops flowing through the tile drains, nitrate concentrations in stream water decrease.

In the same NAWQA study (Fenelon, 1998), tile drains were also found to increase pesticide (atrazine) concentrations in stream water:

> Tile drains "short circuit" the ground water system by intercepting water percolating through the soil and shallow ground water and rapidly transporting it to streams. Because tile drains typically flow during wet periods from late winter to early summer they are able to transport recently applied pesticides to streams (Fenelon, 1998).

In contrast, other studies have shown that atrazine output from drained fields can be less than that from undrained fields (Spaling, 1995). A study in Louisiana (Bengston et al., 1990) found atrazine output from drained plots (23.47 g/ha) was 55% less than atrazine output from undrained plots (51.64 g/ha). In these examples it is hypothesized that the following mechanisms are at work. First, the tile drain network lowers the water table and increases the thickness of the unsaturated zone. The decrease in soil water content in turn leads to exponential decreases in unsaturated hydraulic conductivity (relative to the undrained condition). The combined effect (in Darcy's Law) of lower unsaturated hydraulic conductivity and a greater distance for water to travel through the unsaturated zone is increased residence time of dissolved pollutants in the soil (see Chapter 2). This provides a greater opportunity (as well as aerobic conditions) for decomposition or transformation into more benign chemical compounds relative to undrained fields (with shallow water tables and more frequent anaerobic conditions).

9.3.4 MULTIPLE IMPACTS OF HUMAN ACTIVITY ON AQUATIC BIOTA AND WETLAND BIRDS

Lowell and Culp (1999) examined the cumulative impacts of effluents from pulp mills and sewage plants on the mayfly species *Baetis tricaudatus*. This laboratory study attempts to simulate field conditions that are typical in the province of Alberta, Canada. The authors note that while the toxicity of pulp mill effluent has declined markedly over the last 30 years, it still may contain metals, fatty acids, and chlorinated and nonchlorinated organic compounds. Both pulp mill and sewage effluents add nutrients that stimulate and increase algal and microbial growth. This leads to low dissolved oxygen (DO) concentrations as the additional organic matter decays (increasing biochemical oxygen demand [BOD]). Thus potential adverse impacts may arise in two ways: (1) toxicant effects from chemicals in the effluent and (2) low DO concentrations. At the same time, by stimulating algal and microbial production, nutrients in effluent may provide a more abundant and nutrient-rich food source for benthic macroinvertebrates.

Lowell and Culp (1999) exposed mayflies to four different experimental environments: (1) 1% effluent and high DO (11 mg/L), (2) no effluent and low DO (5 mg/L), (3) 1% effluent and low DO, and (4) no effluent and high DO (control). Survival was reduced by 60% to 90% in the low DO conditions (without effluent) after 2 weeks. The presence of effluent increased mayfly survival (by increasing the nutrient quality of the food supply) in both the high and low DO environments. The survival rate under low DO conditions was lower than the control condition (high DO and no effluent). Lowell and Culp concluded that, at least in the short-term (the 2 week duration of the study), nutrient enrichment and variation in DO were more important than the toxic effects of effluent in determining mayfly survival. Because pulp mills usually operate continuously, except for a 1 week annual maintenance shutdown, it is not clear if these short-term laboratory results accurately replicate field conditions (continuous streamflow, low concentration effluent discharge, and sedentary benthic macroinvertebrates).

Preston and Shackelford (2002) conducted a large-scale study of the Chesapeake Bay and used multiple regression analysis to link biodiversity (as measured by the Shannon-Weaver biodiversity index) with water quality. Water quality indicators included measurements of toxicants (arsenic, cadmium chromium, copper, lead, mercury, zinc, pesticides, PAHs, and PCBs) and other water quality variables (DO, pH, salinity, temperature, and total suspended solids). Benthic macroinvertebrate samples were taken from 1054 fixed and random monitoring stations, resulting in a total of 2596 individual observations. Sediment sampling was done at 353 fixed monitoring stations around the perimeter of the Chesapeake Bay and in most major tributaries and sediment concentrations of 63 toxicants (7 metals, 23 pesticides, 28 PAHs, and 17 PCBs) were analyzed. Temperature, DO, pH, salinity, and total suspended solids data, obtained from the Chesapeake Bay Program, had been collected monthly during the summers from 1984 to 1999 from 112 water quality monitoring stations. Both the combined toxicant variables and the combined water quality variables correlated with benthic biodiversity. The correlation between individual variables and biodiversity was poor (r^2 ranged from 0.01 to 0.16). However, the correlation between biodiversity and all the variables taken together was quite high ($r^2 = 0.73$). Biodiversity was negatively correlated with toxicant concentration. Toxicants as a group accounted for 48% of the variation. Physical variables such as DO, pH, salinity, temperature, and total suspended solids collectively accounted for 30% of the variation; the correlations were low, but positive. While toxicants are clearly anthropogenic in origin, the other water quality parameters may vary naturally or be influenced by human activity. Preston and Shack-elford (2002) concluded that risk assessors must be aware of the many water quality variables that act together to create the aquatic environment at a particular point in space and time. Examining the effect of a single toxic substance is difficult, if not impossible, in an environment with many different toxic substances. In addition, the effect of toxic substances may be exacerbated or ameliorated by other water quality characteristics (DO, pH, salinity, temperature, and total suspended solids).

Harris (1988) described the cumulative impacts that led to the extinction of the dusky seaside sparrow (*Ammospiza maritima nigrescens*), a marsh species in Florida. The bird's natural habitat was quite small, less than a few hundred square kilometers. Marsh impoundment and DDT spraying in the 1940s for mosquito control, marsh drainage for agriculture and residential development, and the construction of an expressway seriously degraded and fragmented habitat for the sparrow. Fire suppression led to a fuel buildup that resulted in large fires in the 1970s. Thirty years of accumulated wetland habitat degradation led to an attempt to capture and breed the birds in captivity, however, the five remaining birds were all males.

9.3.5 LEGACY EFFECTS

Forests now cover large areas of the eastern United States that were once cleared for agriculture. The effects of earlier agricultural land use may still be evident in altered patterns of nutrient cycling (Aber et al., 1998). Changes in the forest floor resulting from an intense fire may increase nitrogen retention more than 100 years after the event (Hornbeck et al., 1997).

Harding et al. (1998) compared the impact of riparian land use and land use across the entire watershed on stream biodiversity in tributary streams of the Little Tennessee and the French Broad Rivers in western North Carolina. Twelve watersheds were chosen in each river basin; of these six were primarily forested and six were agricultural. Land use for all 24 watersheds was assessed from aerial photographs from the 1950s and satellite images from the 1990s. Sampling included both benthic macroinvertebrates and fish. The diversity of benthic macroinvertebrates was signifi-cantly greater in forested streams than in agricultural streams, although the density did not differ. Fish abundance and diversity were greatest in agricultural streams where brown trout (a European predatory species) had not been introduced. Regression analysis showed that whole-watershed land use in the 1950s was the best predictor of diversity in the mid-1990s. Two watersheds that were 92% forested in the 1990s still had species assemblages of macroinvertebrates and fish characteristic of agricultural streams (their condition in the 1950s). Reforestation in the riparian zone had not resulted in the recovery of a macroinvertebrate community that was comparable to the watersheds that were not cleared for agriculture. The authors concluded that "high impact or sustained anthro-pogenic disturbance, such as sustained agriculture, may profoundly alter biotic communities and the effects of this disturbance may be persistent."

The effects of acidic atmospheric deposition and current timber harvesting may be exacerbated by the legacy of past agricultural land use (Hornbeck, 1990; Sidle and Hornbeck, 1991). As discussed in previous chapters, acid precipitation accelerates the loss of calcium from the soil as hydrogen ions (H$^+$) replace calcium ions (Ca^{2+}) on negatively charged soil particles. Calcium ions enter the soil solution and are transported to streams via subsurface flow. Whole-tree harvesting (a method employed in the 1980s in some northeastern forests) removed nutrient capital in mature trees and reduced vegetative uptake and storage immediately following cutting. In whole-tree harvesting, entire trees—boles, branches, and foliage—are removed from the site. Branches are chipped for fuel and pulp. Very little slash is left on site, thus the return of nutrients to the soil from decaying tree parts is minimal (Hornbeck et al., 1990). Net leaching losses during a harvesting rotation were estimated to be 1300 to 2300 kg Ca/ha or 20% to 40% of the calcium capital of the site. In addition, soil calcium levels had already been severely depleted (estimated 1000 kg Ca/ha lost) during the 19th century when these areas were cleared for agriculture and tilled or grazed. While the calcium lost per year due to acid precipitation may have had little effect on the ecosystem, total calcium loss from past land use, whole-tree harvesting, and years of acid precipitation could result in decreased site fertility and forest health (Hornbeck, 1990; Sidle and Hornbeck, 1991).

9.4 SUMMARY AND CONCLUSIONS

9.4.1 MIXED LAND USE PATTERNS AND WATER QUALITY

Most large watersheds in the northeastern United States are characterized by a mix of forest, agricultural, and urban land. Wide-ranging differences in land cover (i.e., relative proportions of forest cover, spatial patterns, seasonal differences, legacy effects, etc.) generate complex ecological effects and daunting watershed management challenges. The studies that are described in this chapter and summarized below are important first steps and examples. The familiar refrain "more research is needed" clearly applies to this subfield of watershed science and management.

Results from mixed land use watershed studies reviewed in this chapter lead to the following conclusions:

1. Increased nitrogen export in rivers is directly related to the rate of nitrogen loading in watersheds. Nitrogen loading is highest in watersheds with the greatest proportion of agricultural and urban land. It also varies in direct proportion to population size and density. In contrast, nitrogen loading is lowest in watersheds with the greatest proportion of forested land area (Boyer et al., 2002).
2. Agriculture is a source of increased phosphorus inputs to waterways. However, urban wastewater can contribute more phosphorus per unit area than agriculture (Fisher et al., 2000a,b; Heisig, 2000; Meals and Budd, 1998; Osborne and Wiley, 1988).
3. High nitrate loadings in streams are associated with residential subdivisions that rely on septic systems (Heisig, 2000).
4. The loading of total coliform, fecal coliform, and fecal streptococcus bacteria in streams and lakes increases substantially in areas with agricultural and residential development. Transport is greater during periods of stormflow (Bolstad and Swank, 1997).
5. Forests are the reference condition against which other land uses should be compared.

9.4.2 CUMULATIVE EFFECTS

Research on cumulative effects focuses on the interaction of human activities and natural disturbances. The magnitude of the effects may vary over time and also in different locations within a watershed. Studies have examined a broad range of topics and used a variety of approaches. These include studying one type of land use, progressive land use change, multiple effects on aquatic biota and wetland species, and legacy effects of previous land uses and disturbances. While the immediate impact of a particular activity—draining one wetland, developing one residential area, cutting one stand of trees—on streamflow and water quality may be small, the combined effects of many individual actions across a watershed landscape can produce substantial changes in hydrologic regime and biogeochemical cycling (Childers and Gosselink, 1990; Gosselink et al., 1990; MacDonald, 2000; Preston and Bedford, 1988):

1. Timber harvesting can cause short-term increases in streamflow as a result of reductions in evapotranspiration and increases in net precipitation. The effect of timber harvesting may be exacerbated by poorly planned roads that reduce the infiltration capacity of soils and divert water directly into streams (Larson, 1984).
2. The progressive loss of forests and forested wetlands—combined with increased nutrient loading from agriculture and channel "improvements"—can lead to increases in peak

discharge and flooding events as well as increased phosphorus concentrations in streams (Childers and Gosselink, 1990).

3. Tile drains transport water, surplus nutrients, and some pesticides to streams in agricultural areas (Fenelon, 1998; Spaling and Smit, 1995). This "short-circuiting" of pathways of flow reduces the storage and residence time of water in soils and alters streamflow regimes.

4. Aquatic organisms are subjected to many pollutants emanating from both agricultural and urban land use, including metals, pesticides, PAHs, and PCBs. Although it is difficult to evaluate the effect of one chemical compound on aquatic biota, the combined or cumulative effect of all chemical compounds has been a significant contributor to decreases in biological diversity (e.g., Chesapeake Bay [Preston and Shackelford, 2002]).

5. Past agricultural land use still influences stream conditions in areas of second-growth forest. The adverse impacts of past agricultural land use are still apparent in stream sediment loads and turbidity (Scott et al., 2002), macroinvertebrate community structure (Harding et al., 1998), and calcium depletion in upland forests (Hornbeck, 1990).

The findings summarized above help to define the broad outlines of the research and development work that will be needed to conserve and protect aquatic ecosystems, water resources, and public health in the 21st century.

REFERENCES

Abbruzzese, B. and Leibowitz, S.G., Environmental auditing: a synoptic approach for assessing cumulative impacts to wetlands, *Environ. Manage.*, 21, 457–475, 1997.

Aber, J.D., McDowell, W., Nadelhoffer, K., Magill, A., Bernston, G., Kamakea, M., McNulty, S., Currie, W., Rustad, L., and Fernandez, I., Nitrogen saturation in temperate forest ecosystems (hypotheses revisited), *BioScience*, 48, 921–934, 1998.

Anderson, R.M., Beer, K.M., Buckwalter, T.E., Clark, M.E., McAuley, S.D., Sams, J.I., III, and Williams, D.R., Water quality in the Allegheny and Monongahela River basins: Pennsylvania, West Virginia, New York, and Maryland, 1996–1998, Circular 1202, U.S. Geological Survey, Reston, VA, 2000, http://pa.water.usgs.gov/almn/almn_pubs.html; accessed July 2006.

Ator, S.W., Blomquist, J.D., Brakebill, J.W., Denis, J.M., Ferrari, M.J., Miller, C.V., and Zappia, H., Water quality in the Potomac River basin, Maryland, Pennsylvania, Virginia, West Virginia, and the District of Columbia, 1992–96, Circular 1166, U.S. Geological Survey, Reston, VA, 1998, http://md.water.usgs.gov/circ/circ1201/; accessed July, 2006.

Ayers, M.A., Kennen, J.G., and Stackelberg, P.E., Water quality in the Long Island–New Jersey coastal drainages, New York, and New Jersey, 1996–1998, Circular 1201, U.S. Geological Survey, Reston, VA, 2000, http://pubs.usgs.gov/circ/circ1201/; accessed July 2006.

Bedford, B.L. and Preston, E.M., Cumulative effects on landscape systems of wetlands: scientific status, prospects, and regulatory perspectives, *Environ. Manage.*, 12, 561–562, 1988.

Bengston, R., Southwick, L., Willis, G., and Carter, C., The influence of subsurface drainage practices on nitrogen and phosphorus losses in a warm, humid climate, *Trans. Am. Soc. Agric. Eng.*, 31, 729–733, 1990.

Bolstad, P.V. and Swank, W.T., Cumulative impacts of land use on water quality in a southern Appalachian watershed, *J. Am. Water Resourc. Assoc.*, 33, 519–533, 1997.

Boyer, E.W., Goodale, C.L., Jaworski, N.A., and Howarth, R.W., Anthropogenic nitrogen sources and relationships to riverine export in the northeastern U.S.A., *Biogeochemistry*, 57/58, 137–169, 2002.

Childers, D.L. and Gosselink, J.G., Assessment of cumulative impacts to water quality in a forested wetland landscape, *J. Environ. Qual.*, 19, 455–464, 1990.

Coburn, J., Is cumulative effects analysis coming of age?, *J. Soil Water Conserv.*, 44, 267–270, 1989.

Council on Environmental Quality, Considering cumulative effects under the National Environmental Policy Act, Council on Environmental Quality, Executive Office of the President, Washington, D.C., 1997.

Fenelon, J.M., Water quality in the White River basin, Indiana, 1992–96, Circular 1150, U.S. Geological Survey, Reston, VA, 1998, http://in.water.usgs.gov/nawqa/wrnawqa.htm; accessed July 2006.

Fisher, D.S., Dillard, A.L., Usery, E.L., Steiner, J.L., and Neely, C.L., Water quality in the headquarters of the upper Oconee watershed, in *Proceedings of the 2000 Georgia Water Resources Conference*, Hatcher, K.J. (ed.), Institute of Technology, University of Georgia, Athens, GA, 2000a.

Fisher, D.S., Steiner, J.L., Endale, D.M., Stuedemann, J.A., Schomberg, H.H., Franzluebbers, A.J., and Wilkinson, S.R., The relationship of land use practices to surface water quality in the upper Oconee watershed of Georgia, *For. Ecol. Manage.*, 128, 39–48, 2000b.

Frey, J.W., Occurrence, distribution, and loads of selected pesticides in streams in the Lake Erie–Lake St. Clair basin, 1996–98, Water Resources Investigations Report 00-4169, U.S. Geological Survey, Reston, VA, 2001, http://oh.water.usgs.gov/nawqa/; accessed July 2006.

Garabedian, S.P., Coles, J.F., Grady, S.J., Trench, E.C.T., and Zimmerman, M.J., Water quality in the Connecticut, Housatonic, and Thames River basins, Connecticut, Massachusetts, New Hampshire, New York, and Vermont, 1992–95, Circular 1155, U.S. Geological Survey, Reston, VA, 1998, http://ma.water.usgs.gov/projects/MA-100/MA-100_Project_Publications.htm; accessed July 2006.

Georgia Department of Natural Resources, http://www.riversalive.org/; accessed July 2006.

Gosselink, J.G., Shaffer, G.P., Lee, L.C., Burdick, D.M., Childers, D.L., Leibowitz, N.C., Hamilton, S.C., Boumans, R., Cushman, D., Fields, S., Koch, M., and Visser, J.M., Landscape conservation in a forested wetland watershed, can we manage cumulative impacts?, *BioScience*, 40, 588–600, 1990.

Grant, G.E. and Swanson, F., Cumulative effects of forest practices, *For. Perspect.*, 1(4), 9–11, 1991.

Grigal, D.F. and Bates, P.C., Assessing impacts of forest harvesting—the Minnesota experience, *Biomass Bioenergy*, 13(4/5), 213–222, 1997.

Groschen, G.E., Harris, M.A., King, R.B., Terrio, P.J., and Warner, K.L., Water quality in the lower Illinois River basin, Illinois, 1995–1998, Circular 1209, U.S. Geological Survey, Reston, VA, 2000.

Harding, J.S., Benfield, E.F., Bolstad, P.V., Helfman, G.S., and Jones, E.B.D., III, Stream biodiversity: the ghost of land use past, *Proc. Natl. Acad. Sci. USA*, 95, 14843–14847, 1998.

Harris, L.D., The nature of cumulative impacts on biotic diversity of wetland vertebrates, *Environ. Manage.*, 12, 675–693, 1988.

Heisig, P.M., Effects of residential and agricultural land uses on the chemical quality of baseflow of small streams in the Croton watershed, southeastern New York, Water Resources Investigations Report 99-4173, U.S. Geological Survey, Reston, VA, 2000.

Hemond, H.F. and Benoit, J., Cumulative impacts on the water quality functions of wetlands, *Environ. Manage.*, 12, 639–653, 1988.

Hornbeck, J.W., Cumulative effects of intensive harvest, atmospheric deposition and other land use activities, FRI Bulletin 159, Ministry of Forestry, Forest Research Institute, Rotura, New Zealand, 1990, pp. 147–154.

Hornbeck, J.W., Bailey, S.W., Buso, D.C., and Shanley, J.B., Streamwater chemistry and nutrient budgets for forested watersheds in New England: variability and management implications, *For. Ecol. Manage.*, 93, 73–89, 1997.

Hornbeck, J.W., Smith, C.T., Martin, Q.W., Tritton, L.M., and Pierce, R.S., Effects of intensive harvesting on nutrient capitals of three forest types in New England, *For. Ecol. Manage.*, 30, 55–64, 1990.

Larson, A.G., Summary of a Washington State study of cumulative effects, in Forest management practices and cumulative effects on water quality and utility, Technical Bulletin 435, National Council of the Paper Industry for Air and Stream Improvements, Research Triangle Park, NC, 1984, pp. 14–23.

Lindsey, B.D., Breen, K.J., Bilger, M.D., and Brightbill, R.A., Water quality in the lower Susquehanna River basin, Pennsylvania and Maryland, 1992–95, Circular 1168, U.S. Geological Survey, Reston, VA, 1998, http://pubs.usgs.gov/circ/circ1168/; accessed July, 2006.

Long, G.R., Chang, M., and Kennen, J.G., Trace elements and organochlorine compounds in bed sediment and fish tissue at selected sites in New Jersey streams—sources and effects, Water Resources Investigations Report 99-4235, U.S. Geological Survey, Reston, VA, 2000.

Lowell, R.B. and Culp, J.M., Cumulative effects of multiple effluent and low dissolved oxygen stressors on mayflies at cold temperatures, *Can. J. Fish. Aquat. Sci.*, 56, 1624–1630, 1999.

Lowenthal, D.M., Introduction to the 1965 reprint edition of *Man and Nature* by George Perkins Marsh (1864), Harvard University Press, Cambridge, MA, 1965.

MacDonald, L.H., Evaluating and managing cumulative effects: processes and constraints, *Environ. Manage.*, 26, 299–315, 2000.

Meals, D.W., Water quality response to riparian restoration in an agricultural watershed in Vermont, USA, *Water Sci. Technol.*, 43(5), 175–182, 2001.

Meals, D.W. and Budd, L.F., Lake Champlain basin nonpoint source phosphorus assessment, *J. Am. Water Resourc. Assoc.*, 34, 251–265, 1998.

Mullaney, J.R. and Zimmerman, M.J., Nitrogen and pesticide concentrations in an agricultural basin in north-central Connecticut, Water Resources Investigations Report 97-4076, U.S. Geological Survey, Reston, VA, 1997.

Myers, D.N., Thomas, M.A., Frey, J.W., Rheaume, S.J., and Button, D.T., Water quality in the Lake Erie–Lake Saint Clair drainages, Circular 1203, U.S. Geological Survey, 2000, http://oh.water.usgs.gov/nawqa/; accessed July 2006.

National Council of the Paper Industry for Air and Stream Improvement, Forestry management practices and cumulative effects on water quality and utility, Technical Bulletin 435, National Council of the Paper Industry for Air and Stream Improvements, Research Triangle Park, NC, 1984.

National Council of the Paper Industry for Air and Stream Improvement, Assessing forest practice rules and managing to avoid landslides and cumulative effects, Technical Bulletin 496, National Council of the Paper Industry for Air and Stream Improvements, Research Triangle Park, NC, 1986a.

National Council of the Paper Industry for Air and Stream Improvement, Papers presented at the American Geophysical Union Meeting on Cumulative Effects, Technical Bulletin 490, National Council of the Paper Industry for Air and Stream Improvements, Research Triangle Park, NC, 1986b.

National Research Council, *Assessing the TMDL Approach to Water Quality Management*, National Academies Press, Washington, D.C., 2001.

O'Brien, A.L., Evaluating the cumulative effects of alteration on New England wetlands, *Environ. Manage.*, 12, 627–636, 1988.

Osborne, L.L. and Wiley, M.J., Empirical relationships between land use/cover and stream water quality in an agricultural watershed, *J. Environ. Manage.*, 26, 9–27, 1988.

Peters, C.A., Robertson, D.M., Saad, D.A., Sullivan, D.J., Scudder, B.C., Fitzpatrick, F.A., Richards, K.D., Stewart, J.S., Fitzgerald, S.A., and Lenz, B.N., Water quality in the western Lake Michigan drainages, Wisconsin and Michigan, 1992–95, Circular 1156, U.S. Geological Survey, Reston, VA, 1998, http://wi.water.usgs.gov/nawqa/; accessed July 2006.

Peverly, J.H., Stream transport of nutrients through a wetland, *J. Environ. Qual.*, 11, 38–43, 1982.

Preston, B.L. and Shackelford, J., Multiple stressor effects on benthic biodiversity of Chesapeake Bay: implications for ecological risk assessment, *Ecotoxicology*, 11, 85–99, 2002.

Preston, E.M. and Bedford, B.L., Evaluating cumulative effects on wetland functions: a conceptual overview and generic framework, *Environ. Manage.*, 12, 565–583, 1988.

Rheaume, S.J., Button, D.T., Myers, D.N., and Hubbell, D.L., Areal distribution and concentrations of contaminants of concern in surficial streambed and lakebed sediments, Lake Erie–Lake Saint Clair drainages, 1990–1997, Water Resources Investigations Report 00-4200, U.S. Geological Survey, Reston, VA, 2001.

Schwab, G.O., Fausey, N.R., and Kopcak, D.E., Sediment and chemical content in agricultural drainage water, *Trans. Am. Soc. Agric. Eng.*, 23, 1446–1449, 1980.

Scott, M.C., Helfman, G.S., McTammany, M.E., Benfield, E.F., and Bolstad, P.V., Multiscale influences on physical and chemical stream conditions across Blue Ridge landscapes, *J. Am. Water Resourc. Assoc.*, 38, 1379–1392, 2002.

Sidle, R.C. and Hornbeck, J.W., Cumulative effects: a broader approach to water quality research, *J. Soil Water Conserv.*, 46, 268–271, 1991.

Spaling, H., Analyzing cumulative environmental effects of agricultural land drainage in southern Ontario, Canada, *Agric. Ecosyst. Environ.*, 53, 279–292, 1995.

Spaling, H. and Smit, B., A conceptual model of cumulative environmental effects of agricultural land drainage, *Agric. Ecosyst. Environ.*, 53, 99–108, 1995.

Sullivan, D.J., Stinson, T.W., Crawford, J.K., and Schmidt, A.R., Surface-water-quality assessment of the upper Illinois River basin in Illinois, Indiana, and Wisconsin—pesticides and other synthetic organic compounds in water, sediment, and biota, 1975–90, Water Resources Investigations Report 96-4135, U.S. Geological Survey, Reston, VA, 1998.

Swank, W.T. and Bolstad, P.V., Cumulative effects of land use practices on water quality, in *Hydrological, Chemical, and Biological Processes of Transformation and Transport of Contamination in Aquatic Environments*, Peters, N.E., Allan, R.J., and Tsirkunov, V.V. (eds.), Publication 219, International Association of Hydrological Sciences, Wallingford, UK, 1994, pp. 409–421.

Vermont Department of Environmental Conservation and New York State Department of Environmental Conservation, Lake Champlain diagnostic feasibility study, Final Report. A phosphorus budget, model, and load reduction strategy for Lake Champlain, Waterbury, VT and Albany, NY, 1997.

Wall, G.R., Riva-Murray, K., and Phillips, P.J., Water quality in the Hudson River basin, New York and adjacent states, 1992–95, Circular 1165, U.S. Geological Survey, Reston, VA, 1998, http://ny.water.usgs.gov/projects/hdsn/; accessed July 2006.

Winter, T.C., A conceptual framework for assessing cumulative impacts on the hydrology of nontidal wetlands, *Environ. Manage.*, 12, 605–620, 1988.

Winter, T.C., The concept of hydrologic landscapes, *J. Am. Water Resourc. Assoc.*, 37, 335–349, 2001.

Wong, C.S., Capel, P.D., and Nowell, L.H., Organochlorine pesticides and PCBs in stream sediment and aquatic biota—initial results from the National Water Quality Assessment Program, 1992–1995, Water Resources Investigations Report 00-4053, U.S. Geological Survey, Reston, VA, 2000.

Zimmerman, M.J., Grady, S.J., Todd Trench, E.C., Flanagan, S.M., and Nielsen, M.G., Water-quality assessment of the Connecticut, Housatonic, and Thames River basins study unit: analysis of available data on nutrients, suspended sediments, and pesticides, 1972–92, Water Resources Investigations Report 95-4203, U.S. Geological Survey, Reston, VA, 1996.

10 Conclusions and Management Implications

Most deeply, an ethic of stewardship is about respecting the beauty and mystery of a natural world we did not create and cannot fully understand, but over which we have acquired dominance. It is about adding a healthy dose of humility as an antidote to our past hubris. And it is about applying the best of our science, policy, and technology not to further manipulate nature but to better adapt ourselves to its time-tested, life-sustaining cycles.

Sandra Postel and Brian Richter, 2003,
Rivers for Life: Managing Water for People and Nature

10.1 INTRODUCTION

Remarkable improvements in water quality occurred in many streams and rivers following implementation of the Clean Water Act in the 1970s. Three decades later, nonpoint source pollution is the central challenge to those concerned with protecting and restoring water quality. Land use effects on the water balance, soil erosion and sediment transport, nutrient cycling, and pollutant transport are a primary cause of water quality degradation.

Most watersheds in New England, the Mid-Atlantic states, and the Midwest contain a variety of land uses in varying proportions: dense urban, industrial, commercial, dispersed residential, agricultural, and forest, plus a network of roads. It is increasingly clear that the pattern of land use throughout an entire watershed determines the condition of the streams. Consequently planning and management for the entire watershed are necessary to protect tributary streams, rivers, lakes, and estuaries. Atmospheric deposition, which is another major source of water pollution, requires policy decisions and management actions on a regional, national, and international scale.

In this concluding chapter, we first review some of the key biophysical processes that alter streamflow and water quality. We then discuss current research challenges and management examples in pollutant mitigation and environmental conservation. Finally, we describe current thinking about the ecological restoration of streams and watersheds. This chapter is derived from current literature and from discussions with the expert panel listed in the preface.

10.2 THE BIOPHYSICAL ENVIRONMENT, STREAMFLOW, AND WATER QUALITY

10.2.1 Natural Variability

Stream channel conditions in undisturbed environments are the product of local and regional climate and the geology, topography, and soils of the watershed. Watershed size, the character and condition of vegetation, and water and sediment delivery to channels determine the geometry of channels and the form and structure of aquatic ecosystems. When natural vegetation is disturbed or removed and soils are compacted or covered over with impervious materials, the compensatory changes in stream channels will vary depending on the topographic, geologic, and edaphic characteristics of the watershed. In other words, some watersheds are more resistant to change than others following natural or anthropogenic disturbance. In addition, the biophysical environment often varies within a watershed from uplands to lowlands. Similarly the response of a stream to management and restoration measures is site specific. An understanding of the biophysical characteristics and hydrologic processes within a watershed is therefore essential to maintaining and improving water quality in receiving waters and to stream restoration (Booth et al., 2003; Driscoll et al., 2003b).

The importance of the biophysical environment of the watershed in determining stream conditions also means that hydrologic and water quality research reflects the ecological conditions in which it is conducted. Hence there is no one study that can predict how much nitrate will be lost from a forest stand when it is clearcut or exactly how a stream channel will change as the area of residential development increases. Rather, research studies help to frame the range of responses that can be expected in different regional and local environments. Regional variability is the norm and should be accepted as the reason why an adaptive management paradigm is needed (Binkley, 2001; Driscoll et al., 2003b).

10.2.2 MASS BALANCE

The mass balance of water and nutrients (and other polluting substances) is the site-specific mathematical representation of the law of conservation of mass. The outflow or output of any component from an ecosystem is the direct result of inflow or inputs minus storage in vegetation, soils, and sediment. In forest ecosystems, nutrients are stored in trees and shrubs and returned to the soil with fallen leaves. In agricultural systems, nutrient inputs are stored in soils and crops. Some of these nutrients are removed when crops are harvested. (As described in Chapters 3 and 5, riparian and other wetland denitrification sites are a major pathway for nitrogen removal.) When pollutant inputs (nutrients, pesticides, and other chemicals associated with human land use) increase and storage capacity decreases (through the removal of perennial vegetation and a decrease in soil porosity), outputs in streamflow must increase. This is evident in changes in stream water chemistry. The mass balance principle also applies to precipitation and streamflow. Streamflow increases when storage is reduced through changes in land use and consequent decreases in soil infiltration. Persistent changes in the quantity and timing of flow and changes in water quality (beyond those attributable to climatic variability) are a clear indication of a change in the balance in a watershed—either increased inputs or a reduction in storage. Examples include increases in water yield and the loss of nutrients (nitrate, calcium, magnesium, etc.) following clearcutting in some forests; increased outflow of nitrogen, phosphorus, pesticides, and microbial pathogens from agricultural areas; and changes in stormflow and water quality in urban areas.

Structural best management practices (BMPs) (both agricultural and urban) and restored riparian buffers can moderate streamflow, slow the movement of nonpoint source pollutants, and reduce pollutant loading to some extent. It is difficult, however, to restore natural streamflow patterns without concerted efforts to increase watershed storage capacity and improve water quality while also reducing pollutant inputs. Legacy pollutants (pollutants stored in the soil, stream sediments, and groundwater during decades of excess application) frequently delay the onset of improved stream conditions. This was evident in the results of the Big Spring Demonstration Project in Iowa (see Chapter 7; Rowden et al., 2000), where groundwater continued to be a source of nitrate and pesticides that had been applied during the 1970s, despite widespread adoption of BMPs designed to reduce the loading of these chemicals. In a 1992 article entitled "Great Ideas in Ecology for the 1990s," the eminent ecologist Eugene Odum concluded that "input management is the only way to deal with nonpoint pollution." Because nutrient pollution is caused by the presence of nutrients in excess of plant requirements, it also is a measure of waste and inefficiency in the ecosystem, enterprise, or household.

10.2.3 POLLUTANT TRANSPORT PATHWAYS FOR NUTRIENTS, PESTICIDES, AND TOXINS

Pollutants may be transported to streams as suspended particles, adsorbed to sediment, or in dissolved form. The primary mode of transport depends upon the chemical characteristics of the substance as well as patterns and rates of water and sediment movement. Controlling erosion can reduce both the loading of sediment and sediment-associated pollutant particles, yet it may not reduce the transport of dissolved substances. This is a particular problem in agricultural areas, where it is necessary to limit both the transport of particulate phosphorus that moves with sediment in overland flow and dissolved nitrate and phosphorus that move with both subsurface flow and overland flow. In the Chesapeake Bay watershed, widespread adoption of no-till cultivation has decreased the transport of sediment-associated phosphorus, but has increased the transport of dissolved biologically reactive phosphorus and nitrate because of the high concentrations of these nutrients in the soil (Boesch et al., 2001; Inamdar et al., 2001; Sharpley, 2003).

Many pollutants—nutrients, pesticides, metals, and organic compounds—are transported locally and regionally through atmospheric deposition. These include nitrates from automobile emissions that contribute to eutrophication in coastal estuaries, pesticides that volatilize from farm fields, and mercury emissions from power plants. The different chemical forms—particulate versus dissolved—and transport pathways of pollutants make it imperative that management strategies focus on source reduction at all the points where pollutants originate.

10.2.4 THE INFLUENCE OF SPATIAL SCALE

Disturbances that degrade receiving waters occur at multiple scales. In forested watersheds, vegetation in riparian areas protects streams from bank erosion and sediment loading (Jones et al., 1999). Forest cover is lost as watersheds are converted to agricultural, urban, or mixed land use. At some point, a threshold is passed and large-scale land use patterns overcome the protection provided by riparian buffers (Hession et al., 2003; Sliva and Williams, 2001). When alterations in groundwater flow, streamflow, and sediment supply are watershed-scale phenomena, small-scale (i.e., stream reach) restoration efforts and localized BMP implementation may not achieve desired or expected results (Booth et al., 2004, Walsh et al., 2005b). Edge-of-stream reductions in nutrient loading, achieved through the restoration of field-scale riparian buffers, may have little influence on stream nutrient concentrations if upstream reaches remain unbuffered and excess loading continues. The Bear Creek, Iowa, riparian buffer demonstration project described in Chapter 5 showed that while buffer restoration resulted in dramatic reductions in sediment, nitrogen, and phosphorus at the site, water quality in the stream was unchanged because upstream loading remained the same (Schultz et al., 2004). There is no escaping the mass balance at the watershed scale.

10.2.5 LAND USE AND HYDROLOGIC FLOW PATHS

Many of the problems, such as changes in the magnitude and frequency of peak flows, channel stability, and water quality, observed in streams are directly linked to increased overland flow and decreased infiltration and storage in the watershed. It may be safely concluded (allowing for variability due to topography, surficial geology, and soils) that the storage and infiltration of precipitation is greatest in forested ecosystems. In these systems, vegetation consists of mature and actively growing trees, leaf litter protects the soil surface beneath a multilayered plant canopy, and soil organic matter concentration, especially in the O layer, is high. This leads to the highest rates (among the various land uses discussed) of evapotranspiration, soil water infiltration, and soil water and groundwater storage. Overland flow is rare in forests. In agricultural ecosystems, removal of the forest canopy (or native grasses), loss of soil organic matter, and excessive soil compaction reduce evapotranspiration, infiltration, and subsurface and groundwater storage. In urban ecosystems, infiltration in open areas is also reduced, due to soil compaction and the virtual absence of soil organic matter. In paved or impervious areas, infiltration and soil water storage capacity are eliminated and all precipitation is directed to overland flow. As emphasized throughout this book, these changes in hydrologic structure and function lead inexorably to a reduction in water quality and aquatic biointegrity.

10.3 CURRENT RESEARCH, CONSERVATION, AND RESTORATION ISSUES

10.3.1 IDENTIFYING POLLUTANT SOURCES

Efficient watershed management requires identification of those areas within a watershed that contribute the greatest proportion of nonpoint source pollutants. Resources may then be directed toward landowners to assist with management and restoration in these critical areas. In Chapter 9 (Box 9.2) we discussed the use of tributary sampling to locate and identify areas within watersheds that contribute disproportionately to pollutant loading. The three examples, described below, use a variety of additional approaches, including geographic information systems (GIS) analysis, soil analysis, and modeling, to address this problem.

10.3.1.1 Lake Mendota, Wisconsin

The watershed surrounding Lake Mendota, Wisconsin, is primarily agricultural (86%) with some developed land (9%). The remaining watershed area consists of wetlands (4%) and forest (1%). Five major streams and two storm sewers deliver water and nonpoint source pollutants to the lake. The lake is eutrophic, a result of decades of phosphorus loading from agricultural and urban sources. Phosphorus concentrations and algal populations in the lake vary from year to year. The watershed is near the Madison metropolitan area and there is growing concern about the potential effect of suburban development on water quality as agricultural land is converted to residential and commercial uses.

Soranno et al. (1996) used Lake Mendota watershed data (topography, hydrography, and land use) and phosphorus concentration and discharge data from the five streams and two storm sewers to develop a model that identified the location of phosphorus sources within a watershed. It estimated the transport of sediment-associated particulate phosphorus in overland flow, while accounting for variability in phosphorus loading caused by variations in annual precipitation. The model used a GIS database—a digital topographic map of the watershed divided into 100 m × 100 m grid cells—and estimated the total phosphorus load from the watershed by adding up the phosphorus contributed from each grid cell. The phosphorus contribution of individual grid cells varied in relation to slope, land use, proximity to a water flow path, and a transmission coefficient (T) that reflected the likelihood that phosphorus from a particular grid cell would reach the lake. The transmission coefficient ranged from zero to one. In urban areas, it was assumed that all the phosphorus that reached a paved surface would be transported to the lake because all precipitation becomes stormwater.

In agricultural areas, small (high frequency) precipitation events might generate little or no phosphorus loading, because some or all of the precipitation can enter the soil and be in a location where sediment does not move. Thus phosphorus loading from agricultural areas decreases during dry years. The variability between years was lower in urban areas, where even small rainfall events produced overland flow. Incorporating the transmission coefficient into the model helped to account for phosphorus storage within the watershed and produced results that matched field data more accurately than simple export models.

The modeling showed that up to half of the watershed generally contributed no phosphorus to the lake. Most of the phosphorus came from riparian areas that varied in width (from 100 m to approximately 6 km) in relation to topography and expansion and contraction of the variable source area. Using the model to predict the effects of future land use changes on phosphorus loading to Lake Mendota showed that "changes in P loading were strongest with conversions of undisturbed vegetated lands, especially riparian areas, to either urban or agricultural uses."

10.3.1.2 Maumee and Sandusky River Watersheds, Northwestern Ohio

The Maumee and Sandusky River watersheds contribute nutrients and sediments to Lake Erie. The relation between land use, primarily agricultural practices, and water quality in this area has been characterized in great detail since the 1970s. The influence of soil types, however, had not been studied as thoroughly. Calhoun et al. (2002) compared 20 years of sediment and nutrient export data from the Maumee and Sandusky Rivers with similar data from studies completed in the 1970s of three intermediate-size (1,000 to 50,000 ha) and three small (40 to 1000 ha) subwatersheds. They also included field-scale and plot measurements of pollutant export from several major soil types of the Maumee watershed. Agriculture is the predominant land use in the region. They found that relationships that were "vague and seemingly contradictory at the large watershed scale" were more fully explained by the smaller scale studies. It became clear that soil type and tile drainage were influential in determining the source areas of pollutants within the larger watersheds.

By comparing the unit area export (kg/ha/yr) of different soils and stream water concentrations, Calhoun et al. concluded that

1. Water from lacustrine or lake plain soils, with high (>50%) clay content and without tile drainage systems, had the highest concentration of suspended solids (6 mg/L). This finding contradicted the conventional wisdom that the greatest source of soil erosion (and particulate P) would originate from steep slopes. The lake plain soils are relatively flat.
2. Tile drained lake plain soils exported more dissolved nutrients (nitrate and soluble reactive phosphorus) than better-drained soils on slopes.

While considerable attention had been given to improving agricultural management on slopes, careful analysis of the available data suggested that the lake plains were a potentially greater source of sediment and nutrients. Although lake plain soils occupied only 5% of the Maumee River basin, they generated 2.5 times more sediment per unit area than soils with greater slopes and lower clay content (approximately 30% clay).

10.3.1.3 Implementing the Phosphorus Index

Various forms of phosphorus indexing based on a framework originally proposed by Lemunyon and Gilbert (1993) have been implemented in 47 states. The phosphorus index is a tool designed to identify critical source areas for phosphorus—areas where excess phosphorus is available and transport potential is high. It is hoped that targeting management assistance to these areas will lead to the most efficient use of resources to protect water quality and allow eutrophic streams, lakes, and coastal estuaries to recover. Phosphorus indexing is one of the tools used by scientists at the USDA Agricultural Research Service Pasture Systems and Watershed Management Research Unit and Pennsylvania State University to assess phosphorus loading to streams in Pennsylvania. Thirty-five percent of the Chesapeake Bay watershed area is located in Pennsylvania. The Susquehanna River, which flows through this region, delivers 50% of the water and 40% of the nutrients entering the bay (Kogelmann et al., 2004; Sharpley and Beegle, 2001; Sharpley et al., 2003).

As discussed in Chapter 7, excess phosphorus accumulates in agricultural soils when the inputs of phosphorus from manure and artificial fertilizers exceed the amount used by crops and removed when crops are harvested. This occurs frequently in areas where intensive animal farming (dairy, poultry, cattle/calves, and swine) is practiced and when manure fertilizers are applied based on crop nitrogen requirements. Because the nitrogen:phosphorus ratio in manure does not match the nitrogen:phosphorus requirements of many crops (corn, for example), applying manure based on crop nitrogen requirements automatically means that excess phosphorus is applied. This leads to an accumulation of phosphorus in surface soils. If those soils are located in areas where overland flow is easily generated, the excess phosphorus can be transported to streams. Phosphorus export is more likely if the phosphorus-enriched soils are in or near riparian areas, although local variability may be related to differences in soil types (e.g., the Maumee and Sandusky River watersheds described earlier). At the same time, it is generally recognized that "most of the annual P export from watersheds occurs from only a small percentage of the land area and during a relatively few large storms" (Sharpley and Beegle, 2001).

The Pennsylvania phosphorus index is a field-scale screening tool designed "to rank and identify areas that have the greatest potential for P loss and to provide flexibility in developing remedial strategies" (Kogelmann et al., 2004). Completion of the phosphorus index involves evaluation of phosphorus sources (soil phosphorus concentration, the type of phosphorus applied, the methods and rate of application) and the phosphorus transport potential of the site (overland flow, soil erosion, subsurface drainage, and proximity to water bodies) (Kogelmann et al., 2004).

Scientists selected sites for phosphorus index screening on the basis of two criteria: (1) a phosphorus soil test greater than 200 mg/L or (2) a site location within 45.7 m (150 ft) of a stream, a distance considered large enough to include most of the variable source area. A GIS analysis was completed to identify potential sites. This involved georeferencing phosphorus soil test data by postal zip code (89,493 soil test phosphorus values in 1,005 zip codes) to create a soil phosphorus map. Land cover data were used to identify the areas of agricultural land within each zip code. This was combined with a streams map to identify the amount of agricultural land within the 45.7 m stream buffer by zip code. Animal density in annualized animal equivalent units (485 kg live weight/acre) was also mapped by zip code. Comparison of these various GIS data layers allowed researchers to examine the relationship between the spatial distribution of agricultural land and soil phosphorus and livestock density and soil phosphorus.

The mapping indicated that, based on soil test data alone, 95% of the agricultural fields in Pennsylvania would not require restrictions on phosphorus use. Three of 67 counties—Lancaster, Schuylkill, and Montgomery—accounted for more than half of the soil tests above 200 mg P/L. All three are located in the southeastern portion of the state, the area with the highest livestock densities and the largest number of confined animal feeding operations (CAFOs). Optimal soil phosphorus concentrations for agricultural cropland range from 30 to 50 mg/L. Nearly half (48%) of the almost 90,000 soil samples had phosphorus concentrations greater than 50 mg/L. This indicates that phosphorus is routinely applied in excess of crop requirements (Kogelmann et al., 2004).

Watershed managers in many states have modified the phosphorus index from its original form to fit local needs and conditions. Within the northeastern region, Vermont has added estimates of flood frequency; Delaware, Maryland, and Pennsylvania have added estimates of the availability or solubility of fertilizer and manure phosphorus; Iowa and Rhode Island have included estimates of BMP effectiveness; and Delaware, Maine, and Maryland have added ranked estimates of the sensitivity of receiving waters. Sharpley et al. (2003) note that the many modifications and accepted versions of phosphorus indices now in use demonstrate the utility and flexibility of the method.

The studies of soil phosphorus concentrations over large areas also highlight the importance of legacy pollutants. It has taken many years of fertilizer and manure application to produce the soil phosphorus concentrations described in these studies. Researchers now predict that it will take 10 to 20 years of corn or soybean production to reverse the process and decrease soil phosphorus concentration from 150 mg/L to 20 mg/L (Carpenter et al., 1998; Kogelmann et al., 2004; Sharpley and Beegle, 2001). Kogelmann et al. (2004) state:

> The P-index has the potential to force a rethinking of how nutrients should be managed on a farm and regional basis as farmers and government officials are compelled to manage nutrients in a way that protects water quality.

10.3.2 Reducing Multiple Sources of Pollutants

While it is important to identify the primary sources of pollutants in order to allocate management resources most efficiently, it is also important to recognize that pollutants may have multiple and diverse sources, all of which may need to be addressed. This is especially true of nitrogen. Primary sources of nitrogen include atmospheric deposition (the result of emissions from the burning of fossil fuels), fertilizer nitrate, wastewater treatment effluent, and discharges from combined sewer overflows (CSOs). Driscoll et al. (2003b) used mathematical models (PnET-BGC and WATERSN) to estimate the reductions in nitrogen loading related to a suite of public policy scenarios in New England and New York, areas that are strongly affected by nitrogen pollution. The scenarios were based on actual or proposed policies, such as the 1990 Clean Air Act Amendments, or on estimates of reductions that might be gained from "aggressive" implementation of newer technologies, such as very low emission vehicles or nitrogen removal at wastewater treatment plants.

The specific management strategies tested included reductions in inputs from atmospheric emissions by regulating the use of fossil fuels by electric utilities, more stringent vehicle emission standards, and reducing agricultural emissions from ammonia gases and ammonium aerosols through improved management of animal wastes at CAFOs. Additional reductions in loading would be achieved through improved wastewater treatment. Adding nitrogen removal to primary and secondary treatment plants can reduce nitrogen loading in effluent by 87% and 67%, respectively.

Using data from the mixed use, 40,744 km² Long Island Sound basin for part of their analysis, Driscoll et al. (2003b) concluded that the removal of biologically available nitrogen at wastewater treatments plants would be the single most effective control of nitrogen input. However, "an integrated management plan, involving reductions from multiple sources, is necessary to achieve the most effective N reduction." They estimated that by removing biologically active nitrogen at all wastewater treatment plants in the Long Island Sound watershed, reducing utility plant emissions by 75% beyond the 1990 Clean Air Act Amendments, converting all vehicles to super-low-emission vehicles, and reducing nitrogen loading from agriculture in accordance with the Clean Water Act, nitrogen loading to Long Island Sound would be reduced by 58%. Combining these actions with reductions in nitrogen loading and protecting areas such as wetlands and riparian areas that are sites for plant uptake and denitrification (nitrogen sinks) would yield an optimal nitrogen reduction strategy.

10.3.3 Forest Cover, Impervious Area, and Stream Degradation

The proportion of forest cover and impervious area in a watershed are frequently used as indicators of watershed condition. In general, studies show that the loss of more than 20% to 25% of the forest cover in a forested watershed will cause measurable, sustained increases in streamflow. These studies relate to timber harvesting and track changes in streamflow immediately after harvesting and through the course of forest regrowth (see Chapter 6). Residential and commercial development replaces forest cover with impervious surfaces and altered areas such as lawns (with reduced infiltration and soil water storage capacity). Simply put, a pollutant and stormwater "sink"—forests—is being replaced by a pollutant and stormwater "source"—impervious area (Figure 10.1).

Prior to 1995, many studies identified the connection between urban land use, total impervious area (TIA), and population density and stream condition, defined by biotic indices and abiotic measures. A TIA threshold of 10% to 15% appeared to exist, after which stream degradation inevitably occurred (Schueler, 1994; Schueler and Claytor, 1996). Recent analyses point out that past studies used a variety of criteria to define and quantify TIA and urban land use, making comparisons among studies difficult (Brabec et al., 2002). In addition, there are other factors to consider: (1) Focusing on TIA tends to ignore the changes in soil characteristics and infiltration in open areas in urban watersheds; (2) Effective impervious area (EIA) (impervious area directly connected to streams by storm drains) appears to be more directly connected with stream degradation than total impervious area (Hatt et al., 2004; Walsh et al., 2005a); (3) The spatial arrangement of development, mature forest, and open pervious areas and the condition and continuity of riparian forest all appear to influence stream condition. Various researchers now conclude that streams in urbanizing watersheds change in a continuous rather than threshold fashion, and that a wide range of stream conditions are possible when urban land occupies 5% to 40% of the watershed (Booth and Jackson, 1997; Brabec et al., 2002; Walsh et al., 2005b). Above 40% to 50% urban land use, stream condition, as measured by biotic indices, consistently declines (Morley and Karr, 2002).

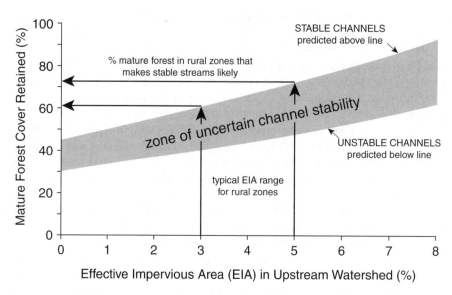

FIGURE 10.1 Conditions of forest cover and impervious area relative to stream stability, from a modeling study by Booth et al. (2002). The range of effective impervious areas (EIA = 3% to 5%) reflects variations in rural land cover conditions; the "zone of uncertain channel stability" reflects uncertainty in the hydrologic parameters. (Booth et al., 2002; reprinted with the permission of the American Water Resources Association.)

Research is needed to more accurately identify the factors that increase stream stability in developing watersheds and to identify those conditions under which restoration is a feasible and efficient use of resources. As discussed in Chapter 8, this would include research on the effects of watershed-scale factors such as drainage designs and the spatial arrangement within the watershed of developed areas and conservation land. The spatial arrangement of forests, wetlands, and developed areas is a reflection of historical and contemporary social, economic, political, and regulatory processes. We now know that these individual and collective decisions can have substantial effects—either positive or negative—on streamflow and water quality.

10.3.3.1 Patterns of Development

During the last 50 years, residential neighborhoods and associated transportation networks and commercial developments have spread over the landscape into many areas that were formerly forested or agricultural land (Johnson and Klemens, 2005) (Box 10.1). This has spurred efforts by the conservation community and others to encourage more concentrated development and reduce automobile use by promoting public transportation and walkable neighborhoods. Recent census data show signs of a modest revival in core urban areas and continuing development at the urban fringe. The central business districts of Chicago, Philadelphia, Boston, Baltimore, Cleveland, Detroit, and Milwaukee all evinced population growth and increased density between the 1990 and 2000 census (a phenomenon labeled "downtown rebound") (Sohmer and Lang, 2003). Recently, efforts to promote residential living in urban centers have been observed in smaller cities such as Franklin, Massachusetts (Urstadt, 2006). This revival appears to be limited to small downtown areas that offer a variety of urban cultural amenities that draw young professionals and retirees. The predominant trend, however, has been one of scattered population decreases in inner suburbs and increases in both population and developed land area further from city centers (Lucy and Phillips, 2003) (Figure 10.2). U.S. census data, released in March 2006, reveal that this trend has not changed. For example, the population of Cook County, Illinois, including Chicago, declined by 1.3% (73,000 residents) between 2000 and 2005. Residents are leaving the city and moving to outlying suburbs of Chicago, and, as

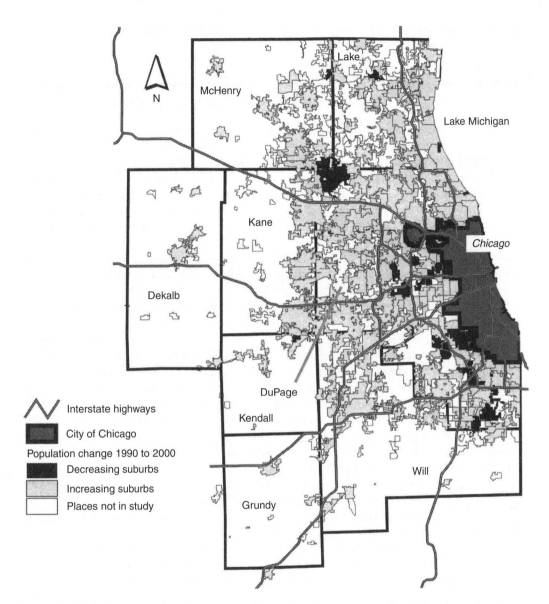

FIGURE 10.2 Chicago, Illinois, primary metropolitan statistical area (PMSA): Population change in suburban places, 1990–2000. (Lucy and Phillips, 2003; reprinted with the permission of The Brookings Institution.)

indicated by national trends, to suburbs in the "sunshine states"—the South, Southwest, and California (Associated Press, 2006). Kendall County, Illinois, 50 miles west of Chicago, was the third fastest growing county in the country, with an increase of 9.4% (6810 residents). Grundy County, adjacent to Kendall County, ranked 11th in the country, with a population increase of 6.5% (2660 residents) (U.S. Census Bureau, 2006).

To evaluate the effects of dispersed development on streamflow and water quality, it is necessary to assess the net result of forest loss and impervious area increases. Booth et al. (2002) found that aquatic resources in the Puget Sound lowlands were degraded by low density rural development (5 acre zoning). Although the area of impervious surface was relatively small, the forest was cleared in patches to create large lawns, pastures, or hobby farms. This resulted in a fragmented forest cover in which cleared areas comprised up to 60% of the watershed. In addition to land clearing, new homeowners frequently cleared riparian areas and removed large woody debris from stream channels on their property. The forest conversion and fragmentation and channel alteration led to significant changes in streamflow and water quality.

Box 10.1   Urban Sprawl and Water Quality in Pennsylvania

A study in Pennsylvania (Interlandi and Crockett, 2003) examined the effect of suburban development in the Philadelphia metropolitan area on water quality in the Schuylkill River (the largest tributary in the Delaware River basin). The Schuylkill River watershed has an area of 4903 km^2 in southeastern Pennsylvania and is home to approximately 3 million people. Between 1980 and 2000 the Philadelphia metropolitan area lost 10% of its population (Brookings Institution, 2003). During the same time period, the surrounding counties were growing by as much as 20% per decade. Suburban population gains led to an increase in the total area of developed land in the watershed (from 21.5% in 1982 to 28.3% in 1997). Most of this development took place on agricultural land. Here, the biggest change was not the loss of forest cover, but the increase in suburban roadways and residential and commercial areas, that is, an increase in impervious area and traffic volume.

There was a marked improvement in water quality in the Schuylkill River after passage of the Clean Water Act and reductions in industrial point sources. Now the major threat is from nonpoint source pollution caused by suburban development (Schuylkill Watershed Conservation Plan, 2001). Interlandi and Crockett (2003) found that alkalinity, conductivity, and mean annual concentrations of sodium and chloride in river water have increased during the last 10 years. Instantaneous peak chloride discharge in the Schuylkill River increased from 4 kg/sec in 1990 to 23 kg/sec in 1999. Similarly the peak transport of sodium, less than 3 kg/sec in 1990, had risen to almost 15 kg/sec in 1999. After examining the effects of climate and streamflow rates during the same period, Interlandi and Crockett concluded, "increases in solute transport in the Schuylkill River in recent decades appear to be the direct result of modern suburban development in the watershed."

It may be argued that increases in sodium and chloride concentrations (the constituents of road salt) are the result of increased application rates (or more frequent application) by town, county, and state highway departments rather than an increase in the total mileage of roads. Without examining the highway department records, it is impossible to determine which explanation is more accurate. In any case, two additional points should be considered. First, residential development on agricultural land requires the construction of additional roads. Second, increased commuter traffic volume and average speeds on secondary roads usually leads highway departments to increase the application rate or application frequency of road salt. This change is made to reduce accidents and maintain safe conditions for emergency vehicles. In either case, the conversion of agricultural land to residential use is a direct or indirect cause of water quality degradation. A statistically significant decline in ammonium concentration was probably related to a decrease in livestock in the watershed and is not necessarily an indication of an overall decrease in nutrient loading. Other water quality parameters (e.g., total nitrogen, total phosphorus, nitrate, phosphate) not measured in the Interlandi and Crockett (2003) study are needed to construct an accurate mass balance.

Americans have mixed opinions about sprawl. A poll conducted in 1998 for the National Trust for Historic Preservation (Coffin and Elder, 2005) found that many Americans associated the word sprawl with "traffic, congestion, concrete, pollution, ugly subdivisions, and strip malls." To others, sprawl suggested "a way out of current congestion, the ability to spread out, to move to a quiet, peaceful area, and economic growth." Allowing this type of dispersed development was viewed as giving people choices about where to live and providing affordable housing and good schools for families with young children. Many Americans also wanted to have larger backyards and the opportunity to be close to nature. Robert Bruegmann (2005), a professor in the departments of art history and architecture and urban planning at the University of Illinois–Chicago, contends that sprawl as a development pattern is neither recent, nor uniquely American. He describes sprawl as a consequence of economic development that "afforded many people greater levels of mobility, privacy, and choice than they were able to obtain in the densely settled large cities that were the norm through the end of the nineteenth century." His contention that "at low densities, it would be possible to trap and treat almost all of the water onsite or nearby and do without centralized wastewater facilities altogether" is diametrically opposed to the research results summarized in this volume.

In contrast, David Owen, a staff writer for *The New Yorker*, argues that concentrated development protects the environment—*an assertion with which we agree*:

> Because densely populated urban centers concentrate human activity, we think of them as pollution crisis zones. Calculated by the square foot, New York City generates more greenhouse gases, uses more energy, and produces more solid waste than most other American regions of comparable size. On a map depicting negative environmental impacts in relation to surface area, therefore, Manhattan would look like an intense hot spot, surrounded at varying distances by belts of deepening green.

> If you plotted the same negative impacts by resident or by household, however, the color scheme would be reversed. My little town (in Connecticut) has about four thousand residents spread over 38.7 thickly wooded square miles, and there are many places within our town limits from which no sign of settlement is visible in any direction. But if you moved eight million people like us, along with our dwellings and possessions and current rates of energy use, into a space the size of New York City, our profligacy would be impossible to miss, because you'd have to stack our houses and cars and garages and lawn tractors and swimming pools and septic tanks higher than sky scrapers. (Conversely, if you made all eight million New Yorkers live at the density of my town, they would require a space equivalent to the land area of the six New England states plus Delaware and New Jersey.) Spreading people out increases the damage they do to the environment, while making the problems harder to see and to address (Owen, 2004).

10.3.4 CONSERVATION AND RESTORATION

It is not possible for all the streams, rivers, lakes, and estuaries in the northeastern United States to be restored to the condition discovered by European settlers in the 1600s. The constraints imposed by land use in agricultural, urban, and mixed use watersheds and a growing population are too pervasive. The effects of past land use and atmospheric deposition also limit recovery potential, even in forested areas (Driscoll et al., 2003a; Harding et al., 1998; Likens et al., 1996). Hence the goal of management, conservation, and restoration activities should be to achieve the best possible conditions given the practical limits imposed by a human-dominated landscape. An understanding of these limits also helps to direct efforts and funding to those sites where they will be most effective and yield the greatest marginal benefit (Booth et al., 2004; Hess and Johnson, 2001).

It is reasonable to predict that streams in the best condition are found in forests that are protected from development or sustainably managed. Figure 10.3 depicts the range of hydrologic effects (per unit area) of the conversion of forests to agricultural and open land, and urban and suburban development. The hydrologic reference condition in the diagram is deliberately shown as a broad gray area. This is because there is no single or universal description of baseline water quality, but rather a range of acceptable attributes that vary across ecological regions. The flow regime and water quality constituents used to define the reference condition for a forested watershed in the Adirondack Mountains would be substantially different from sites in coastal New England, the Ohio River Valley, or northern Wisconsin. The reference conditions for each site reflect differences in topography, geology, soils, climate, and forest type.

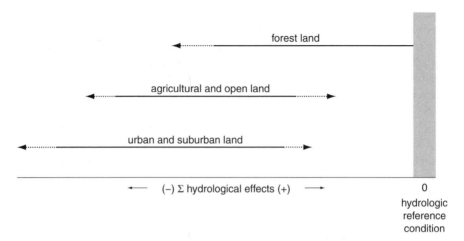

FIGURE 10.3 The range of hydrologic effects (per unit area) of the conversion of forests to agricultural and open land, and urban and suburban development.

Each horizontal bar (forest, agricultural, and urban) represents a range of stream conditions that may occur, depending on land management practices within the watershed. Moving toward the reference condition (vigorous and diverse forest) is a positive change; moving away from the reference condition is a negative change because it increases streamflow instability and pollutant loading. Thus small-scale forestry operations (removing less than 20% of the forest canopy), using BMPs to protect riparian areas and mitigate the effects of roads, keep streams on forested land close to the reference condition. The water balance and associated biogeochemical cycling is restored to the reference condition as the forest regrows. As the scale and intensity of timber harvesting increases, the departure from the reference condition also increases, along with the time required for recovery. Similarly, in agricultural areas where BMPs are diligently applied, nutrient and pesticide inputs are minimized, integrated pest management is effective, wetlands and riparian areas are protected or restored, and forests are conserved, the departure from the reference condition is minimized. In contrast, as the scale and intensity of animal and row crop agriculture increases—especially if agricultural chemicals are applied in excess and riparian areas and wetlands are not effectively protected—the watershed and stream condition will steadily decline.

In each of these land use scenarios, it is the cumulative effect of many different land use and management decisions that determines the magnitude and direction (either positive or negative) of movement along the range indicated by each arrow. The conservation of forests, wetlands, and riparian areas is vital to minimizing the departure from the reference condition. Pollution mitigation and ecosystem restoration (upland, riparian, or stream channel) are needed to move any undesirable existing condition back to the right.

Box 10.2   The Bronx River Restoration

The Bronx River originates in wetlands located below the Kensico Reservoir in upper Westchester County, New York. From there, it flows south, a total distance of 38.6 km, through White Plains, Yonkers, and the Bronx, where it joins the East River, not far from Manhattan (Bitner, 1980). In 1998, land use in the 87 km² Bronx River watershed was estimated to be 54% urban, 25% forest, and 21% cropland or open land (e.g., golf courses and cemeteries) (Olson, 1998). Programs to restore the river are notable because of (1) the involvement of a wide range of participants from the community, government agencies, and educational institutions; (2) the long history of public concern about the condition of the river; and (3) the commitment to ecological monitoring of the river and riparian areas.

Urban development in the Bronx River watershed began in the early 19th century. The New York Central Railroad system built tracks along the river channel in 1844. The land near the tracks and the river became a dumping ground for nearby towns. In the 1890s, state officials were concerned about the "noxious, odiferous, and unsanitary conditions in the river" (Olson, 1998). The river was described as an "open sewer" and the lakes in Bronx Park were declared to be a "positive menace to public health." Restoration efforts began with the establishment of the 268 ha Bronx Park in 1888. This park, which totaled 291 ha in 2001, extends 24.9 km, from the river headwaters to the Bronx Zoo. Thus, while passing through one of the more densely populated areas in the Northeast, the Bronx River is protected by a continuous riparian buffer for nearly two-thirds of its length. Below the park, the river flows through dense urban areas, dissected by highways (Bronx River Alliance, 2006).

The Bronx River Restoration Project, Inc. was formed in 1974, a period of community deterioration in the South Bronx. The project succeeded in removing much of the trash and debris from the riverbanks, including "refrigerators, tires, and even a wine press." The Bronx Riverkeeper program was established in 1996. The Bronx River Working Group (established in 1997) brought in many groups from the community, government agencies, schools, and businesses to aid in the restoration effort. In 2001 the Bronx River Alliance was formed to build on these previous efforts. Current participants include the Wildlife Conservation Society, National Oceanic and Atmospheric Administration (NOAA), National Park Service, New York Botanical Garden, New York State Department of Environmental Conservation, U.S. Army Corps of Engineers, Bronx Zoo, Lehman College, and Youth Ministries for Peace and Justice, and many others (approximately 70 organizations are now involved). In 2005 Congressman José Serrano obtained $12 million for restoration projects from NOAA's community-based restoration program. These funds will be directed toward restoration of salt marshes, tidal flats, floodplains, and river banks, and public access and education in the South Bronx, the most degraded section of the river (http://www.wcs.org/sw-wcs_in_new _york/communityaffairs/bronxriver; accessed July 2006).

The alliance supports a wide variety of programs designed to achieve restoration goals. Community volunteers have removed litter, trash, and invasive species and have worked to stabilize the riverbanks with native vegetation. The Bronx River Conservation Crew conducts a full-length reconnaissance of the river once a week to detect problems and ensure safe public access. The Alliance Education Program works with teachers in local schools, developing educational materials based on the river and its watershed. The education program also sponsors walking tours, canoe trips, and educational presentations. The alliance works to ensure that construction within the park follows urban BMPs to "minimize erosion, buffer sensitive natural areas, capture water onsite, and maximize open space." The alliance also coordinates scientific studies conducted by participating groups. These include a watershed soil survey (USDA), a pesticide discharge study, and a study of animal waste discharge associated with the Bronx Zoo. Extensive species lists have been developed for vegetation, birds, amphibians, and mammals that reside in riparian areas, as well as for fish and macroinvertebrates.

A baseline biological assessment of the Bronx River within the Bronx Park was completed in 1999. This included fish studies, macroinvertebrate studies, and an assessment of discharges to the river and its tributaries (Bode et al., 1999). Results indicated that water quality was good in the headwaters area, but declined to fair at the next sampling station, approximately 7 miles downstream in White Plains. The decrease in water quality was associated with organic wastes (sewage) and municipal and industrial discharges upstream of the White Plains stations. As a result, Bode et al. recommended that the identification and remediation of discharges upstream of White Plains be given the highest priority. Water quality improved slightly at sampling stations downstream of White Plains. Future biological sampling will assess the effectiveness of the many restoration efforts now being implemented.

10.3.4.1 Stream Corridor Restoration

Successful stream restoration should result in dynamic (variable) streams with improved ecological condition. Ecological conditions include the biological, hydrological, and geomorphic aspects of streams channels and riparian areas (Palmer et al., 2005). In addition to improving local conditions, stream restoration is one of the primary ways to improve and protect the condition of larger receiving waters such as the Great Lakes, East Coast estuaries, and the Gulf of Mexico.

There is increasing recognition and consensus that successful stream restoration means creating stable channels in which the processes of erosion and sedimentation are in balance. This requires the restoration of more natural flow regimes and channel migration (meandering). Within the stream corridor and riparian areas, this implies the restoration of riparian and floodplain processes to the greatest extent possible (see Chapter 5, Figure 5.1) (Poff et al., 1997; Woltemade, 1994). While replanting riparian vegetation is vital to any stream restoration plan, it is seldom the only action required to achieve measurable and lasting changes in stream condition (Booth et al., 2004; Petersen, 1999; Schultz et al., 2004).

This is an exciting time in the field of ecological stream restoration. There is increased scientific understanding and public awareness of stream and floodplain characteristics and billions of dollars are being spent on restoration projects throughout the United States (Bernhardt et al., 2005; Malakoff, 2004; Palmer et al., 2005). At the same time, ecological stream restoration is a relatively new area of research and investigation. Scientists and managers are just beginning to develop evaluation techniques and acquire examples of both success and failure that can produce well-tested and practical recommendations for managers.

Many recent publications address the topic of stream restoration. *Stream Corridor Restoration: Principles, Processes, and Practices* (Federal Interagency Stream Restoration Working Group [FIS-RWG], 1998) provides a wealth of detailed information about the scientific principles and processes along with practical guidelines for organizing, designing, and implementing stream restoration, monitoring, and evaluation. It is the result of a collaborative effort among scientists from 15 federal agencies and partners, including the U.S. Department of Agriculture (Agricultural Research Service, Cooperative Extension, Forest Service, and Natural Resources Conservation Service), EPA, Department of Commerce, Army Corps of Engineers, and Department of the Interior. The World Wildlife Fund coordinated input and review from nongovernmental organizations.

Stream Corridor Restoration describes steps involved in restoration projects, including initial community organization, analysis of stream corridor condition (including hydrologic and geomorphic processes and chemical and biological characteristics), development of a restoration design, and implementation, monitoring, and management. Other integrative sources include the proceedings of a symposium on stream protection and restoration published by the American Society of Civil Engineers (ASCE) (2003). Journal articles are numerous (Asbridge, 2004; Booth, 2005; Booth et al., 2004; Gillilan et al., 2005; Palmer et al., 2005; Ward et al., 2001).

In addition to comprehensive manuals, such as the FISRWG (1998) publication cited above, a large group of scientists involved in stream and river restoration are working together to develop standard criteria for evaluating the success or failure of restoration projects (Bernhardt et al., 2005; Palmer et al., 2005). They have also established an online database, the National River Restoration Science Synthesis (NRRSS) database (containing information on nearly 40,000 projects). An additional resource for coastal areas is the National Oceanic and Atmospheric Administration (NOAA) Office of Response and Restoration website, which provides links to restoration resources and examples of current restoration projects.

Palmer et al. (2005) have proposed five criteria for evaluating the success of restoration efforts.

1. "The design of an ecological river restoration project should be based on a specified guided image of a more dynamic, healthy river that could exist at the site."
   The image of the restored stream can be derived from historical records or from examination of relatively undisturbed streams of similar size located in a similar topographic

and geologic setting. Physical and biological restoration goals need to account for the limitations that may be imposed by the geologic setting and local geomorphic processes (Booth et al., 2003). They also should reflect what is possible, given the constraints imposed by the current condition of the stream corridor and watershed. In many urbanized settings, ecological stream restoration is unattainable. If buildings, highways, and private lands confine stream corridors or the proportional area of watershed urbanization is very high, true ecological restoration may be functionally impossible or prohibitively expensive (Booth et al., 2004; Hess and Johnson, 2001). Stream rehabilitation under these conditions may still be a worthwhile goal for public amenities.

2. "Ecosystems are improved: the ecological conditions of the river are measurably enhanced."

   This includes improvement of habitat conditions through the stream corridor, a more natural streamflow and sediment delivery regime, hydraulic connections to riparian wetlands, and improved water quality. These changes should increase the diversity and abundance of aquatic biota (plants and animals).

3. "Resilience is increased: the river ecosystem is more self-sustaining than prior to restoration."

   In undisturbed settings, river ecosystems are both "self-sustaining and dynamic." Seasonal and interannual variations in rain and snowmelt lead to changes in streamflow, channel scour, sediment distribution and the distribution of woody debris. Some habitat may be temporarily destroyed during storm events, while new habitat is created. Populations may suffer periodic declines, but, in general, biotic abundance and diversity across the ecosystem is stable. Ecosystem resilience is the ability to recover from natural disturbances. Increased resilience in restored streams is derived from more reliance on natural stream processes. Streams with enhanced ecosystem resilience should require less maintenance and repair. For example, mature riparian trees that provide a continuing source of large woody debris preserve the resilience of stream channels to future stormflow events. Woody debris artificially added to stream channels may temporarily create more habitat diversity and stream channel stability, but must be replaced if it is washed away.

4. "No lasting harm is done: implementing the restoration does not inflict irreparable harm."

   The restoration project should not cause the loss of native riparian vegetation or interrupt seasonal biological activity such as fish spawning.

5. "Ecological assessment is completed: some level of both pre- and postproject assessment is conducted and the information made available."

   Ecological stream restoration projects need to be systematically evaluated. Information about these projects should be shared nationally and internationally. If restoration goals are not achieved, it is important to understand why and to learn from mistakes. Gillilan et al. (2005) note that restoration project assessment has frequently been limited to the evaluation of physical attributes of the restored stream. Biological indices are the more demanding measures of success.

10.3.4.2 Watershed Restoration

Beyond the stream corridor, restoring natural streamflow regimes, patterns of sediment delivery, and water quality also must include efforts to (1) increase infiltration, (2) reduce overland flow, soil erosion, and sediment transport, (3) reduce inputs of nonpoint source pollutants, and (4) improve wastewater treatment. This can be achieved by conserving forests and wetlands wherever possible, with consistent use of forestry, agricultural, and urban BMPs, and by siting new development in the most benign areas of the watershed with the consistent use of low impact design practices in new development and construction. Watershed-scale (and regional) actions include identifying sources and reducing inputs of nitrogen and phosphorus, limiting pesticide use, and increasing the care with which we use and dispose of the many potential pollutants in our homes and workplaces.

An example of this type of comprehensive approach is described in the Chesapeake 2000 Agreement. A broad range of improvements in wastewater treatment and nonpoint source pollution mitigation implemented from the 1970s to the present have reduced phosphorus and nitrogen loading, but have yielded only slight improvements in ecological metrics. There have been modest reductions in nutrient concentrations and algal biomass, but no significant improvement in dissolved oxygen concentrations in the Chesapeake Bay. One positive development is the expansion of seagrass beds from a low of approximately 18,000 ha in 1984 to approximately 28,000 ha in 1999. Boesch et al. (2001) summarized the five goals for the program.

1. Linking eutrophication and living resources.
   The 2000 agreement calls for the development of water quality standards that reflect the conditions necessary for a healthy biotic community. Standards (e.g., dissolved oxygen, chlorophyll, and water clarity) will be used to develop nutrient load allocations.
2. Reducing atmospheric deposition.
   The Chesapeake Bay receives about one-fourth of its nitrogen load from atmospheric deposition, both from local and regional (coal-burning power plants in the Ohio Valley) sources.
3. Enhancing nutrient sinks.
   Additional effort is needed to increase the area and capacity of terrestrial nutrient storage and processing sites in addition to protection of wetlands and restoration of riparian forests.
4. Limiting suburban sprawl.
   A proposed 40% reduction in nutrient loading is part of a commitment to maintain caps on nutrient loading in perpetuity. This must be done despite population growth. Of particular concern is the pattern of development. Boesch et al. (2001) note that "the conversion of forested and agricultural lands to development has grown at a rate 2 to 3 times greater than population growth."
5. Predicting and preventing harmful algal blooms.

10.4 BUILDING PUBLIC SUPPORT

Scientific understanding of the effects of human land use on streamflow and water quality has grown exponentially throughout the 20th century, especially the last 5 to 10 years. The task is now to interpret and communicate this information to the public at large and to build support for programs that will restore and protect streams, rivers, lakes, and estuaries in the future. This is already happening in many areas through the noteworthy efforts of schools, museums (Museum of Science, Boston; American Museum of Natural History, New York, Peggy Notebaert Nature Museum, Chicago; and Academy of Natural Sciences, Philadelphia, to name just a few), and conservation organizations. Local watershed associations are now common. These efforts must be supported and enabled to grow and continue. Many of the actions described in this chapter—conserving forests, limiting the area of development, restoring stream corridors—require comprehensive local and regional planning. They also require that individuals within a community work together to protect the natural systems that preserve healthy streams and clean water. In fact, public opinion research has shown that people are most likely to support restrictions on development if they believe that development will destroy wetlands and forests that protect water quality (Coffin and Elder, 2005).

10.5 SUMMARY AND CONCLUSIONS

The goal of this chapter has been to summarize and highlight the key lessons from the scientific and watershed management literature. Three themes emerge.

10.5.1 BIOPHYSICAL PATTERNS AND PROCESSES

Practitioners, policy makers, and the public are often frustrated by the complexity of watersheds and the seeming evasiveness or noncommittal nature of scientists. Miscommunication, such as "100 year floods" that occur in successive years (in fact, a peak discharge that has a 1% chance of being equaled or exceeded in any given year), and the apparent inability or unwillingness of accomplished scientists to specify thresholds or boundary conditions (i.e., $x\%$ impervious surface or $y\%$ forest loss causes water quality degradation, etc.) undermine public confidence in what is known, let alone established with certainty. This information or knowledge gap, exacerbated by the differences in perspective that sometimes separate scientists, policy makers, and the public, is a major impediment to progress in watershed management and restoration.

We believe that the following generalizations are well supported by the literature and, though broad in scope, can guide our efforts to more effectively manage watersheds. They can be dismissed as "academic" or applied in practical and useful ways to solve "real-world" problems.

- Site-specific consideration and characterization of surficial and bedrock geology, topography and terrain features, soils, climate, and hydrologic regime are needed. Variability is the rule, not the exception.

The laws of nature are universal, yet their expression is strongly influenced by watershed characteristics. It stands to reason that the hydrologic regime of a small watershed high in the Adirondack Mountains—with shallow, stony soils derived from glacial till, crystalline bedrock, a forest comprised of red spruce, balsam fir, and heath shrubs, and a severe alpine climate—will be drastically different from a watershed of equal size in the Piedmont of Maryland, the lacustrine soils of the Ohio Valley, or the karst region of southeastern Minnesota. It also follows that the differences can be anticipated and explained in relation to the hydrogeologic setting, topography, soils, and climate of each site.

- Use the mass balance to identify, design, and implement watershed management strategies to avoid, prevent, or, at last resort, mitigate nonpoint source pollution.

The mass balance—the working version of the law of conservation of mass—frequently demonstrates that human activities exceed the assimilative capacity of upland, riparian, or aquatic ecosystems. Essential nutrients become unwanted pollutants when total input exceeds total storage capacity or overwhelms the transformation processes that convert them to relatively immobile or benign forms. When storage capacity is exceeded, the conversion of excess inputs to unwanted outputs becomes inevitable and the train of downstream effects is set in motion. This inescapable law reminds us that watershed management strategies that focus on mitigating unwanted outputs, rather than avoiding or preventing unneeded inputs, are, at best, a delaying action.

- Focus on changes in the pathways of water flow to assess the effects of various land uses or site conditions.

Because water is both the universal solvent and the transport mechanism for nonpoint source pollutants, seemingly minor changes in the pathway and rate of flow can quickly lead to an escalating chain of events. When trees, understory plants, groundcover plants, the litter layer, and the organic (O) soil horizon of a forest are replaced or altered by a new land use, overland flow replaces infiltration and subsurface flow and the proportion of baseflow and stormflow is correspondingly altered. As described in scores of studies that were summarized in the preceding chapters, this change in flow path and rate initiates other ecological changes and structural interventions that can culminate in watershed-scale problems and impacts of daunting proportions.

- Stream channel instability can be caused by reach-scale disturbance or alteration (e.g., encroachment, armoring, etc.) and watershed-scale changes (e.g., reduced infiltration and increased stormflow). Both contribute to the cumulative effects of land use and to the complexity and cost of mitigation and restoration efforts.

In open channel hydraulics, kinematic wave or translatory wave calculation procedures are commonly used to quantify or forecast the routing of flood flows through watersheds (Henderson, 1966). A kinematic or translatory wave is also an apt description of land use effects on streamflow, channel stability, and, consequently, water quality and aquatic biointegrity. When increases in stormflow volume and velocity are also regarded as increases in the hydraulic energy (mass and shear stress, respectively) available to scour and realign stream channels, the typical causes of channel instability are clear. Stream channels, riparian areas, flowing water, and suspended sediment exist in a relatively fragile dynamic equilibrium that is subject to immutable physical laws. The laws of motion (simply paraphrased: 1. Objects at rest tend to stay at rest; objects in motion tend to stay in motion. 2. Force equals mass times acceleration. 3. For every action there is an equal and opposite reaction.) forecast the adverse effects of inappropriate land use and related hydrologic changes on streams and rivers. They also underscore the need for coordinated watershed management and restoration efforts that extend from the headwaters to the outlet.

10.5.2 IDENTIFICATION OF PRIMARY POLLUTANT SOURCES

- Determining the proportional contribution of various pollutant sources to total loading is the central challenge of watershed science and management.

Many of the studies summarized in this and earlier chapters show that a large proportion of the total pollutant load in a watershed often emanates from a relatively small area. In fact, the relation of pollutant load to source area may differ by an order of magnitude (i.e., 50% of the pollutant load originates from less than 5% of the area). Therefore, it is critically important to focus mapping, modeling, and field assessment methods on the detection and effective management of these areas. Variations in the effects of land use at the watershed scale trace back to the wide range of combinations and permutations in location, spatial arrangement, timing, climate (seasonal and interannual), legacy effects, management practices, and many other attributes. This also underscores the importance and potential of spatially distributed mass balance models that are calibrated and verified with long-term monitoring data. In addition, careful consideration of the location and timing of samples is needed to ensure key questions can be answered. As noted in several earlier chapters, event-based sampling on tributary streams often yields more useful information than an equal number of fixed-frequency (e.g., weekly or monthly compliance monitoring) measurements on a transect along the main stem of a river.

- Many pollutants have multiple sources, so assessment and management is unavoidably complex.

It follows that focusing watershed management efforts on one or two sources for a pollutant with many more local and regional sources and potential forms is not likely to be effective at the watershed scale. Nitrogen, for example, may enter water from a wide range of sources (e.g., fertilizer, manure, wastewater, atmospheric deposition, etc.). In this case, riparian area restoration to mitigate agricultural sources, even if completely effective, can be overshadowed by wastewater discharge and atmospheric deposition. In addition, the store of accumulated nitrogen in the watershed may take years or decades to dissipate even if all new inputs are greatly reduced. Centuries of environmental alteration and damage cannot be reversed in a year or two.

- The conversion of forest land to impervious or substantially altered areas replaces a stormwater and pollutant sink with a stormwater and pollutant source and amplifies the ecological effect of the land use change.

This is the two-edged sword of sprawl. Even when total impervious area (TIA) and effective impervious area (EIA) are minimized, population density is maximized, and low impact development practices are scrupulously followed, some impact is inevitable. The relative magnitude and watershed-scale consequences of this impact are then determined by where it occurs and how well the design fits the site. Basic attributes such as slope and aspect, proximity to water or the riparian area, and soil depth and permeability strongly influence the net result of forest conversion.

10.5.3 WATERSHED MANAGEMENT PRINCIPLES AND PRACTICES

- Conserve forests, wetlands, and riparian areas—they are the ecological counterweights to undesirable environmental change.

It is difficult to overstate the intrinsic ecological and hydrological values and benefits of forests, riparian areas, and wetlands. It is, however, relatively easy to overestimate their assimilative capacity when they are battered by inappropriate land use. Forests, wetlands, and riparian areas form the foundation of sustainable watershed management programs. They are the first barrier in a multiple barrier approach to source water protection (Barten and Ernst, 2004).

A century or more ago, when the U.S. population was less than 100 million people, forest conversion or degradation was obvious, abrupt, and often alarming—as were the results (erratic streamflow, sedimentation, contamination of water supplies, extirpation of fish and wildlife, etc.). Today, as the U.S. population passes 300 million people, forest loss is more likely to take place at the scale of a building lot. One this year, two the next, one on the east side of town, two on the west, the change is subtle and seemingly benign. Even if the scale of development is larger (e.g., a residential subdivision, a shopping mall, etc.), there always seems to be more forest. In addition, second- or third-growth forest (the result of farm abandonment or natural regeneration of cutover land in the early 1900s) is still growing in stature.

A recent large-scale, long-term analysis presents a more accurate and sobering view. Stein et al. (2005) linked land cover, land ownership, the 2000 U.S. census, and 2030 development projection data in a GIS analysis of private forest lands across the United States. For the period 1982 to 1997, they documented a net loss of forest land in 26 states in the conterminous U.S. The national rate of forest conversion ranged from approximately 280,000 ha/yr to 400,000 ha/yr (an area the size of Catskill Park in New York). What will be the ultimate ecological and economic cost of this transformation of the American landscape? How would people and nongovernmental organizations respond to a proposal to convert in one year the entire 162,000 ha Green Mountain National Forest (Vermont) to a four-season resort and retirement community for about 100,000 people? Recall that this magnificent forest, and virtually all of the national forest land east of the Mississippi, was purchased in the early 1900s as a result of the Weeks Act of 1911 in order to protect the headwaters of navigable streams. What is the fate of the middle and lower reaches of navigable streams a century later?

- Restore wetlands and riparian areas that have been degraded by earlier land and resource use whenever possible.

Wetlands and riparian areas occupy key landscape positions and thus provide opportunities for restoration of the hydrologic regime and the amelioration of water quality. At the same time, wetland and riparian restoration at the reach scale cannot overcome the effects of reduced infiltration and pollutant loading beyond the range of natural variation at the watershed scale. Hence reducing stormwater inflows and pollutant loads to a range that is compatible with the wetland or riparian ecosystem is the necessary precursor to successful and sustained restoration of ecological function.

- Reduce the input of potential pollutants.

A recurrent theme in the studies summarized in this book is the unavoidable need to reduce inputs or loading of nitrogen, phosphorus, and other ubiquitous pollutants. The notion that BMPs, created wetlands, and other after-the-fact approaches will alone be sufficient is not supported by science or common sense. There are many highly successful examples of point source pollution management, such as the pioneering 3P (Pollution Prevention Pays) program of the Minnesota Mining and Manufacturing (3M) Corporation. Once pollution is regarded as a (1) direct cost (not an externality); (2) measure of inefficiency; (3) waste of labor, materials, and energy; (4) health risk; and (5) legal liability, progressive managers immediately find ways to limit the financial damage to the balance sheet and the future of the firm. As Dr. Joseph Ling, the founder of the 3P program in 1975 stated: "Pollution is waste, and waste today leads to shortages tomorrow." In 30 years, the program has reduced 3M's pollutant emissions by 1 billion kg and saved at least $810 million (http://www.3m.com/; accessed July 2006). Similarly, whole-farm planning and precision farming focuses on managing soil erosion, soil fertility, and the use of pesticides and fertilizers in a comprehensive manner. Inputs contribute to crop yields, not pollutant outputs. Optimizing production efficiency maximizes profits and environmental protection. Nonpoint source pollution prevention pays.

- Diligently implement BMPs on forest, agricultural, and developed (urban and suburban) land.

While the form and effectiveness of BMPs varies widely, they bear the name for a reason. Research and development work, operational experience, new materials and technologies, and performance monitoring can lead to improvements, but at any given time they are the *best* management practices at our disposal. When BMPs are diligently implemented and maintained, they can help to reverse the sign (from negative to positive) of the cumulative effects of land use on streamflow and water quality. In many cases, a simple and inexpensive action can have far-reaching and valuable effects. Consider, for example, fencing livestock out of riparian areas and the sediment, nutrient, and pathogen loading that is avoided. Also note that the ability of the protected stream reach (adjacent to the pasture) to assimilate pollutants from upstream sources is not forfeited by localized damage.

- Provide socially and economically viable alternatives to sprawl; namely, reinvest in urban areas to encourage high density use and energy conservation.

Sprawl and forest conversion have been accepted by some as inevitable or at least as the price of progress. Acquiescence to earlier forms of environmental degradation was refuted by the conservation movement in the first half of the 20th century and the environmental movement in the second half. At the beginning of the 21st century, innovative thinkers like David Owen (2005) and our respected colleague, Rutherford Platt (2004a,b)—building on the work of William H. Whyte (1917 to 1999) and others—provide a new vision and viable alternatives to the unsustainable status quo.

In *Rivers for Life: Managing Water for People and Nature*, Postel and Richter (2003) present a compelling vision for the 21st century. Coupling this approach with the ecological cities paradigm, a renewed commitment to forest and wetland conservation, and whole-farm planning could substantially improve the effectiveness of watershed management efforts and give new meaning to the term "sustainable development." As Professor Otis Graham (2000) puts it: "Fifty or a hundred years ago, when a now forgotten word was viable, this would have been called Planning—for a different and better future than the stressful one dead ahead."

- Employ low impact development principles and practices for all new construction and urban redevelopment.

Increasing public awareness of the long-term costs and consequences of conventional development and stormwater management is increasing the demand for and interest in alternative approaches to residential, commercial, and industrial design. The ecological and financial success of many prototype projects is very encouraging.

- Use systematic monitoring to guide implementation and foster continuous improvement.

This familiar refrain is included in the recommendations of every publication about watershed or environmental management. It also is the first item to be cut, or at least to be rendered ineffective, when the project budget is limited (as always). This well-intentioned and seemingly rational diversion of monitoring resources into planning, BMPs, and treatment technologies breaks the adaptive management loop. It consigns learning and continuous improvement efforts to anecdotal information and subjective impressions of performance and cost. Watershed management programs that claim to employ the adaptive management paradigm must include a well-designed, sustained monitoring component or refrain from self-congratulatory rhetoric.

- Extend awareness of and participation in watershed management as broadly as possible through education, outreach, and demonstration projects.

In the last chapter of his seminal biography of George Perkins Marsh, Professor David Lowenthal (2000) describes how we can benefit from Marsh's insights and sterling example 140 years after the publication of *Man and Nature*. He discusses (1) the inevitability of impact, (2) the primacy of the unexpected, (3) the singularity of human intention, (4) the necessity of stewardship, and (5) the primacy of the amateur (meaning a person who is not an expert or specialist). As Aldo Leopold said in 1935 (Meine and Knight, 1999:162): "Relegating conservation to government is like relegating virtue to the Sabbath. [It] Turns over to professionals what should be the daily work of amateurs."

The primacy of the amateur is the reason why we wrote this book. We hope this volume will provide the integration and synthesis of the primary literature needed to advance the work of people with an abiding commitment to the conservation and stewardship of forests and water.

REFERENCES

American Society of Civil Engineers, *Proceedings of the Protection and Restoration of Urban and Rural Streams Symposium*, World Water and Environmental Resources Congress, Philadelphia, PA, June 24–26, 2003, American Society of Civil Engineers, Reston, VA, 2003.

Asbridge, G.M., Using large woody debris to help restore stream channels and floodplains, in *16th International Conference, Society for Ecological Restoration, August 24–26, 2004, Victoria, Canada*, Society for Ecological Restoration, Tucson, AZ, 2004.

Associated Press, Suburbs grow while big cities shrink, June 20, 2006.

Barten, P.K. and Ernst, C.E., Land conservation and watershed management for source protection, *J. Am. Water Works Assoc.*, 96(4), 121–135, 2004.

Bernhardt, E.S., Palmer, M.A., Allan, J.D., Alexander, G., Barnas, K., Brooks, S., Carr, J., Clayton, S., Dahm, C., Follstad-Shah, J., Galat, D., Gloss, S., Goodwin, P., Hart, D., Hassett, B., Jenkinson, R., Katz, S., Kondolf, G.M., Lake, P.S., Lave, R., Meyer. J.L., O'Donnell, T.K., Pagano, L., Powell, B., and Sudduth, E., Synthesizing U.S. river restoration efforts, *Science*, 308, 636–637, 2005.

Binkley, D., Patterns and processes of variation in nitrogen and phosphorus concentrations in forested streams, Technical Bulletin 836, National Council for Air and Stream Improvement, Research Triangle Park, NC, 2001.

Bitner, R., The diversity of the benthic macroinvertebrates of the lower Bronx River, submitted in partial fulfillment of the requirements for the degree of Bachelor of Arts, SUNY Purchase, New York, 1980.

Bode, R.W., Novak, M.A., Abele, L.E., and Carlson, D., Biological stream assessment: Bronx River, Bronx and Westchester Counties, New York, Stream Biomonitoring Unit, Bureau of Watershed Assessment and Research, Division of Water, New York State Department of Environmental Conservation, Albany, New York. 1999.

Boesch, D.F., Brinsfeld, R.B., and Magnien, R.E., Chesapeake Bay eutrophication: scientific understanding, ecosystem restoration, and challenges for agriculture, *J. Environ. Qual.*, 30, 303–320, 2001.

Booth, D.B., Challenges and prospects for restoring urban streams: a perspective from the Pacific Northwest of North America, *J. North Am. Benthol. Soc.*, 24, 724–737, 2005.

Booth, D.B. and Jackson, C.R., Urbanization of aquatic systems: degradation thresholds, stormwater detection, and the limits of mitigation, *J. Am. Water Resourc. Assoc.*, 33, 1077–1090, 1997.

Booth, D.B., Hartley, D., and Jackson, R., Forest cover, impervious surface area, and the mitigation of stormwater impacts, *J. Am. Water Resourc. Assoc.*, 38, 835–846, 2002.

Booth, D.B., Haugerud, R.A., and Troost, K.G., The geology of Puget lowland rivers, in *The Restoration of Puget Sound Rivers*, Montgomery, D.R., Bolton, S., Booth, D.B., and Wall, L. (eds.), University of Washington Press, Seattle, 2003, pp. 14–45.

Booth, D.B., Karr, J.R., Schauman, S., Konrad, C.P., Morley, S.A., Larson, M.G., and Burges, S.J., Reviving urban streams, land use, hydrology, biology, and human behavior, *J. Am. Water Resourc. Assoc.*, 40, 1351–1364, 2004.

Brabec, E., Schulte, S., and Richards, P.L., Impervious surfaces and water quality: a review of current literature and its implications for watershed planning, *J. Plan. Lit.*, 16, 499–514, 2002.

Bronx River Alliance, http://www.bronxriver.org/; accessed July 2006.

Brookings Institution, Philadelphia in focus: a profile from census 2000, 2003, http://www.brookings.edu/es/urban/livingcities/philadelphia.htm; accessed July 2006.

Bruegmann, R., *Sprawl*, University of Chicago Press, Chicago, 2005.

Calhoun, F.G., Baker, D.B., and Slater, B.K., Soils, water quality, and watershed size: interactions in the Maumee and Sandusky river basins of northwestern Ohio, *J. Environ. Qual.*, 31, 47–53, 2002.

Carpenter, S.R., Caraco, N.F., Correll, D.L., Howarth, R.W., Sharpley, A.N., and Smith, V.H., Nonpoint pollution of surface waters with phosphorus and nitrogen, *Ecol. Applic.*, 8, 559–568, 1998.

Coffin, C. and Elder, J., Building public awareness about the effects of sprawl on biodiversity, in *Nature in Fragments: The Legacy of Sprawl*, Johnson, E.A. and Klemens, M.W. (eds.), Columbia University Press, New York, 2005, pp. 335–348.

Driscoll, C.T., Driscoll, K.M., Roy, K.M., and Mitchell, M.J., Chemical response of lakes in the Adirondack region of New York to declines in acid deposition, *Environ. Sci. Technol.*, 37, 2036–2042, 2003a.

Driscoll, C.T., Whitall, D., Aber, J.D., Boyer, E.W., Cronan, C.S., Goodale, C.L., Groffman, P., Hopkinson, C., Lambert, K.F., and Lawrence, G.B., Nitrogen pollution in the northeastern United States: sources, effects, and management options, *BioScience*, 53, 357–375, 2003b.

Federal Interagency Stream Restoration Working Group (FISRWG), Stream corridor restoration: principles, processes, and practices, Publication 0120-A, SuDocs A 57.6/2:EN 3/PT.653, U.S. Government Printing Office, Washington, D.C., 1998.

Gillilan, S., Boyd, K., Hoitsma, T., and Kauffman, M., Challenges in developing and implementing ecological standards for geomorphic river restoration projects: a practitioner's response to Palmer et al. (2005), *J. Appl. Ecol.*, 42, 223–227, 2005.

Graham, O.L., Jr., Epilogue: A look ahead. Special Issue: Environmental Politics and Policy, 1960s–1990s, *J. Policy Hist.*, 12, 157–176, 2000.

Harding, J.S., Benfield, E.F., Bolstad, P.V., Helfman, G.S., and Jones, E.B.D., III, Stream biodiversity: the ghost of land use past, *Proc. Natl. Acad. Sci. USA*, 95, 14843–14847, 1998.

Hatt, B.E., Fletcher, T.D., Walsh, C.J., and Taylor, S.L., The influence of urban density and drainage infrastructure on the concentrations and loads of pollutants in small streams, *Environ. Manage.*, 34, 112–124, 2004.

Henderson, F.M., *Open Channel Flow*, Macmillan, New York, 1966.

Hess, A.J. and Johnson, P.A., A systematic analysis of the constraints to urban stream enhancements, *J. Am. Water Resourc. Assoc.*, 37, 213–221, 2001.

Hession, W.C., Charles, D.F., Hart, D.D., Horwitz, R.J., Johnson, T.E., Kreeger, D.B., Marshall, B., Velinsky, D.J., Newbold, J.D., and Pizzuto, J.E., Final report. Riparian reforestation in an urbanizing watershed: effects of upland conditions on instream ecological benefits, EPA Grant R825798, U.S. Environmental Protection Agency, Washington, D.C., 2003, http://cfpub.epa.gov/ncer_abstracts/index.cfm/fuseaction/display.abstractDetail/abstract/182/report/F; accessed July 2006.

Inamdar, S.P., Mostaghimi, S., McClellan, P.W., and Brannan, K.M., BMP impacts on sediment and nutrient yields in the Coastal Plain region, *Trans. Am. Soc. Agric. Eng.*, 44, 1191–1200, 2001.

Interlandi, S.J. and Crockett, C.S., Recent water quality trends in the Schuylkill River, Pennsylvania, USA: a preliminary assessment of the relative influences of climate, river discharge and suburban development, *Water Res.*, 37, 1737–1748, 2003.

Johnson, E.A. and Klemens, M.W. (eds), *Nature in Fragments: The Legacy of Sprawl*, Columbia University Press, New York, 2005.

Jones, E.B.D., III, Helfman, G.S., Harper, J.O., and Bolstad, P.V., Effects of riparian forest removal on fish assemblages in southern Appalachian streams, *Conserv. Biol.*, 13, 1454–1465, 1999.

Kogelmann, W.J., Lin, H.S., Bryant, R.B., Beegle, D.B., Wolf, A.M., and Petersen, G.W., A statewide assessment of the impacts of phosphorus-index implementation in Pennsylvania, *J. Soil Water Conserv.*, 59, 9–18, 2004.

Lemunyon, J.L. and Gilbert, R.G., The concept and need for a phosphorus assessment tool, *J. Prod. Agric.*, 6, 483–496, 1993.

Likens, G.E., Driscoll, C.T., and Buso, D.C., Long-term effects of acid rain: response and recovery of a forested ecosystem, *Science*, 272, 244–246, 1996.

Lowenthal, D., *George Perkins Marsh: Prophet of Conservation*, University of Washington Press, Seattle, 2000.

Lucy, W.H. and Phillips, D.L., Suburbs: patterns of growth and decline, in *Redefining Urban and Suburban America*, Katz, B. and Lang, R.E. (eds.), The Brookings Institution, Washington, D.C., 2003, pp. 117–136.

Malakoff, D., Profile: David Rosgen, the river doctor, *Science*, 305, 937–939, 2004.

Meine, C. and Knight, R.L. (eds.), *The Essential Aldo Leopold: Quotations and Commentaries*, University of Wisconsin Press, Madison, 1999.

Morley, S.A. and Karr, J.R., Assessing and restoring the health of urban streams in the Puget Sound basin, *Conserv. Biol.*, 16, 1498–1509, 2002.

National Oceanic and Atmospheric Administration, Office of Response and Restoration, http://response.restoration.noaa.gov/topic_catalog.php?RECORD_KEY%28topics_chosen%29=topic_id&topic_id(topics_chosen)=4; accessed July 2006.

National River Restoration Science Synthesis, http://www.restoringrivers.org; accessed July 2006.

Odum, E.P., Great ideas in ecology for the 1990s, *BioScience*, 42, 542–545, 1992.

Olson, C., Results of a benthic macroinvertebrate study conducted at four sites on the Bronx River on September 20–21, 1997, New York Department of Environmental Protection, Albany, NY, 1998.

Owen, D., Green Manhattan, everywhere should be more like New York, *New Yorker*, October 18, 2004, pp. 111–123.

Owen, D., Green Manhattan: why New York is the greenest city in the U.S., 2005, http://www.greenbelt.org/downloads/resources/newswire_11_04GreenManhattan.pdf; accessed July 2006.

Palmer, M.A., Bernhardt, E.S., Allan, J.D., Lake, P.S., Alexander, G., Brooks, S., Carr, J., Clayton, S., Dahm, C.N., Follstad Shah, J., Galat, D.L., Loss, S.G., Goodwin, P., Hart, D.D., Hassett, B., Jenkinson, R., Kondolf, G.M., Lave, R., Meyer, J.L., O'Donnell, T.K., Pagano, L., and Sudduth, E., Standards for ecologically successful river restoration, *J. Appl. Ecol.*, 42, 208–217, 2005.

Petersen, M.M., A natural approach to watershed planning, restoration, and management, *Water Sci. Technol.*, 39, 347–352, 1999.

Platt, R.H., *Land Use and Society: Geography, Law, and Public Policy*, Island Press, Washington, D.C., 2004a.

Platt, R.H., Toward ecological cities: adapting to the 21st century metropolis, *Environment*, 46(5), 10–27, 2004b, http://www.umass.edu/ecologicalcities/; accessed July 2006.

Poff, N.L., Allan, J.D., Bain, M.B., Karr, J.R., Prestegaard, K.L., Richter, B.D., Sparks, R.E., and Stromberg, J.C., The natural flow regime: a paradigm for river conservation and restoration, *BioScience*, 47, 769–784, 1997.

Postel, S. and Richter, B., *Rivers for Life: Managing Water for People and Nature*, Island Press, Washington, D.C., 2003.

Rowden, R.D., Liu, H., and Libra, R.D., Results from the Big Spring water quality monitoring and demonstration projects, *Hydrogeol. J.*, 9, 487–497, 2000.

Schueler, T., The importance of imperviousness, *Watershed Protect. Techniques*, 1, 100–111, 1994.

Schueler, T. and Claytor, R., Impervious cover as a urban stream indicator and a watershed management tool, in *Effects of Watershed Development and Management on Aquatic Ecosystems*, Roesner, L.A. (ed.), American Society of Civil Engineers, Reston, VA, 1996, pp. 513–531.

Schultz, R.C., Isenhart, T.M., Simpkins, W.W., and Colletti, J.P., Riparian forest buffers in agroecosystems—lessons learned from the Bear Creek watershed, central Iowa, USA, *Agrofor. Syst.*, 61, 35–50, 2004.

Schuylkill Watershed Conservation Plan, 2001, http://www.schuylkillplan.org/plan.html; accessed July 2006.

Sharpley, A.N., Soil mixing to decrease surface stratification of phosphorus in manured soils, *J. Environ. Qual.*, 32, 1375–1384, 2003.

Sharpley, A.N. and Beegle, D., Managing phosphorus for agriculture and the environment, Publication CAT UC162, Pennsylvania State University, College of Agricultural Sciences Research and Cooperative Extension, University Park, PA, 2001.

Sharpley, A.N., Weld, J.L., Beegle, D.B., Kleinman, P.J.A., Gburek, W.J., Moore, P.A., Jr., and Mullins, G., Development of phosphorus indices for nutrient management planning strategies in the United States, *J. Soil Water Conserv.*, 58(3), 137–153, 2003.

Sliva, L. and Williams, D.D., Buffer zone versus whole catchment approaches to studying land use impact on river water quality, *Water Res.*, 35, 3462–3472, 2001.

Sohmer, R.R. and Lang, R.E., Downtown rebound, in *Redefining Urban and Suburban America*, Katz, B. and Lang, R.E. (eds.), The Brookings Institution, Washington, D.C., 2003, pp. 63–74.

Soranno, P.A., Hubler, S.L., Carpenter, S.R., and Lathrop, R.C., Phosphorus loads to surface waters: a simple model to account for spatial pattern of land use, *Ecol. Applic.*, 6, 865–878, 1996.

Stein, S.M., McRoberts, R.E., Alig, R.J., Nelson, M.D., Theobald, D.M., Eley, M., Dechter, M., and Carr, M., Forests on the edge: housing development in America's private forests, PNW-GTR-636, USDA Forest Service, Washington, D.C., 2005, http://www.fs.fed.us/projects/fote/reports/fote-6-9-05.pdf; accessed July 2006.

Urstadt, B., Smart thinking, *Boston Globe Magazine*, February, 12, 2006, pp. 54–61.

U.S. Census Bureau, Population estimates program, 2006, http://www.census.gov/popest/estimates.php; accessed July 2006.

Walsh, C.J., Fletcher, T.D., and Ladson, A.R., Stream restoration in urban catchments through redesigning stormwater systems: looking to the catchment to save the stream, *J. North Am. Benthol. Soc.*, 24, 690–705, 2005a.

Walsh, C.J., Roy, A.H., Feminella, J.W., Cottingham, P.D., Groffman, P.M., and Morgan, R.P., II, The urban stream syndrome: current knowledge and the search for a cure, *J. North Am. Benthol. Soc.*, 24, 706–723, 2005b.

Ward, J.V., Tockner, K., Uehlinger, U., and Malard, F., Understanding natural patterns and processes in river corridors as the basis for effective river restoration, *Regul. Rivers Res. Manage.*, 17, 311–323, 2001.

Woltemade, C.J., Form and process: fluvial geomorphology and flood-flow interaction, Grant River, Wisconsin, *Ann. Assoc. Am. Geogr.*, 84, 462–479, 1994.

Index

A

Acid precipitation, 52, 64–67, 75, 276
Acidity, *See* pH
Adirondack Long-Term Monitoring Program, 66
Adirondack Mountains, 21, 56–57, 66, 131, 140, 293, 299
Adopt-a-Stream, 270
Agricultural best management practices (BMPs), 195, 206
  case studies, 196–201
  effectiveness, 196, 197, 201
  livestock exclusion, 199–200
Agricultural land abandonment, forest regeneration and, 2, 175–176
Agricultural land use, 176–177, *See also* Mixed land use
  aquatic ecosystems and, 195
  biotic community and, 179
  channel form impacts, 187–188
  contour plowing, 182
  cumulative effects, 273–276, 278
  current management issues, 202–205
  drainage systems, 174–175, 273–275, 278, 287
  erosion and sediment deposition, 184–187, *See also* Sediment loading; Soil erosion
  estuary or marsh conversion, 172–173
  forest conversion, 171, 173–175, 205–206
  historical patterns, 171, 173–175
  identifying pollutant sources, 286–288
  legacy effects, 276, 278
  nutrient management approaches, 204–205, *See also* Nutrients
  pesticides and, *See* Pesticides
  regional variation, 258
  soil compaction effects, 23–24, 179, 287
  streamflow impacts, 178–183
  tillage practices, *See* Tillage effects
  water quality effects, 188–191, 206
  water temperature effects, 188
watershed management challenges, 206–207
  whole farm planning, 203, 207
Agricultural research, 206
  Agricultural Research Service, 177–178
  Natural Resources Conservation Service, 177
  NOAA, 178
A horizon, 23
Alachlor, 61, 119, 191, 193–194, 229
Algal growth, 68, 71, 72, 75, 196
Alkalinity, 47
Allochthonous material, 84
Aluminum, 50, 64–65, 66–67
Ammonium, 47, 50–51, 64
Animal wastes, 171, *See also* Manure; Microbial contamination
  agricultural BMPs and, 198–199

Oconee watershed analysis, 269
  pathogens, 194
  riparian buffers and, 120
Anoxic/hypoxic conditions, 72–73, 94
Antibiotics, 235
AP horizon, 179
Appalachian geology, 3, 5
Aquatic biota, 83–84, 100, *See also* Benthic macroinvertebrates; Fish
  agricultural inputs and, 195
  biological monitoring, 94–100
  carbon dynamics and community composition, 86–88
  cumulative effects of human activities, 275–276, 278
  eutrophication effects, 70–71
  habitats, *See* Stream habitat
  logging effects, 154–155
  pesticide toxicity, 192
  salt tolerance, 231
  streamflow variability and, 89–91
  temperature effects, 92–93
  trophic classifications, 85–86
  urban development effects, 235–240
  urban water quality and, 228
Aquatic food webs, 84–88, 100, 108
Aquatic macrophytes, 83
Aroclor, 235
Arsenic, 63
Atmospheric deposition, 64–67, 75, 285, *See also specific pollutants*
  Clean Air Act, 65
  comprehensive watershed restoration approach, 298
  cumulative effects, 276
  metals, 64, 233
  nitrates, 11, 64–65, 75, 227
  PAHs, 234
  pesticide residues, 60
  policy and management issues, 283
  reducing multiple pollutant sources, 288–289
  water quality trends, 65–67
  wind patterns and, 8
Atrazine, 60, 61, 119, 191–194, 197–198, 203, 229, *See also* Herbicides
Aufwuchs, 83
Autochthonous material, 84
Autotrophs, 83

B

B horizon, 23
Baltimore Ecosystem Study (BES), 216
Bank erosion, 222, 223, 236, 246, *See also* Stream channel form
  forest management effects, 132, 150